Applications of Cryogenic Technology
Volume 10

Applications of Cryogenic Technology
Volume 10

Edited by

J. Patrick Kelley
SURA/CEBAF
Newport News, Virginia

Plenum Press • New York and London

Proceedings of the sessions sponsored by the Cryogenic Society of America, Inc., at the CRYO '90 conference, held June 17–23, 1990, in Binghamton, New York

Library of Congress Catalog Card Number 68-57815 (ISSN 0093-8815)

ISBN 0-306-43892-5

FOREWORD

Applications of Cryogenic Technology, Vol. 10, is the proceedings from the portion of the conference CRYO-90 sponsored by the Cryogenic Society of America (CSA). CRYO-90, held on the campus of the State University of New York, Binghamton, New York, was an unusual interdisciplinary event, drawing from the life sciences as well as the physical science and engineering areas of the low temperature community. Co-sponsoring CRYO-90 with CSA were the Society for Cryobiology and the Symposium on Invertebrate and Plant Cold Hardiness. These latter two organizations brought an exciting developing field to the conference, a field whose exploration will lead to the betterment of all mankind through improved cryosurgical and organ preservation techniques in addition to improved agricultural and herd yields under extreme conditions. Specific goals of the cryobiological community are cryopreservation, the arrest and recovery of living processes of cells, tissues and organs; and cryosurgery - the local cryodestruction of diseased cells while preserving the healthy surrounding tissue. These goals present great technological challenges. The technological requirements of the cryobiologist include the ability to cool tissues at rates of 10^6 degrees per second (vitrification), to thaw frozen tissue without damaging the delicate cells, to freeze dry tissue using molecular distillation (vacuum) drying, to supercool cell structures below 0°C without freezing, and to successfully store the preserved tissues and organs for any required length of time. CSA, with its broad based experience in the development, application and manufacturing of cryogenic instrumentation, refrigeration, storage systems and heat transfer, is ideally suited to interface with the cryobiologist and overcome these technological hurdles. The selection of the CSA session topics was driven by the cryobiologist's requirements.

CRYO-90 was held June 17-24, 1990. There were 285 registered participants. During the conference, 181 oral papers were presented in addition to 56 poster papers. CSA-sponsored sessions ran from

Wednesday morning, June 20, through Friday afternoon, June 22. There were 27 oral papers presented at the CSA sessions. These papers provided a mixture of overview/review articles, which were intended as a mechanism with which to familiarize the cryobiologist (and others) with the existing technology, and articles describing recent advances. The sessions were well attended, each with audiences of between 20 and 50 people.

The papers presented in this volume, with the exception of the invited papers, have been peer reviewed.

I would like to thank John Baust and the Society for Cryobiology for providing CSA with the opportunity to participate in CRYO-90. It was obvious to all involved that the tremendous success of the conference was primarily due to John's efforts. Much deserved credit also goes to the local organizing committee, the SUNY, Binghamton Cryobiology graduate students. I know that all who attended the conference were favorably impressed by the courtesy and hard work of this incomparable group of women and men. I would also like to thank the CSA Board of Directors for their support, especially Laurie Huget for her steadying influence and Stan Augustynowicz for his ideas and organizational help. Thanks are also due to the CSA technical chairmen for their fine organizational work and for overseeing the peer review process. It is through their efforts, and those of their reviewers, that we are able to offer a credible contribution to the cryogenic literature. Special recognition goes to my secretary, Linda Williams, for her time and energy, which she contributed cheerfully to all aspects of CSA's preparations for the conference and these proceedings. Finally, I would like to thank my Fiancee, Laura Parrish, for her patience, understanding and support during the preparation of this volume.

J. Patrick Kelley

ACKNOWLEDGEMENTS

The organizers of CRYO-90 are deeply grateful for the support received from the following organizations:

American Magnetics, Inc.
Cell Systems, LTD
Cryogenic Services, Inc.
CryoLife
Cryomech, Inc.
CryoMed
Cryomedical Sciences, Inc.
Energy Beam Sciences
Forma Scientific, Inc.
FTS Systems, Inc.
R.G. Hansen & Associates
Hoffer Flow Controls, Inc.
Keller-PSI/Pressure Systems, Inc.
Koch Process Systems, Inc.
Lakeshore Cryotronics, Inc.
Leybold Vacuum Products, Inc.
LifeCell Corp.
LOX Equipment Company
Meyer Tool and Manufacturing
MVE Cryogenics
Neslab Instrument, Inc.
Perkin Elmer Corp.
Revw Scientific, Inc.
Queue Systems
Scientific Instruments, Inc.
Tempshield, Inc.
TS Scientific
Virtis Company, Inc.

CONTENTS

PLENARY LECTURES
Chair: J.C. Chato

SHORT AND LONG TERM STORAGE SYSTEMS
Chair: I.V. LaFave

CRYOGENICS IN THE MEDICAL FIELD
Chairs: S.A. Wilchins and S. Augustynowicz

GENERAL SESSION
Chair: R.W. Fast

PROGRESS IN CRYOCOOLERS[*]

Ray Radebaugh

National Institute of Standards and Technology
325 Broadway
Boulder, CO   80303 USA

ABSTRACT

A significant number of advances have been made during the last few years in a variety of cryocoolers. This paper discusses some of these advances in Brayton, Joule-Thomson, Stirling, pulse tube, Gifford-McMahon, and magnetic refrigerators. Reliability has been a major driving force for new research areas. This paper reviews various approaches taken in the last few years to improve cryocooler reliability. The advantages and disadvantages of different cycles are compared, and the latest improvements in each of these cryocoolers is discussed.

INTRODUCTION

The term cryocooler has generally been used for refrigerators of small and intermediate size, which are capable of reaching temperatures below about 120 K.[1] We follow this definition here and exclude the large refrigerators, which are generally used for commercial liquefaction systems. The increased use of cryogenics in both commercial and military applications has brought about a considerable demand for cryocoolers over the last decade. Many times the lack of a cryocooler adequate for the job hinders further use of cryogenic phenomena in a particular application. The problem is now recognized and a sizeable effort is underway in the U.S. as well as in other countries to improve the performance and reliability of cryocoolers.

---

[*]Contribution of NIST, not subject to copyright.

APPLICATIONS

<u>Infrared sensors</u>

By far, the largest application for cryocoolers is the military use of cooling infrared sensors for night vision and missile guidance. Most of these sensors use HgCdTe, InSb, or PtSi as the detector material and require a refrigeration capacity of 0.1 to 1.0 W at a temperature of about 80 K. Several manufacturers supply this market, which amounts to several thousand cryocoolers per month. Ten years ago, these tactical applications used either closed-cycle Stirling cryocoolers, with lifetimes of a few hundred hours, or open-cycle Joule-Thomson coolers, which had to be recharged with high pressure gas after each test. The demand for higher reliability has led, in the last few years, to the introduction of Stirling cryocoolers with lifetimes of 2000 to 3000 hours. These units are often replacing the older open-cycle Joule-Thomson coolers.

Recently, this infrared technology has found increased use in civilian applications such as police work and border patrol, as well as thermographic scanning in medical diagnosis.[2]

Applications of infrared sensors in space by NASA and the military have greatly expanded the requirements placed on the cryocoolers for these sensors. Over sixty Stirling cryocoolers may be needed for the NASA Earth Observation System (EOS) for cooling infrared sensors aboard satellites to temperatures of 60 to 80 K. These flight-qualified coolers cost about one-half million dollars each. Observation of colder targets or better temperature resolution demands greater sensitivity, which has been achieved by cooling the sensors to lower temperatures and by using other sensor materials. Temperatures of 30 to 40 K, which require two stages of cooling, are now sometimes needed. In some cases, one-stage devices are called upon to provide refrigeration in the range of 60 to 70 K. Temperatures of 10 K and refrigeration powers of 0.2 to 3 W are generally required for long-wavelength surveillance sensors of Si:As. The cooling of bolometers to 0.1 K is of interest to NASA for infrared astronomy. This application has led to the development of adiabatic demagnetization refrigerators, although studies regarding the use of He^3-He^4 dilution refrigerators in zero gravity are also underway.

The lifetimes demanded of cryocoolers for these space applications range from 2 to 10 years. Lifetimes above 5 years are desired but have never been demonstrated. Because of the limited power available on spacecraft, high efficiency is also required. Finally, vibration amplitudes must be kept very low at the sensor to eliminate vibrational noise in the signal output. In addition, the vibration amplitudes in remotely situated compressors must also be kept low to prevent disturbances to the accurate orientation of the satellite and to prevent structural resonances.

<u>Cryopumps</u>

The largest commercial application for cryocoolers is the cooling of cryopumps, which are used by the semiconductor industry to achieve the clean vacuum needed for the production of high-density VLSI circuits. Approximately 4000 to 5000 cryopumps per year are being manufactured in the U.S. The cryocoolers used for this application are the two-stage Gifford-McMahon refrigerators, operating at about 20 K. Refrigeration powers at that temperature vary from about 1 to 15 W. Reliability, low maintenance, and low cost are the driving forces in this application. The efficiency of the Gifford-McMahon refrigerator is relatively low, but for ground-based applications, this is of little concern.

<u>Other applications</u>

There are many other applications of cryocoolers, although the number of cryocoolers manufactured for these purposes is quite small in comparison with those made for infrared and cryopump applications. In the last few years, there has been considerable interest in cryocoolers for magnetic resonance imaging (MRI) magnets. In most cases, a two-stage Gifford-McMahon (GM) refrigerator is used to cool the shields of storage dewars for liquid helium and thereby greatly reduce the boil-off rate of the liquid helium. In some cases, a Joule-Thomson (JT) stage is added at the low end to achieve 4.2 K and reliquefy the boiloff helium in the MRI cryostat. GM/JT cryocoolers have also been used for many years to cool the maser amplifiers used to communicate with satellites in deep space. The parametric amplifiers used to communicate with satellites in earth orbit are cooled to 20 K with two-stage GM cryocoolers.

In the last few years, the computer industry has been studying cooled semiconductor computers. Temperatures of about 80 K lead to an increase in operating speed by a factor of two. So far only large computers have been considered for cryogenic operation and the cooling powers are in the kW range at 80 K. A few such units have been manufactured in the U.S. The use of small Joule-Thomson refrigerators has recently been considered for the cooling of single chips on a printed circuit board.

When high-temperature superconductors become a practical reality, there will be a need for cryocoolers operating in the temperature range of 40 to 60 K to cool such devices to about one-half their critical temperature. These devices operate in the right temperature range to be combined with cooled semiconductor components.

TYPES OF CRYOCOOLERS

Cryocoolers are usually divided into two types: recuperative and regenerative. Recuperative cryocoolers use recuperative heat exchangers exclusively. Common recuperative cryocoolers are the Joule-Thomson and the Brayton devices, as shown in Figure 1. A regenerative cryocooler has at least one regenerative heat exchanger, or regenerator. Figure 2 shows schematics of four types of regenerative cryocoolers. In these figures, the displacer operates synchronously with the compressor piston but leads it in phase by about 90°. A fifth regenerative cryocooler, known as the pulse tube refrigerator, is similar to the Stirling cycle but it does not use a displacer. This refrigerator will be discussed in detail later.

<u>Operating Ranges</u>

Figure 3 shows the typical operating ranges of the various cryocoolers. Certainly the boundaries are not rigid, but are shown to indicate general trends. Some applications may use a regenerative cycle, such as the Stirling cryocooler, up in the kilowatt range at 80 K. However, in these large sizes the turbo-Brayton systems are particularly well suited because they are relatively efficient in those sizes and their gas bearings offer high reliability. As the size decreases, their efficiency drops considerably because of the practical difficulty in making efficient expansion turbines less than a few millimeters in diameter and because of the geometric scaling problem in the conduction loss along the shaft between the turbine and the warm brake mechanism.

The low temperature limit of about 10 K for regenerative systems is due to a lack of matrix heat capacity in the coldest

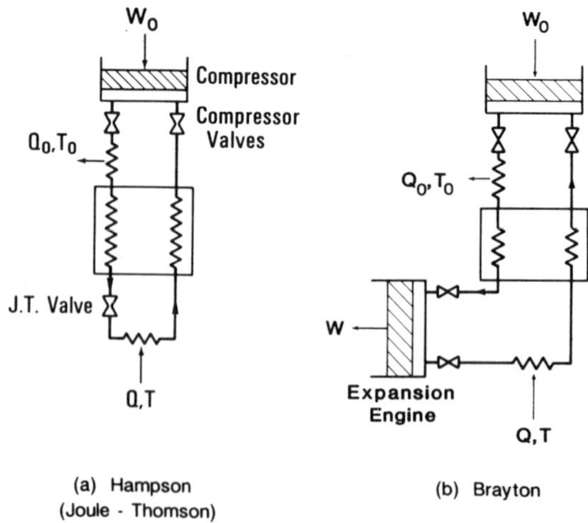

Figure 1. Various types of recuperative cryocoolers

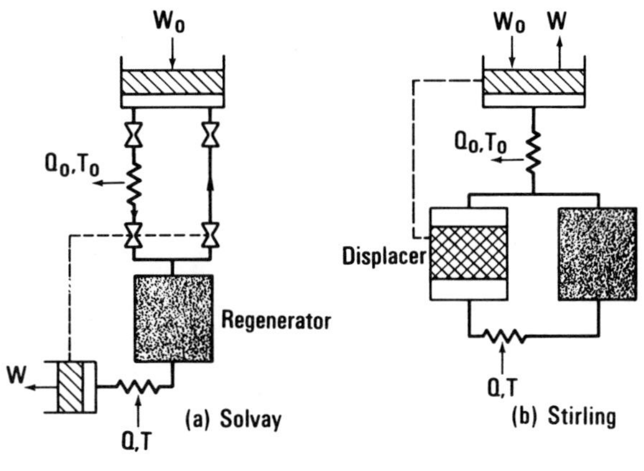

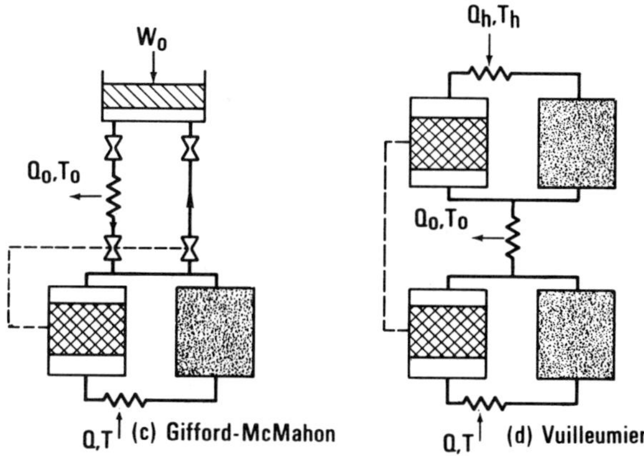

Figure 2.   Various types of regenerative cryocoolers

regenerator.   Temperatures of 4 K can now be reached[3,4] with the use of magnetic materials, like GdRh and Er_3Ni, in the regenerator, but the system efficiency is still quite low.   Joule-Thomsom cryocoolers that use helium for the last stage can reach 4 K, but reliable compressors for such a systems are still under development.   Thus, the region below 10 K and below a few watts of refrigeration has not been handled well with any refrigerator.   So far, that region is usually handled with a two-stage Gifford-McMahon cryocooler with a Joule-Thomson stage added at the bottom.   Magnetic refrigerators are being studied quite extensively for this same region since they could be used in place of the Joule-Thomson stage.

Advantages and Disadvantages

Table 1 shows advantages (+), disadvantages (-), and neutral status (0) of the recuperative and regenerative cryocoolers regarding

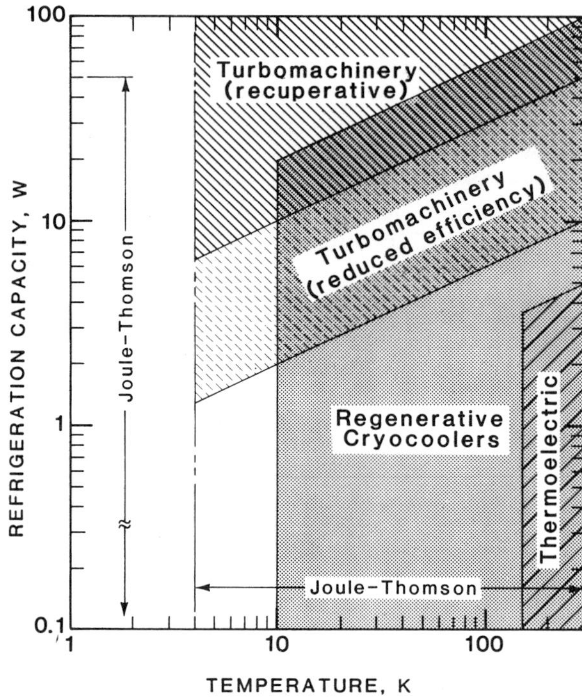

Figure 3. Typical operating region of various cryocoolers

various categories. Because void volumes do not affect the
recuperative systems, they can transport the refrigerant over long
distances. The recuperative heat exchangers are more complex and
larger than regenerators. Because of the few moving parts in the
regenerative cryocoolers, they are sometimes more reliable than
recuperative systems. That aspect is discussed further in the next
section.

Table 1. Advantages and disadvantages of cryocooler type

	Recuperative	Regenerative
Reliability	0-	0+
Efficiency	-	+
Vibration	+	-
Transport refrigerant	+	-
Simple Heat Exchanger	-	+

RELIABILITY

The marketplace demand for higher reliability has been the driving force behind much of the research and development work for cryocoolers. Space applications require cryocooler lifetimes of at least 3 to 5 years. Over the last ten years, the U.S. Defense Department and NASA have made significant progress in reaching these goals. Table 2 lists typical lifetimes (mean time to failure) achieved with various cryocooler types.

Table 2. Typical lifetimes achieved with mechanical cryocoolers

	Unsupported	Supported
crank-driven (rubbing seals)	200-2000 h (Stirling) 20,000 h (GM)	
linear drive (clearance seals)	1000-4000 h	2-5 y
turbo-systems (gas bearings)		2-5 y

<u>Unsupported Systems</u>

All the early cryocoolers used crank driven compressors and expanders with rubbing seals. For Stirling cryocoolers the system had to be oil-free, which limited the lifetimes even more. Most of the small 80 K Stirling cryocoolers for cooling infrared detectors are of this type. Their lifetimes have usually been in the range of 200-500 hours, although recently these lifetimes have been pushed into the 1000-2000 hour range. However, special efforts have been made in some cases to extend these lifetimes even further. In 1978, four Stirling refrigerators were flown in space for approximately two years to cool gamma-ray detectors.[5] Their performance slowly deteriorated due to a slow leak of the helium gas in the system.

The Vuilleumier refrigerator uses a thermal compressor instead of a mechanical one (see fig. 2d), and as a result, there is less force on the compressor seals. The few Vuilleumier refrigerators built did have somewhat longer lifetimes than their Stirling counterparts. In fact, after considerable effort in selecting the right seal materials and designs, an aerospace contractor with Air Force sponsorship was successful in building six Vuilleumier refrigerators that show potential lifetimes of 3 to 5 years. One successfully demonstrated a lifetime of 2-1/2 years at twice the normal speed. These units are three-stage devices capable of providing 150 mW of refrigeration at 10 K. Three of them are flight qualified and are the only flight-qualified cryocoolers for the 10 K range. At present, there is no planned mission for them.

The Gifford-McMahon refrigerators use oil lubricated compressors, which increase the lifetimes. Because dead volume before the expander valves has no effect on the performance, oil removal filters can be install there at the expense of additional weight. Extensive operating experience with the GM refrigerator for cryopumps has helped to increase their reliability. One manufacturer now accumulates 1-1/2 million hour of operating experience per month with its systems in the field.

In the last couple of years, many manufacturers have been making linear-drive compressors for the Stirling cryocoolers. Because most side forces are eliminated, the lifetimes have been increased to the range of 1000 to 4000 hours. In addition, most of these linear drive devices use clearance seals instead of rubbing seals. The piston and displacer are not supported in any manner so there is still some rubbing between parts.

<u>Supported Systems</u>

In order to reliably achieve the lifetimes of 3 to 5 years needed for satellite applications, some means of supporting the piston and displacer (expander) inside the cylinder must be used. Turbo-Brayton systems have used gas bearings for many years. Another Brayton system, the rotary reciprocating refrigerator (R³) uses a reciprocating compressor and expander which also rotates to provide a continuous gas bearing support. Extensive life tests have not begun on that device yet. The first Stirling refrigerator to use supported drive elements was the magnetic bearing cooler made in 1983 for NASA/Goddard. This one-stage cooler produces 5 W at 65 K and has operated for five years with no mechanical failure. In fact, the components limiting the life of this refrigerator may very well be the complicated electronics for the magnetic bearing system.

A simpler support system, first used in the early 1980's for Stirling cryocoolers by G. Davey of Oxford University, consists of flexure diaphragm springs which hold the piston or displacer drive rods rigidly in the radial direction.[6] Figure 4 depicts these springs and how they are used. One of these 1 W, 80 K Oxford cryocoolers ran for 2-1/2 years before a bushing in the displacer worked loose. A similar unit is to be used for cooling an infrared radiometer on the NASA Upper Atmosphere Research Satellite (UARS), scheduled for deployment in 1990. These Stirling cryocoolers are available in flight qualified form from a British aerospace company. The U.S. companies have been somewhat slow to pick up this new technology, although at this time there are about three or four U.S. companies actively developing or studying these Oxford cryocoolers.

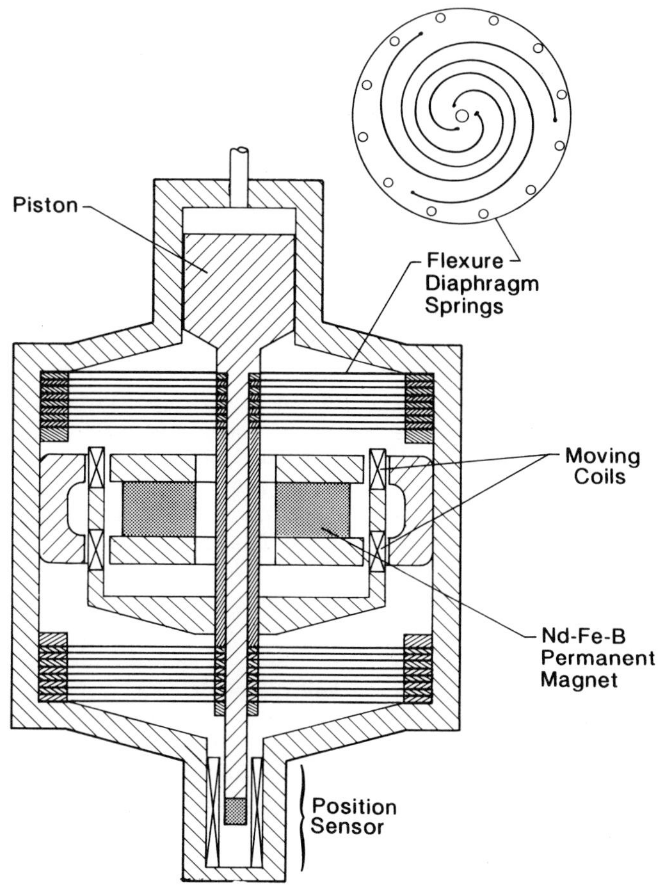

Figure 4. Oxford split Stirling cryocooler (compressor only)

DEVELOPMENTS IN RECUPERATIVE CYCLES

Joule-Thomson Cryocoolers

These coolers have been used for many years in open-cycle systems where a high-pressure tank of nitrogen or argon provides the input gas. The rapid cooldown (2-5 seconds) possible with these cryocoolers makes them suited for cooling infrared guidance systems on missiles. Miniature finned tubing is used for the heat exchangers. About 10 years ago, W.A. Little developed a photolithographic process for making the heat exchanger, expansion capillary and liquid container.[7] In this process, the gas channels are abrasively etched in thin, planar glass substrates. The substrates are then fused together to yield a laminated structure containing the channels of the complete refrigerator. Because of the heat capacity of the glass substrates, these refrigerators, in general, do not have the fast cooldown of the miniature finned-tubing devices. However, the manufacturing process is ideal for mass production. These glass refrigerators are normally used in university and industrial R & D laboratories for a variety of experiments. A lack of a reliable compressor for these miniature JT refrigerators has limited their application to laboratory use. Within the last year, the manufacturer of the glass JT refrigerators has teamed with a Japanese manufacturer of compressors to offer an oil-free compressor for use with their refrigerators. The lifetime of the compressor appears to be over 1 year.

There has been a surge of interest, in the last few years, regarding the use of mixtures of nitrogen and hydrocarbon gases in JT refrigerators, although the idea was proposed by Alfeev of the Soviet Union in 1971.[8] Alfeev showed that a mixture of 30% methane, 20% ethane, and 20% propane gave a factor of 10 increase in the system efficiency compared with that using nitrogen alone. Interest in the capability of these mixtures was stimulated a few years ago when Little[9] showed that the addition of a few percent of CF_3Br renders the mixture nonflammable. These mixtures are capable of achieving temperatures in the 80-100 K range. For lower temperatures mixtures of N_2-Ne and N_2-He are now being investigated.

Another possible solution to the JT compressor problem is the use of sorption compressors, which operate with heat input rather than mechanical work input. A chemisorption system with H_2 and $LaNi_5$ was developed by JPL several years ago for refrigeration at 20 K. More recently, promising physisorption and chemisorption systems for other stages in a JT cryocooler have been identified.[10] A praseodynium-cerium-oxide/oxygen chemisorption system shows promise for an 80 to 90 K stage, and systems based on carbon/krypton or carbon/methane physisorption show promise for the range of 110 to 140 K. To reduce the input power, schemes are being studied where heat rejected from one stage is used to help heat another stage, and where various parts of a single-stage compressor are heated and cooled regeneratively. Predicted power inputs in such schemes for a 1 W, 80 K refrigerator are now about 40 to 50 W.

Brayton Cryocoolers

The Brayton cycle has been chosen by the military to provide a few watts of cooling at 10 K for cooling infrared focal plane arrays in space. Mission lifetimes are to be 5 years. For this program, two cooler developments are being funded. One is the R^3 system discussed earlier. The other system is the turbo-Brayton refrigerator.

The turbo Brayton cycle is also used in a development by another manufacturer to provide a 5 W, 65 K cryocooler for

NASA/Goddard. Because there are no reciprocating parts, the vibration is extremely low.

DEVELOPMENTS IN REGENERATIVE CYCLES

<u>Stirling Cryocoolers</u>

Many of the new developments in this field are taking place in the U.K. with the development of the Oxford split Stirling cryocooler.[6] Within the U.S., NASA and JPL have plans to use these cryocoolers on various satellites. Extensive testing of the units is planned in the near future. These cryocoolers provide about 0.7 W of cooling at 80 K with a total input power of only 30 W of electrical power. The specific power input is 43 W/W and the relative efficiency is 6.4% of Carnot.

Significant developments in Stirling cryocoolers have been the use of linear motor drives, clearance seals, and demand control to increase the lifetimes and reduce the power input for the coolers used in tactical applications. With demand control, the cryocooler can cool down quickly, but then the power input is reduced to maintain a fixed temperature of 80 K independent of the heat load. The reduced power input slows down the device and increases the operating lifetime. Within the last few years, some microminiature integral Stirling cryocoolers have been developed for cooling hand-held infrared sensors. The smallest of these units requires only 3.5 W of power input to give 130 mW of cooling power at 80 K. The cryocooler mass is only 0.3 kg with a maximum dimension of 85 mm. Such low input powers can be provided by batteries.

<u>Pulse Tube Refrigerators</u>

Five years ago, workers at the National Institute of Standards and Technology (NIST) showed[11] it was possible to achieve 60 K in a one-stage orifice pulse tube refrigerator, shown schematically in Figure 5. Since then, there has been world-wide interest in this new cryocooler with only one moving part. The intrinsic efficiency is between that of a Joule-Thompson and a Stirling cycle.[12] Recently, a temperature of 39 K was obtained in a one-stage orifice pulse tube refrigerator at NIST. Dual inlet pulse tubes are discussed by Zhu et al[13] and offer the potential for further improvements to pulse tube refrigerators. Work is underway at NIST and other laboratories to optimize the refrigeration power and efficiency of both one- and two-stage devices. A temperature of 26 K was reported for a two-stage pulse tube refrigerator developed by Tward et al.[14] In a joint program between NIST and Los Alamos National Laboratory, a heat-driven thermoacoustic driver (TAD) is used to power an orifice pulse tube refrigerator (OPTR). The combined TADOPTR provides refrigeration with no moving parts. The first model constructed reached a low temperature of 90 K. It is the first cryogenic

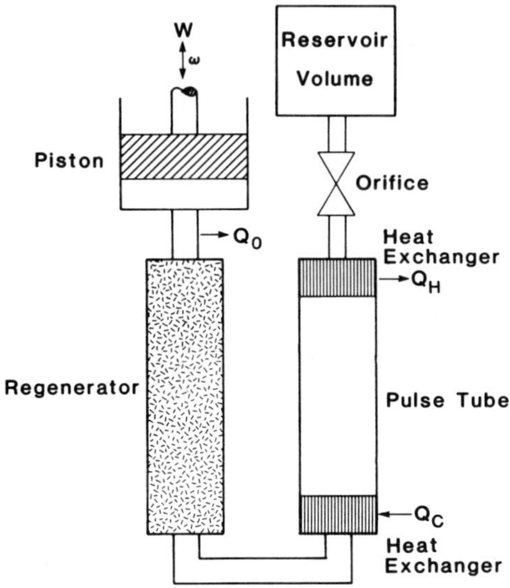

Figure 5. Orifice pulse tube refrigerator

refrigerator with absolutely no moving parts. Problems to be
overcome are the viscous losses in the resonant tube and the
operation of the OPTR at pressure ratios as low as 1.1.

Gifford-McMahon Refrigerators

Two-stage Gifford-McMahon refrigerators usually reach
temperatures of about 8-10 K with no load. Lower temperatures are
normally achieved by adding a Joule-Thomson stage. The 8-10 K limit
of the GM refrigerator is caused by the low specific heat of the lead
spheres used in the second stage regenerator as the temperature
decreases below about 15 K. In the last one or two years, there has
been considerable effort in Japan in the use of materials with low
temperature magnetic transitions to obtain high specific heats.
Experiments with GdRh and Er_3Ni powders in the second-stage
regenerator have enabled the GM refrigerator to liquefy helium at
4 K.[4,15]

DEVELOPMENTS IN MAGNETIC REFRIGERATORS

Most of the work underway on magnetic refrigerators has been
for spanning the range of 1.8 to 4 K and 4 to 20 K. The material
most commonly used is gadolinium gallium garnet (GGG). A wheel made
of plates of GGG rotates in and out of a stationary magnetic field
and exchanges heat with flowing helium gas at either 4 K or 20 K.[16]
Clearance seals are used between the rotating GGG and the stationary
frame to prevent the two helium gas flows from mixing with each
other. These refrigerators generally follow a Carnot cycle. Most

of the U.S. work on the systems for 4-20 K is now being done under
the direction of John Barclay. New materials are being investigated
for this temperature range as well as for higher temperatures.
Active magnetic regenerators using ferromagnetic materials show the
most promise for magnetic refrigerators operating above about 20 K.[17]

CONCLUSIONS

The amount of research and development work on cryocoolers has
increased considerably in the last few years due to the greater use
of cryogenics for various applications. Improved reliability has
usually been the most important demand placed on new cryocoolers.
As this reliability improves, more applications begin to appear
attractive. Space-based applications are now demanding lifetimes of
5 years for cryocoolers with low vibration. Considerable new work
is underway on Brayton, Joule-Thomson, Stirling, pulse tube, Gifford-
McMahon, and magnetic systems. Each has its own unique set of
advantages and disadvantages for a particular application.

REFERENCES

1. G. Walker, "Cryocoolers," Plenum Press, New York (1983).
2. G. Walker, "Miniature Refrigerators for Cryogenic Sensors and
    Cold Electronics," Oxford Press, Oxford (1989).
3. H. Nakashima, K. Ishibashi, and Y. Ishizaki, Developments of a
    4-5 K cooling Stirling cycle refrigerator, in: "Proceedings of
    the 4th Int'l Cryocooler Conference," G. Green, G. Patton, and
    M. Knox , ed., Bethesda (1986), p. 263.
4. M. Nagao, T. Inaguchi, H. Yoshimura, T. Yamada and M. Iwamoto,
    Helium liquefaction by a Gifford-McMahon cycle cryocooler, in:
    "Advances in Cryogenic Engineering," vol. 35B, R. W. Fast, ed.,
    Plenum Press, New York (1990), p. 1149.
5. L. G. Naes and T. C. Nast, Long life orbital operation of
    Stirling cycle mechanical refrigeration, in: "Proceedings of the
    8th Int'l Cryogenic Engineering Conf.," IPC Science and
    Technical Press, Surrey, England (1980), p. 204.
6. G. Davey, Review of the Oxford cryocooler, in: "Advances in
    Cryogenic Engineering," vol. 35B, R. W. Fast, ed., Plenum Press,
    New York (1990), p. 1423.
7. W. A. Little, Advances in Joule-Thomson cooling, in: "Advances
    in Cryogenic Engineering," vol. 35B, R. W. Fast, ed., Plenum
    Press, New York (1990), p. 1305.
8. V. N. Alfeev, et al., Great Britain Patent No. 1,336,892 (1973).
9. W.A. Little, U.S. Patent application, Serial #919,699 (1985)
    Patent pending.
10. J. A. Jones, Sorption cryogenic refrigerator-status and future,
    in: "Advances in Cryogenic Engineering", vol. 33, R. W. Fast,
    ed., Plenum Press, New York (1988), p. 869.

11. R. Radebaugh, J. Zimmermann, D. R. Smith, and B. Louie, A comparison of three types of pulse tube refrigerators: New methods for reaching 60 K, _in_: "Advances in Cryogenic Engineering," vol. 31, Plenum Press, New York (1986), p. 779.

12. R. Radebaugh, Pulse tube refrigeration - a new type of cryocooler, _Japanese J. Appl. Phys._, vol. 26, suppl. 26-3: 2076 (1987).

13. S. Zhu, P. Wu, and Z. Chen, Double inlet pulse tube refrigerators: an important improvement, _Cryogenics_, 30: 514 (1990).

14. E. Tward, C. K. Chan, and W. W. Burt, Pulse tube refrigerator performance, _in_: "Advances in Cryogenic Engineering," vol. 35B, R. W. Fast, ed., Plenum Press New York (1990), p. 1207.

15. T. Kuriyama, R. Hakamada, H. Nakagome, Y. Toka M. Sahashi, R. Li, O. Yoshida, K. Matsumoto, a T. Hashimoto, High efficient two-stage GM refrigerator with magnetic material in the liquid helium temperature region, _in_: "Advances in Cryogenic Engineering, " vol. 35B, R. W. Fast, ed., Plenum Press New York (1990), p. 1261.

16. J. A. Barclay, Magnetic refrigeration: a review of a developing technology, _in_: "Advances in Cryogenic Engineering," vol. 33, R. W. Fast, ed., Plenum Press, New York (1988), p 719.

17. E. Schroeder, G. Green, and J. Chafe, Performance predictions of a magnetocaloric refrigerator using a finite element model, _in_: "Advances in Cryogenic Engineering," vol. 35B, R. W. Fast, ed., Plenum Press, New York (1990), p. 1149.

LINEAR DRIVE STIRLING COOLER TECHNOLOGY

Peter J. Kerney

CTI-CRYOGENICS

Waltham, Ma

ABSTRACT

Stirling cycle cryocoolers are used for cooling IR detectors in a number of military of military and commercial applications including missile guidance systems and night vision security systems. There is an ever increasing demand for longer and longer maintenance free operation of these coolers.

A critical component in the development of a long life, highly reliable Stirling cooler is the drive motor. Present state-of-the-art production 77K Stirling cryogenic coolers utilize rotary drive motors in a variety of configurations, brush/brushless, with AC and DC input power. In existing packaging concepts, these motors tend to be a source of particulate and gaseous contamination which reduces the overall system reliability. Linear drive technology offers the potential for significantly reducing the life limiting elements in the cooler design. Operational flexibility and low mechanical vibration are promising added features of this drive motor approach.

This paper reviews the status of linear drive technology developments at CTI-CRYOGENICS and describes mechanical designs and cryogenic performance for several sizes of Stirling coolers.

INTRODUCTION

Stirling cycle cryogenic coolers have historically found applications in cooling infrared (IR) imaging systems for night vision and missile guidance (1). Although military requirements have been the primary driver in the development of these cryocoolers, applications in the commercial marketplace for security systems and sensitive instrumentation are beginning to emerge.

Typical performance specifications include refrigerator capacities from 0.25W to 1.5W at 77K with input power from 20W to 80W. Total system weight ranges from 1.0 kg to 2.3 kg with mean time to failure (MTTF) rates of 500 to 2500 hours. The earliest systems were of the integral configuration wherein the cold end is mechanically connected through a common drive to the compression piston. A typical production unit is the U.S. Army HD 1033 Common Module cryocooler which is driven by a 400 HZ, single phase, 115 volt AC motor through a 16.0 to 1.0 reduction gear assembly. The original design of this cooler utilized 3 contacting lip seal assemblies, one on the compressor piston and two on the displacer/regenerator. The compressor seal isolates the helium working gas volume from the crankcase volume and the displacer seals prevent gas by-passing of the regenerator.

These contacting seals were one of the life-limiting components in the early version of the integral Stirling cryocoolers. The lip seals were often spring loaded to ensure positive contact with the cylinder wall and produced particulate contamination during operation resulting in premature failure of the cryocooler. A rotary brush motor was the original drive system of choice and it was another life limiting component due to rapid brush wear in a dry helium environment.

In an effort to extend cryocooler lifetimes, spring loaded lip seals have been replaced by hard on hard and hard on soft clearance seals (2). By definition a contacting seal is designed to prevent gas leakage by eliminating the annual gap between the piston and cylinder wall over a relatively short seal length. The seals most frequently used contain a soft seal jacket and a metallic expander. The soft seal jacket material wears and eventually the seal/spring force becomes insufficient to restrict leakage flow and a blow-by condition results. Also the frictional drag created by the contacting seal is a source of cooler power consumption.

In contrast the clearance seal is designed to restrict gas leakage by maintaining a very small annular gap over the entire length of the piston. To minimize wear induced by side loads the clearance seal materials have a very high hardness. The combination of high hardness, and low frictional drag have been shown to produce negligible wear over very long periods of time. The low frictional characteristics of clearance seals result in reduced mechanical loading, significantly reduced seal wear and lower input power. Since clearance seals generate minimum of particulate wear debris in comparison to their fully contacting counterparts, consistent performance over long operating periods is achieved.

The second major life limiting component in the Stirling cryo-

cooler is the drive motor. Conventional Stirling cryocoolers use
rotary electric motors and a crank mechanism to obtain reciprocating
motion of the piston. A relatively large number of moving and
contacting mechanical parts are required for converting rotary to
linear motion. In addition bearing grease and motor constituents
are a primary source of helium gas contamination.

     Utilization of linear drive technology can eliminate many of
the life limiting features of rotary drive systems. Several imple-
mentations of linear drive technology can isolate motor coils and
eliminate the necessity of lubricants thus extending cryocooler life
by reducing contamination of the helium gas (3). A low radial side
load associated with linear drive motors will also decrease seal
wear in comparison to a conventional rotary crank mechanism configuration.

     A variety of motor types have been used in designing linear
drive motors including moving iron, moving magnet and moving coil
configurations. The moving iron approach is the simplest and lowest
cost approach with low leakage flux but with high weight for a given
output and efficiency. Also, because of the small air gap inherent
in this approach the resulting high inductance leads to a low power
factor.

     The moving coil motor (4) has very low eddy current losses and
low weight when operated at high frequencies. A major disadvantage
of this approach is the inability to isolate motor coils from the
helium gas and potential failure of the moving electrical connections.
The moving magnet design has the highest output and efficiency with
the lowest weight and volume. Its disadvantages include high
hysteresis and eddy current losses and high leakage flux from
exposed magnets.

     Moving iron and moving magnet type motors have been applied to
single and double opposed piston designs. In order to achieve dynamic balancing a single piston approach requires a passive balancer
tuned to the right operating frequency. Dual opposed piston compressors are inherently balanced at all frequencies which provides added
operating flexibility.

     In the split Stirling system configuration wherein the compres-
sor is separated from the cold expander by a gas transfer line, linear
drive motor technology offers additional operating benefits. Conventional split systems utilize the helium gas to pneumatically drive
the cold end displacer. The reciprocating motion produces mechanical
vibrations in the cold end with a resulting deleterious effect on
the IR detector. Specifically the mechanical vibrations interact
with the IR detector leads which can produce currents that are in-
distinguishable from IR signal currents a phenomenon called
"microphonics". By employing a linear drive motor in the cold end (5)

to assist the pneumatic drive the motion of the displacer can be carefully controlled and modulated to significantly reduce these mechanical vibrations and the resultant microphonics.

SUMMARY

Linear drive motors are the drive system of choice for emerging applications utilizing Stirling cycle cryocoolers. Linear drive motors offer potential for significant systems enhancements. These include longer life through minimization of outgassing components and use of noncontacting seals. Inherent dynamic balancing leads to low vibration and low acoustic noise operation. Cold end stroke control for fast cooldowns and minimum microphonics.

REFERENCES

1. F. Eberth, F. Chellis, "Comparing Closed Cycle Cryocoolers", <u>Electro-Optical System Design</u>, November 1979.

2. F. W. Pirtle, P. K. Bertsch, P. J. Kerney, "Cryogenic Cooler Seals Investigation", AFWAL-TR-87-3046, Air Force Wright Aeronautical Laboratories, September 1987.

3. P. J. Kerney, G. J. Higham, J. Vigilante, "Linear Resonance Split Cryocooler/Design and Component Development Program", Final Report, U.S. Army Contract DAAK 70-82-C-0222, February 1985.

4. F. Stolfi, A. K. de Jonge, "Stirling Cryogenerators with Linear Drive", <u>Phillips Technical Review</u>, Volume 42, No. 1.

5. N. O. Young, R. Henderson, P. J. Kerney, "Refrigeration System with Linear Motor Trimming of Displacer Movement", U. S. Patent 4,475,346, October 1984.

COOLING OF GeZn DETECTORS BY COMMERCIALLY AVAILABLE CLOSED-CYCLE CRYOGENIC REFRIGERATORS

Robert G. Hansen

R. G. Hansen & Associates
631 Chapala Street
Santa Barbara, CA 93101

ABSTRACT

In the late 1960's, zinc doped germanium (GeZn) detectors were developed for broad range detection in the near infrared spectrum spanning a spectral range of 2 to 40 micron wavelength units with high detectivity.

For optimum use of this detector material, performance data indicated operation at 4.2 Kelvin; i.e., at liquid helium temperature. This being the case, the use of this detector was solely dependent upon the availability of liquid helium, thus limiting or restricting its use both nationally and internationally. It could only be used for applications where factors other than economics warranted its use.

This paper presents a unique and innovative technique for using this broad range, high performance detector by integrating it with a commercially-available, closed-cycle refrigeration system, thus broadening both its application and utility.

INTRODUCTION

In the early 1960's, considerable advancement was achieved in the development of infrared detectors and electro-optic technology. Organizations such as the Santa Barbara Research Center made commercially available a variety of doped germanium and doped silicon detectors for use in both commercial and military applications. One of the most exciting series of detectors was the doped germanium materials such as GeCu, GeHg and GeZn. This technology is described in detail in Puttey's pioneer paper on "Solid State Devices for Infrared Detection."[1] These specific detectors exhibited a broad range detectivity that required correspondingly lower cooling temperatures to support their performance.[2]

Specifically, GeHg required operation at temperatures below 28 K, GeCu required temperatures equal to or below 16 K and GeZn required temperatures below 10 K. As Figure 1 indicates, the need for broad range detectivity was solely dependent upon the availability of liquid helium.[2] Figure 1 shows a plot of detectivity versus temperature for various germanium detectors. Figure 2 illustrates detectivity versus wavelength for various detectors.[2]

As a matter of historical note, liquid helium was both a controlled and extremely expensive commodity from the mid-1960's through 1970. Availability was restricted to limited distribution throughout the United States and heavily restricted for export due to both logistical and economic considerations. As a result, the infrared technology within the research and commercial arena was limited to detectors and technology supported by either liquid nitrogen cooled sensors or conventional non-cooled sensors.

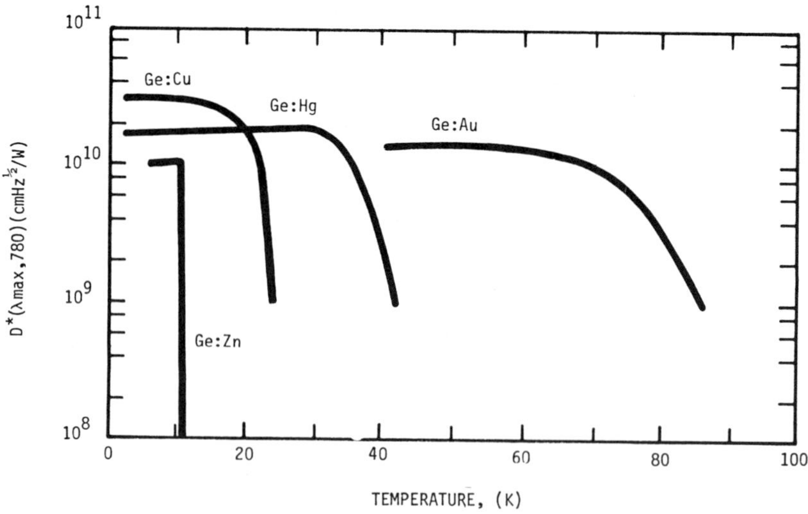

Figure 1. Performance Versus Operating Temperature
of Selected Infrared Detectors

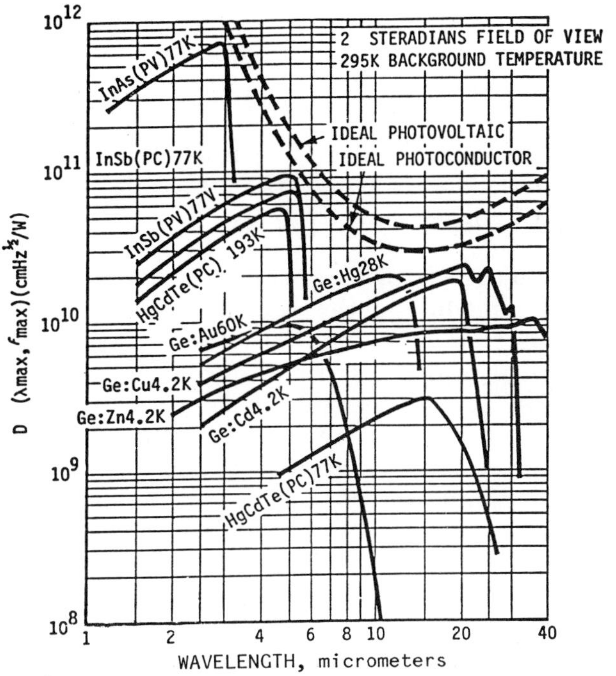

Figure 2. Performance Characteristics Versus
Spectral Response of Selected Infrared Detectors

In 1970, organizations such as Texas Instruments, Hughes Aircraft, Aerojet General and Honeywell commenced fabrication of forward-looking infrared systems (FLIRs) for military applications. These systems required cooling by airborne, closed-cycle refrigeration systems which were expensive and limited in lifetime. Typical cost were ~$13,000 with mean time between maintenance of ~1,000 operating hours.

With the implementation of FLIR systems within the military inventory, a new market was created for night vision and thermal imaging technology. These systems placed emphasis on alternative materials which operated at higher operating temperatures; i.e., mercury cadmium telluride operating at 77 K, indium antimonide operating at 77 K and most recently, platinum silicide operating at 77 K.

The basic research community, as well as the semiconductor industry, learned that detectors such as GeZn and GeCu detectors could be used very effectively for the analysis and determination of impurity levels in semiconductor materials. Specifically, the broad range sensitivity of GeZn would allow the presence of carbon, oxygen, phosphorous and boron to be determined both quantitatively and qualitatively. This established quality levels for use in very large scale integrated circuitry. Additionally, the advent of Fourier Transfer Infrared Radiation (FTIR) spectrophotometers within the semiconductor industry has created the necessity for broader range infrared sensors with higher sensitivity.

One of the objectives of this paper is to acquaint the reader with a technique for modifying the conventional FTIR to utilize the superior performance of a GeZn detector. This technique may extend the capability and performance of conventional FTIR spectrophotometers. It also permits use of the GeZn detector in laser and conventional spectroscopy. It is not the purpose of this paper to describe in detail, or discuss the electrical and/or optical details and specifications of GeZn detectors other than on an application orientated basis.

DISCUSSION

In 1986, in conjunction with the Ethyl Corporation's Process Development Center in Baton Rouge, Louisiana,[3] a specially modified FTIR system was designed which utilized a helium cooled GeZn detector. Performance levels were achieved which exhibited state-of-the-art analysis and determination of carbon, oxygen, phosphorous and boron levels within bulk silicon materials.

As a further means of enhancing this system, a developmental program was initiated which resulted in the cooling of the GeZn detector with a conventional closed-cycle refrigeration system. After two years of operation with the closed-cycle refrigerator, this particular FTIR system is currently exhibiting a 98% up time.

In addition to the cooling of an infrared detector with a closed-cycle refrigeration system, a specially designed cryogenic system was fabricated which cooled two heads simultaneously; i.e., the first for the cooling of an infrared sensor, and the second for the cooling of a variety of samples or specimens. In the specific case of Ethyl Corporation, a quantity of 12 samples could be cooled simultaneously to temperatures at or below 12 K.

As Figure 3 indicates, the cryogenic support system can be easily adapted to conventional spectrophotometers such as FTIRs.

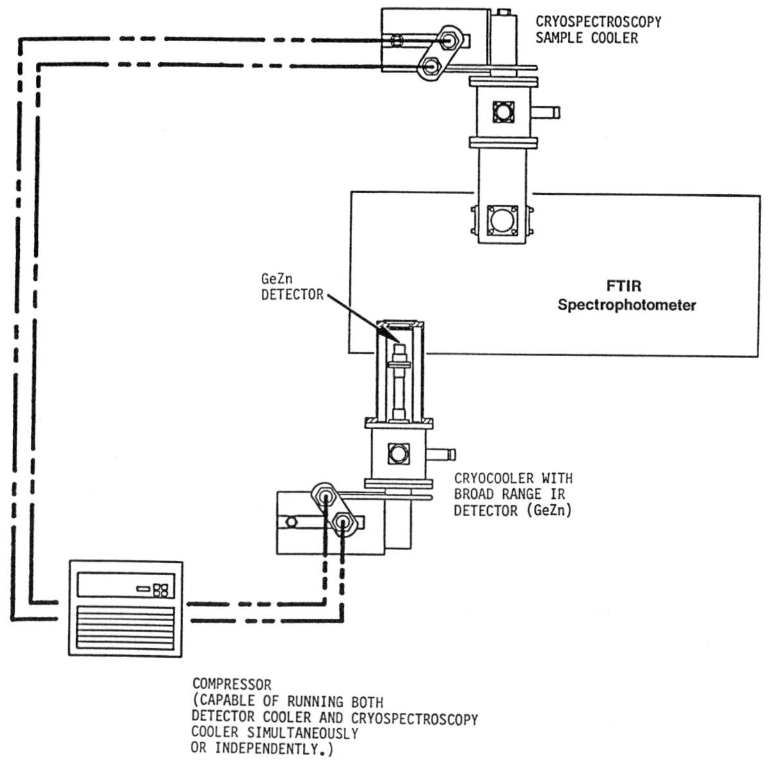

Figure 3. Typical Cryogenic Support System for FTIR

In integrating closed-cycle refrigeration systems into spectroscopy applications, considerations such as optical access, vacuum shrouds, and mechanical stability must be addressed. In most cases, the inherent microphonics typical of closed-cycle refrigerations systems has been found to be compatible with sample cooling requirements.

With regards to the cooling of infrared detectors, which are temperature and noise dependent, it is important to be aware of microphonics and their effect on detector performance. Microphonics, in this particular instance, may be characterized as the linear displacement inherent to the dynamic force of the displacer piston within a cryogenic refrigeration system. In addition, the minute thermal cycling ($\sim\pm0.10$ K at 10 K) inherent to the thermal cycling of the refrigerator affects microphonics. Microphonics directly affects the stability, detectivity and performance of a closed-cycle refrigerated detector or device. It is imperative that these issues be successfully addressed.

A technique was established by R. G. Hansen & Associates to substantially reduce, to an acceptable level, both linear displacement and thermal cycling through a microphonic dampening network. Resulting systems have exhibited temperature stabilities of better than ±0.01 K with limited linear displacement or mechanical motions demonstrating that microphonics do no affect performance objectives. The technique used for successful reduction of microphonics is a thermal mass, usually consisting of lead, coupled with microphonic dampening network of indium spacers, separated with aluminized Mylar and attached with an insulating and/or mechanically integrating fastener. This technique is easily accommodated on the end of a closed-cycle refrigeration system in an envelope of less than 2.54 cm (1") diameter and approximately 3.8 cm (1.5") in length.

With the above considerations addressed, we believe that much broader utility of mechanically cooled GeZn detector materials can be enjoyed by both the research and process control community. For applications requiring broad range detectivity and state-of-the-art performance, GeZn should be reviewed as a viable alternative.

The following sketch, Figure 4, demonstrates a typical GeZn detector cooled within a conventional CTI[4] closed-cycle refrigeration system. A wide variety of cold-shield configurations, apertures, window materials, et cetera, can be integrated into the suggested system.

The researcher should be aware that the system is orientation independent; i.e., can operate in the horizontal or vertical position. The system can be operated for extremely long periods of time without maintenance or support. Typically, the mean time before maintenance on a closed-cycle refrigeration system assembly is 12,000 operating hours.

In designing a support vacuum shroud and/or housing assembly for the cryocooler detector interface, it is imperative to address the outgassing of the internal surfaces. The author has found that electron-beam welded stainless steel, electro-polish processed internally, is the preferred material for dewar manufacturing. This results in low outgassing characteristics, good high-vacuum practice and the ability to reduce the cryocollection of water and hydrocarbon vapors on the detector surface. Typically, hygroscopic materials such as aluminum can result in the cryocollection of contaminants on cooled detector surfaces.

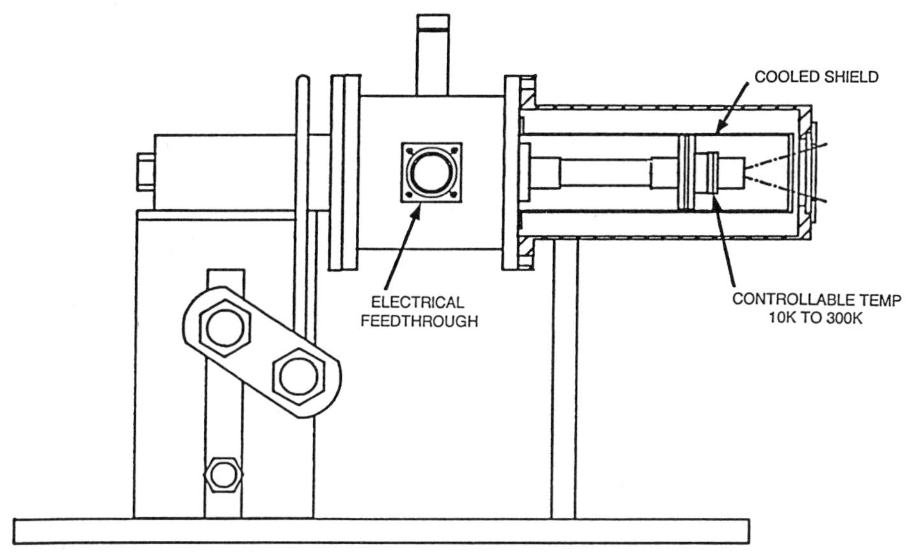

Interfaces for:
GeCu	InSb	Platinum Silicide
GeZn	SiAs	CCD's
HgCdTe	PbSe	Arrays
GeHg		PbSnTe

Figure 4. Typical GeZn Detector Mounted on a
CTI Refrigerator

CONCLUSION

In summary, we believe that GeZn detectors, cooled by commercially available closed-cycle refrigeration systems, can successfully, conveniently and performance-wise be considered as the option of choice for state-of-the-art spectral analyses. Misconceptions regarding the operating temperature of GeZn detectors and their performance at elevated temperatures, such as 9 to 11 K, have been demonstrated to be unfounded. Current prices for commercially available closed-cycle refrigeration systems operating in the 9-11 K range make this option not only technically, but also economically feasible.

REFERENCES

1. E. H. Puttey, "Solid States Devices for Infrared Detection", <u>J. Scient. Instr.</u>, 43:857-868 (1966).

2. Santa Barbara Research Center, "Infrared Components Brochure," Ed. 16. Cover chart, Figure 16 (1982).

3. The author gratefully acknowledges the pioneering development efforts of Dr. Don Imhoff, Senior Scientist with Ethyl Corp.'s Process Development Center, Baton Rouge, LA for his contribution to this paper and this technology.

4. CTI is a trademark for a series of closed-cycle refrigerators manufactured by Cryogenic Technology, Inc. of Waltham, MA.

MAGNETIC REFRIGERATION: A LARGE COOLING POWER

CRYOGENIC REFRIGERATION TECHNOLOGY

Stephen F. Kral and John A. Barclay

Astronautics Corporation of America
Madison, Wisconsin 53716-1387

INTRODUCTION

While magnetic refrigeration has long been used for refrigeration below 1 K in research laboratories, use of the technology for refrigeration above 1 K is relatively recent. In 1966, van Guens[1] provided the first discussion of magnetic refrigeration above 1 K. Beginning in 1976, with publications by Brown[2,3] on the use of ferromagnetic materials near their Curie points as refrigerants, magnetic refrigeration from 1 K to near 300 K has been the focus of activity in several laboratories throughout the world. Recent reviews provide an excellent discussion of progress from these efforts.[4,5,6]

A Survey of Refrigeration Applications

In the lowest temperature range of interest, from approximately 1.5 K to about 15 K, large growth in refrigeration applications has occurred in the past decade. Probably of greatest significance has been the commercial development of medical applications of magnetic resonance imaging; these systems demand about 0.25 W cooling power at the temperature of liquid helium plus shield cooling at 20 K and 70 K. This development has legitimized low temperature superconductivity as a commercial technology and, by association, has begun consideration for commercial use of other low temperature systems requiring cryogenic refrigeration.

Other, non-commercial, new applications requiring refrigeration in this temperature range include particle accelerators such as the Superconducting Super Collider (SSC) and Superconducting Magnetic Energy Storage (SMES). These applications demand very large cooling powers, of the order of several kilowatts at liquid helium temperatures.

The cooling of Long Wavelength Infrared Sensors (LWIR) for scientific and military applications has also influenced considerably the development of low temperature cryogenic refrigeration. These LWIR systems make modest cooling power demands, e.g., a fraction of a Watt near or below 10 K plus higher stage cooling. They make extremely challenging demands on reliability and lifetime, especially for space-based applications.

Applications of Cryogenic Technology, Vol. 10
Edited by J.P. Kelley, Plenum Press, New York, 1991

At somewhat higher temperatures, from approximately 15 K to approximately 100 K, growth in the demand for refrigeration comes from several sources. Cryopumps are probably the single biggest commercial application of cryocoolers, while medium wave-length infrared system coolers make up the bulk of the military market. Hydrogen liquefiers, air separators, and LNG plants all require large cooling power refrigeration systems. The discovery of high temperature superconductivity, with the need for cooling in the 40 K to above 100 K temperature range, is expected to significantly increase the demand for refrigeration above about 40 K.

Refrigeration needs from 100 K to room temperature are more commonplace. Included here are home refrigerators and freezers, commercial refrigeration, air conditioners and heat pumps. Recent government requirements for the reduction and eventual elimination of the use of chloroflourocarbons, coupled with requirements for increased efficiency, produces an unprecedented need for innovative room-temperature refrigeration technologies.

Many refrigeration technologies are available for cooling in all of these temperature ranges. Aerospace applications have led to the development of many refrigerators, which, for reasons of cost and very special application constraints, are not of commercial interest. Excluding use for aerospace applications, the list of available technologies becomes interestingly brief. For cooling to a few fixed temperatures, the normal boiling point, the melting point and possibly the sublimation point of common materials are used. Included here are water, ice, solid carbon dioxide, liquid nitrogen, liquid hydrogen and liquid helium. These liquids and solids are limited in their temperature applications, often require special handling, and can be expensive and difficult to obtain, particularly in small lots.

Figure 1 shows the cooling power and cold temperature of common, commercially available, closed-cycle refrigerators. For applications above about 40 K, the single stage Gifford-McMahon (GM), Stirling, or Brayton cycle refrigerators provide cooling over the widest cooling power range. For lower cooling powers, systems using free expansion through a Joule-Thompson (JT) valve are used. At higher cooling power, systems using reciprocating expansion engines and a JT valve at a final stage are commonly used.

Below about 40 K, two-stage GM, Stirling and Brayton cycle refrigerators are used, but these are limited to temperatures of about 10 K because of the inadequate regenerator effectiveness below that temperature. Recent developments using rare-earth metals for the regenerator matrix suggest that GM or Stirling refrigerators might be useful to temperatures below 4 K.[7,8,9] Commercial refrigerators combining GM refrigerators with a final JT stage and operating to below 4 K with cooling powers of one Watt or more are available. For higher cooling powers, refrigerators with reciprocating expansion engines, and, at the highest cooling powers, turbine expanders, are available. The borders of regions in Figure 1 are approximations and with new developments in each type of refrigeration, overlap is increasing.

This brief survey identifies the various refrigeration technologies available for commercial applications. Magnetic refrigeration is only now becoming commercially available and will be an increasingly used alternative mode of refrigeration as the technology matures. The purpose of this paper is to begin to define the region of cooling power-load temperature space where magnetic refrigerator systems best apply.

<u>Present State of Magnetic Refrigeration</u>

Magnetic refrigeration is usually thought of as a technology that can provide only small cooling powers at very low temperatures. The work presented below should dispel that perception.

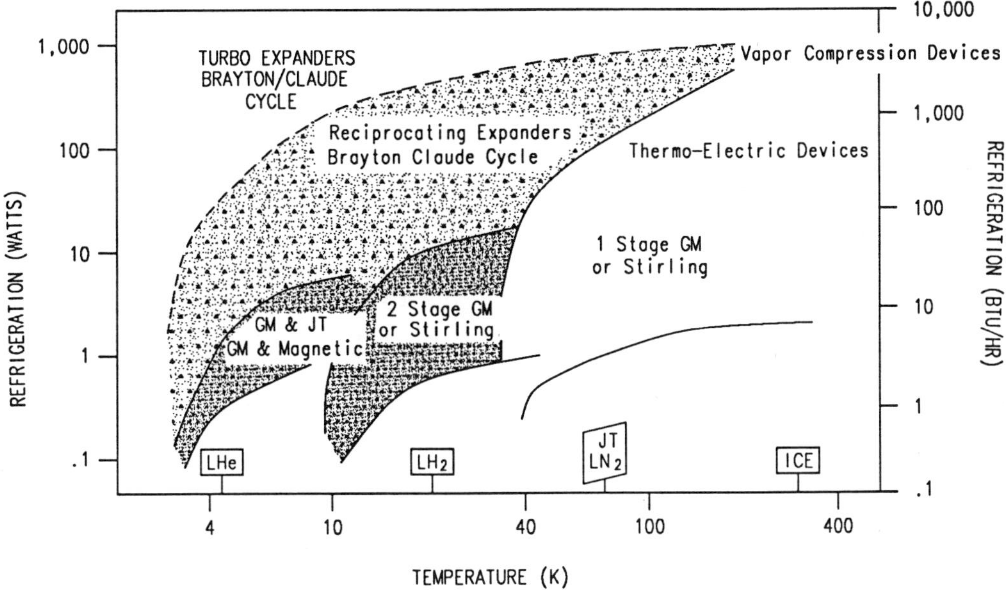

Figure 1. Cooling power - load temperature space for commercial refrigeration systems.

Many large-scale cooling applications have been studied with interesting results. Table I provides performance data and design predictions for a wide range of magnetic refrigerators. This table is adapted from Lacaze[6]. Information made available since 1986 has been added. Much of the more recent data are results from design studies we have pursued over the past four years. Of interest are the larger cooling powers of the recent designs and the higher operating temperature ranges of some of them. Included are 10 W and 100 W magnetic refrigerators operating between 1.8 K and about 5 K, a proof-of-principle design for an Active Magnetic Regenerator refrigerator (AMR), a 50 kW magnetic slushifier operating between 12 K and 21 K, an approximately 10 kW magnetic liquefier operating between 15 K and 85 K, and a room temperature magnetic refrigerator with about 50 kW cooling power operating between 265 K and 320 K.

These higher cooling power designs have been analyzed for all the important qualities promised by magnetic refrigeration: efficiency, size, potential reliability and long lifetime.

INTRODUCTION TO MAGNETIC REFRIGERATION

The Magnetocaloric Effect

The operating principles of magnetic refrigerators are discussed in a review by Barclay[4] and references therein. The entropy change, dS, experienced by a constant volume magnetic material undergoing a change in temperature, T, or external magnetic field, H, is:

$$dS = (C_H/T)\, dT + \mu_o V\, (\delta M/\delta T)_H\, dH \tag{1}$$

where C_H is the field dependent heat capacity, μ_o is the permeability of free space, V is the volume, and M is the magnetization.

Table I

Compilation of Magnetic Refrigerator Data

	Cycle Type	Material Used	Magnetic Field (T)	Tc-Th (K)	Cooling Power (W)	Frequency (Hz)	Ref.
NASA	Carnot Static	Gd_2SO_4	0-3	0.2-1.5	~ 0	~ 0	10
Hitachi	Carnot Static	GGG	0-3	1..7-4.2 1.8-4.2	0 0.5	0.2 0.2	11
Hitachi	Carnot Rotating Magnets	GGG	0-3	1.9-4.2 2.1-4.2	0 1.5	0.067 0.067	12
Toshiba	Carnot Static	GGG	0-5	4.2-16	6	0.005	13
Los Alamos	Carnot Rotating	Gd_2SO_4	0-2.1	2.1-4.2 2.75-4.2	0 0.53	- -	14
Los Alamos	Carnot Rotating	GGG	0-4	(4.2-15)	-	0.2	15
CEA Grenoble	Carnot Recipr.	GGG	1-4	1.38-4.2 1.8-4.2 2.1-4.2	0 1.35 2.35	0.5 0.8 0.6	16
JPL	Carnot Recipr.	GGG	0-7	(4.2-20)	-	-	17
Los Alamos	Ericsson Rotating	Gd	0-3.5	289-298 290-297	0 400	0.05 0.2	18
Los Alamos	Ericsson Recipr.	Gd_2SO_4	0-2.5	3.6-4.2	0.052	0.017	19
NASA	Ericsson Recipr.	Gd	0-7	254-334	6		20
Los Alamos	Active Regen.	Gd	0.5-4.5	287-301 292-296	0 3	0.003 0.006	18 21
MIT	Active Regen.	GGG	0-4	(4.2-10)	-	-	
CEA Grenoble	Active Regen.	GGG	0-3	(4.2-15)	-	-	22
Hitachi	Carnot Rotating	GGG	0-3	1.8-4.2	1.81	.33	23
Toshiba	Carnot Static	GGG	0-5	4.2-15 1.8	.4 0	.04	24 25
Kurchatov Institute	Carnot Static	GGG	0-2.9	1.8-4.2	.5	.25	26
DTRC	AMR Static	Gd	0-7	250-290	0	.011	27
Tokyo Inst. Tech.	Ericsson Recipr.	$DyAl_{22}$	0-5	48-59	0	.0026	28

Table I (cont.)

	Cycle Type	Material Used	Magnetic Field (T)	Tc-Th (K)	Cooling Power (W)	Frequency (Hz)	Ref.
ACA	Carnot Rotating	GGG	0-6	4.2-15	1.0	.33	29
ACA	Carnot Rotating	GGG	0-4.5 0-2.4	1.8-4.2 3.0-4.5	0.5 0	.17 .02	30
ACA	Carnot Rotating	GGG	0-5	1.8-4	10	.08	31
ACA	Carnot Rotating	GGG	0-7	13-20	1000.	.25	32
ACA	AMR Recio. Mag.	ErGdAl	0	7-20	1.0	.25	33
ACA	AMR Recip.	GGG	2.5-5.5	1.8-4.7	100.	1.0	34
ACA	AMR Rotating	ErGdAl$_{22}$	0-8	20-77 in 3 stages	8000 @20 +smaller loads -higher temps.	1.0	35 36
ACA	AMR Rotating	Gd	0-7.5	255-305	50,000.	4.2	36
NASA Ames	Carnot Static	GGG	0-9	2-10	.10	.0014	37
JPL	Carnot	DAG	0-6	4-20	.5	.1	38

The temperature change experienced by a thermally isolated magnetic material when the material undergoes a change in magnetization induced by a change in external magnetic field is called the adiabatic temperature change. Under adiabatic conditions $dS = 0$, and equation 1 yields

$$dT = -(T/C_H)\mu_o V \, (\delta M/\delta T)_H \, dH. \tag{2}$$

Another quantity important to magnetic refrigeration following from Equation 1 is the material's isothermal entropy change, the entropy change experienced at constant temperature in the presence of a changing field, given by

$$dS = \mu_o V (\delta M/\delta T)_H \, dH. \tag{3}$$

Magnetic Refrigeration Cycles

All magnetic refrigerators cool through the use of the entropy changes associated with magnetic field changes. Table I indicates that the most common design cycle for magnetic refrigerators is the Carnot cycle. As larger operating temperature ranges and very large cooling powers are sought, regenerative cycles such as the Ericsson, Brayton or Active Magnetic Regenerator cycles are being used. These cycles are illustrated in Figures 2, 3, 4 and 5.

The Carnot cycle, shown in Figure 2, is the simplest refrigeration cycle, but its use is limited by the adiabatic temperature changes that can be achieved by magnetic materials. The lattice heat capacity of the magnetic refrigerants used in these refrigerators increases rapidly with temperature. The adiabatic temperature change is limited by the increasing heat capacity as temperature rises, restricting Carnot cycle refrigerators to use at temperatures below about 20 K. At temperatures above 20 K, and, for many applications below 20 K, regenerative cycles are used.

Regenerative cycles allow much larger temperature spans and potentially much higher cooling powers than Carnot cycles. Figure 3 presents a regenerative magnetic Ericsson cycle. In this cycle, for high efficiency, the heat transferred from the regenerator to the magnetic material as it warms on path AB must be exactly balanced by the heat transferred to the regenerator as the magnetic material cools along path CD. Because the heat transferred during regeneration is simply the area between the entropy axis and the path followed during regeneration, CD or AB, for an ideal Ericsson system, the magnetic material constant field curves in the T-S plane must be parallel straight lines.

In both the magnetic Carnot and magnetic Ericsson cycles, heat transfer occurs isothermally while the material is being magnetized or demagnetized. For an ideal cycle, without irreversibilities that distort the cycle, the cooling power is given by

$$Q_{ideal} = T_c \, \Delta s_T \, f \, m \tag{4}$$

where T_c is the cold source temperature; Δs_T is the isothermal entropy change per unit mass for the demagnetization occurring at the cold sink temperature; f is the cycle frequency; and m is the mass of the magnetic material.

The magnetic Brayton cycle is another possible regenerative cycle. In the ideal magnetic Brayton cycle, heat is absorbed at a constant magnetic field. The cooling power can be calculated easily for the ideal case, shown in Figure 4, because the heat absorbed during each cycle is the area of the trapezoid AA'BB'. If ΔT_s is the adiabatic temperature change experienced by the material during the demagnetization, and Δs_T is the isothermal entropy change (per unit mass) that would occur if the material underwent an isothermal demagnetization between points E and B, the cooling power is seen to be

$$Q_{ideal} = T_c \, \Delta s_T \, f \, m - (1/2) \, \Delta s_T \, \Delta T_s \, f \, m \tag{5}$$

The cooling power for an ideal Brayton cycle is seen to be reduced from that of an ideal Carnot or Ericsson cycle operating under similar conditions.

Another important cycle in magnetic refrigeration is the Active Magnetic Regenerator (AMR) refrigerator cycle.[39] For this cycle, in the T-S plane (Figure 5), the magnetic material ideal isofield lines are lines that are not parallel, but rather straight lines that intersect at T = 0 K.[40] This cycle also requires a second medium to effect heat transfer. In an ideal AMR, each element of the magnetic material is carried through a small ideal Brayton cycle which spans only part of the total temperature range spanned by the refrigerator. Over a cycle, upon magnetization, all of the elements are raised to their highest temperature. The second medium, typically fluid helium, is warmed from the cold source temperature to slightly above the hot sink temperature by heat absorption from the magnetic material.

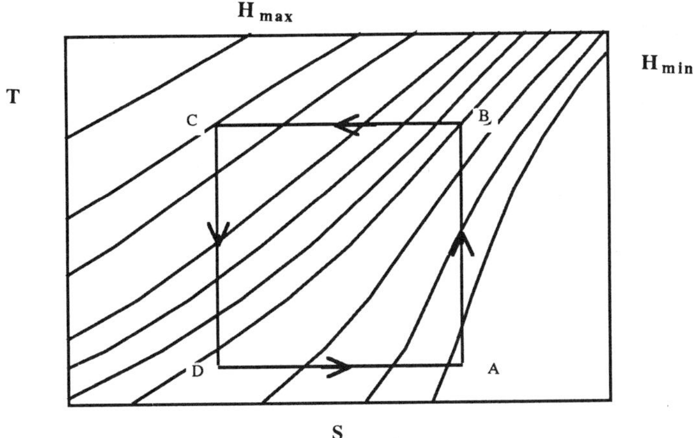

Figure 2. Magnetic Carnot Cycle with magnetic material isofield lines.

The second medium then rejects heat to the hot sink. All the elements are then cooled to their lowest temperature by demagnetization. The second medium is cooled to slightly below the cold source temperature as it rejects heat to the magnetic material elements. Heat is then absorbed from the cold source by the second medium and the cycle begins again. In this way, the heat rejected by the second medium as it cools is absorbed by the elemental Brayton cycle refrigerators when they are at the low temperature part of their cycles. Heat is rejected to the second medium when the elemental Brayton cycle refrigerators are at the high temperature part of their cycles. In a superficial way, this process resembles internal regeneration; the process is more precisely understood as the action of an ensemble of elemental Brayton cycle refrigerators acting in parallel to warm and cool a second medium to effect heat transfer from the cold source to the hot sink. The cooling power of an AMR depends directly on the capacity of the second medium to absorb heat from the cold source. If m is the mass of the second

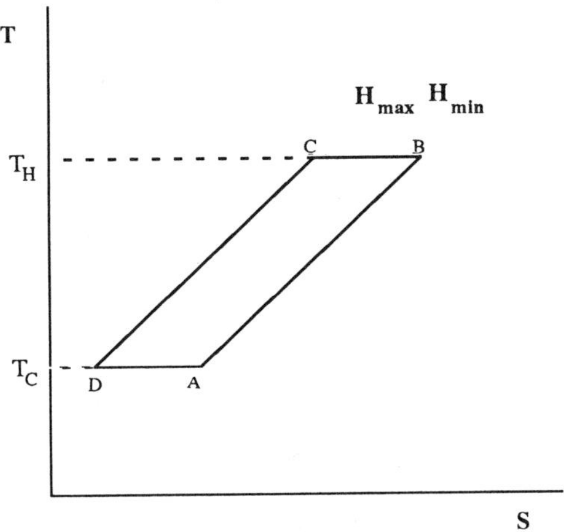

Figure 3. Magnetic Ericsson Cycle: isothermal heat transfer and isofield regeneration.

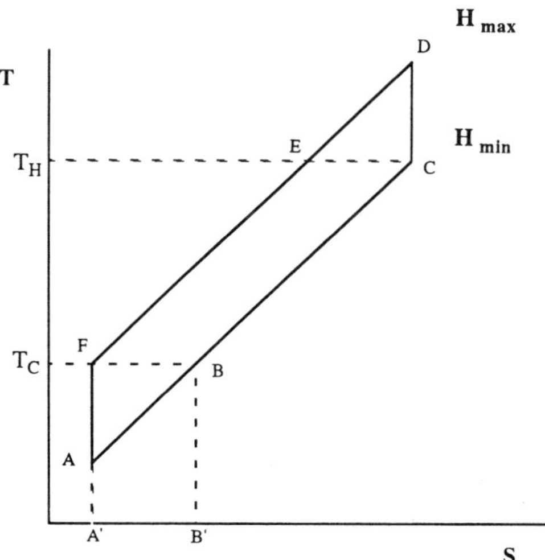

Figure 4. Magnetic Brayton Cycle: isofield heat transfer with isofield regeneration.

medium, c_p is its specific heat, f is the cycle frequency and ΔT is the temperature difference between the cold source and the second medium, the cooling power is given by

$$Q = mc_p f\, \Delta T. \tag{6}$$

We note that because the second medium is cooled by Brayton cycle refrigerators, the AMR is intrinsically limited to efficiencies less than those of a Carnot cycle.

The description of the AMR refrigeration cycle given above is correct under conditions where the thermal mass of the magnetic material is much greater than that of the second medium. When the thermal mass of the second medium is greater than that of the magnetic material, the ideal isofield lines in the T-S plane are parallel, to ensure balanced regeneration. This case is discussed by DeGregoria.[34]

Magnetic Refrigerator Scaling

The cooling power of a magnetic refrigerator depends on the isothermal entropy of the magnetic working material. The two curves in Figure 6 permit a comparison of the molar isothermal entropy change of helium gas and magnetic materials. The helium gas curve is drawn assuming the gas undergoes a 10:1 compression. The curve for magnetic materials is drawn through a series of discrete points. Each point in the series represents the molar isothermal entropy change experienced by a magnetic material near its Curie temperature when it experiences a magnetic field change of 8.5 T. Materials with Curie temperatures ranging from 4 K to 300 K are included.

Figure 6 shows that the molar entropy change of magnetic materials is considerably less than that of helium at all temperatures except at the lowest temperatures. Because the isothermal entropy changes are expressed in molar form, they conceal a very important difference between magnetic material refrigerants and helium gas refrigerant. This difference can be appreciated by comparing the molar volumes of magnetic materials and helium. A mole of helium at STP occupies 22.4 liters while,

34

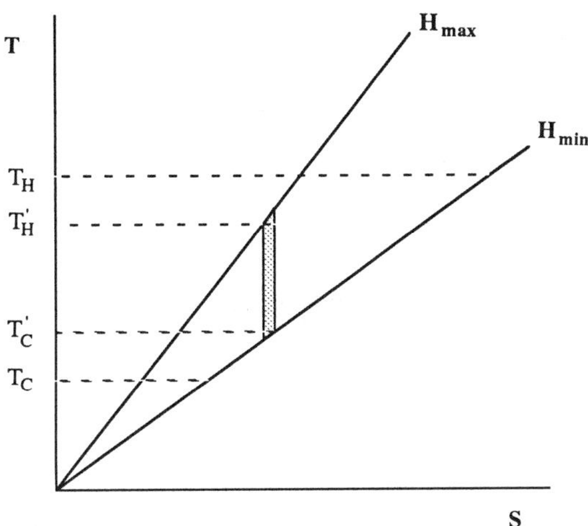

Figure 5. AMR Cycle: Elemental Brayton cycle operation between T_H' and T_C' is shown.

typically, a mole of magnetic material occupies about 0.3 liters. Thus, the volume of a mole of magnetic material is about one-thousandth the volume of a mole of helium. When the difference in the isothermal entropy change is taken into account, a refrigerator operating by cycling 300 K helium gas between, say, 0.1 MPa and 1 MPa will require a refrigerant volume about 400 times larger than a comparable magnetic refrigerator! Magnetic refrigerator designs of extremely large cooling power are remarkably compact. For recent Astronautics' designs, the material volume per Watt of cooling power is between 30 cm^3 and 100 cm^3, for field changes of about 5 T and operating frequencies of about 0.25 Hz.

While the very high volumetric isothermal entropy change of the magnetic materials used as refrigerants leads to compact designs, it also has some unexpected consequences when refrigerators of small cooling power are considered. To reduce the cooling power of an optimized refrigerator design, the volume of refrigerator material would be reduced. The design of a small cooling power magnetic refrigerator would seem to require only the reduction in size of all other refrigerator components to match the reduction in magnetic material volume.

However, not all components can be efficiently reduced in scale. Consider scaling the superconducting magnets. Figure 7 shows the ratio of the bore volume of a superconducting solenoid to the total magnet volume as a function of the bore radius. The bore radius is expressed as a fraction of the solenoid length. For small bore radii, the bore volume of the magnet (where a magnetic material would reside to experience maximum field) decreases much more rapidly than the total magnet volume. Since the magnet volume outside the bore consists primarily of the superconductor winding volume, another way of expressing this result is that the cost of the bore volume of the magnet increases rapidly, as the bore volume decreases. This property of superconducting solenoids precludes the cost-effective construction of small cooling power magnetic refrigerators. It is this property of superconducting solenoids that leads to the conclusion that magnetic refrigeration is most appropriately a high cooling power technology. While the result has been argued only for solenoids, the argument is correct for other superconducting magnet configurations.

There are other problems resulting from the need for physically large magnets in small cooling power refrigerators. To support the attractive forces between the magnets, large structures are required. The support structures contribute to the internal heat leak which reduces both the cooling

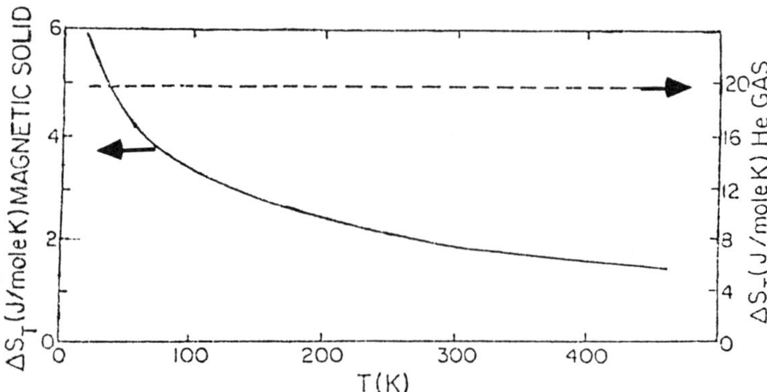

Figure 6. The molar isothermal entropy change of a series of magnetic solids or helium gas as a function of temperature.

power available to users and the system efficiency. As smaller cooling power designs are attempted, this internal leak will grow disproportionately, because of the still necessary large magnet structure.

The other components of a magnetic refrigerator, the heat exchangers, pumps, instrumentation, dewars, etc., are less significantly affected by the system compactness. These components either scale linearly with material volume or are uninterestingly independent of material volume.

Thus, we conclude that there is a lower limit for the cooling power of magnetic refrigerators, determined primarily by the properties of superconducting magnets. There is also a greatest size. Two effects lead to this conclusion. First, large cooling power magnetic refrigerators will have large magnets. Ultimately, the capacity of the magnetic-force supporting members of the refrigerator will limit designs that achieve high cooling power by means of very large magnetizing fields.

A second reason for an upper limit on cooling power results from the need to transfer heat from the solid magnetic refrigerant. The specific area, a_v, of a volume of material is defined as the surface area of the material divided by the material volume. The heat flux into or out of the solid magnetic material due to a temperature difference ΔT is given by

$$Q = ha_v V \, \Delta T \tag{7}$$

where h is the heat transfer coefficient, V is the total material volume, and ΔT is the temperature difference for heat transfer. Typical heat transfer coefficients for heat transfer mechanisms used in existing magnetic refrigerators are given in Table II. To accomplish efficient heat transfer, i.e. heat transfer with small temperature differences, very large specific areas for the small volumes of magnetic material used are required. In the most demanding case, this means configuring the material into particle beds. While particle beds yield large specific area, they can introduce large pressure drops in the fluid flows used to transfer heat from the magnetic solids.

Also, increased cooling powers are achieved by increasing the refrigerator's operating frequency. At larger operating frequencies, the mass flow rate of the heat transfer fluid must be increased so that the heat transferred between the magnetic material and the heat transfer fluid for each

36

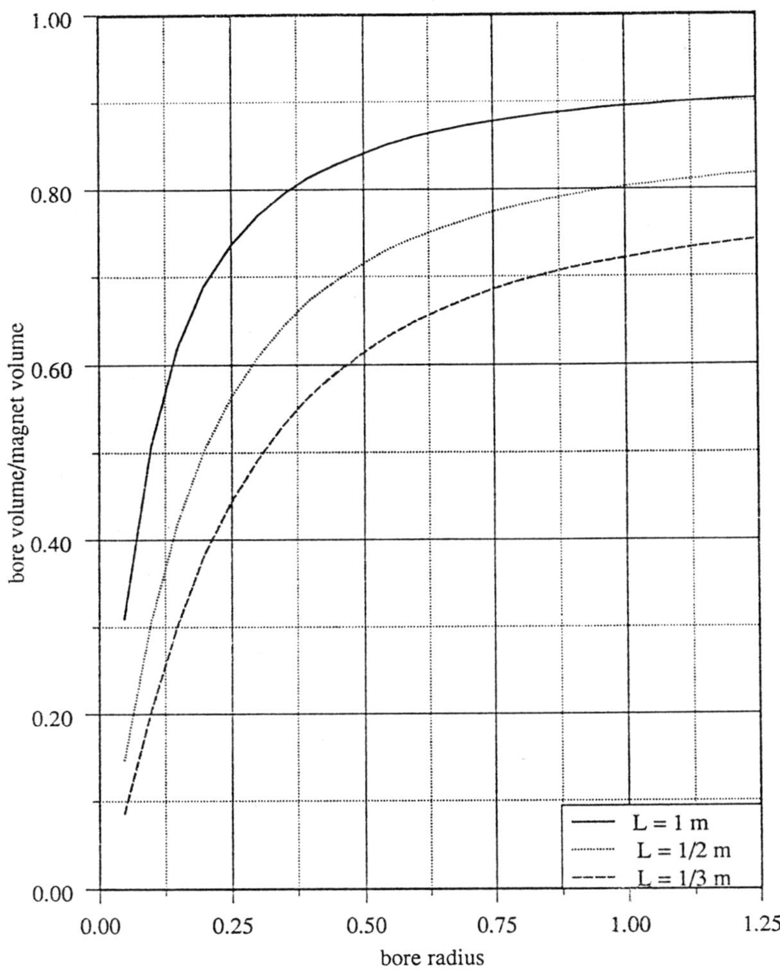

Figure 7. Ratio of superconducting solenoid bore volume to total volume as a function of solenoid bore radius. Bore radius is expressed as a fraction of solenoid length.

cycle is maintained. At large fluid mass flow rates, heat transfer rates are limited and at sufficiently high operating frequency, the cooling power of the refrigerator no longer grows with increased frequency. The peak in the cooling power of the AMR refrigerator, described in Figure 8, results from this effect.

The existence of a maximum cooling power for particular designs is not a severe restriction. When the limit is reached, modular designs that use two or more of these refrigerators achieve even higher cooling powers. Because these refrigerators are very compact, the physical size of these modular refrigerators will still be modest compared to common, extremely large, gas-cycle refrigerators.

Table II

Approximate Values For Various Heat Transfer Coefficient
Mechanisms For Magnetic Refrigeration

Heat Transfer Mechanisn	$h(W/cm^2 \cdot K)$
Conduction across 0.01 cm He filled gap at 4.2 K	0.02
Laminar forced convection	0.2
Nucleate boiling of He	2.0
Kapitza resistance to He II at 1.8 K	0.1 - 1.0

The lower limit imposed on the magnetic refrigerator cooling power described above is also not an absolute restriction. It is possible to build magnetic refrigerators using permanent magnets, thereby achieving cost-effective small cooling power designs. A permanent magnet magnetic refrigerator requires magnetic materials capable of large isothermal entropy changes under the influence of the relatively small magnetic field changes available using permanent magnets. For special purpose applications, refrigerators of this type can be built.

CONCLUSION

Figure 9 is a plot in cooling power-load temperature space of the magnetic refrigerator data presented in Table I. As can be seen, the cooling powers of the magnetic refrigerators now being designed, fabricated and tested are extremely large, some with cooling powers in the tens of kilowatt range. These designs are all extremely compact. In one design unit a cooling power of approximately 10 kW at 20 K has the refrigerant confined to a annular cylinder less than 2 m in diameter and less than 0.5 m in height![36] Modeling predictions for this design also indicate a higher efficiency than competing gas systems. This result, while remarkable, is not unique. All the large systems described in Table I have this property.

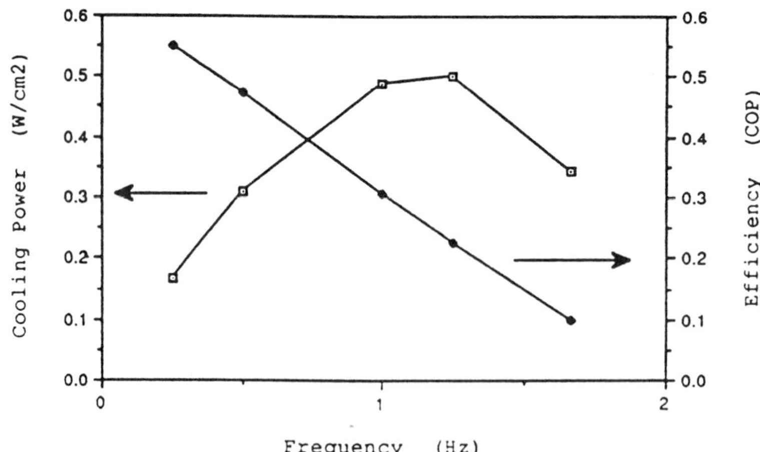

Figure 8. Cooling power per unit bed cross section and efficiency as a function of frequency for the AMR.

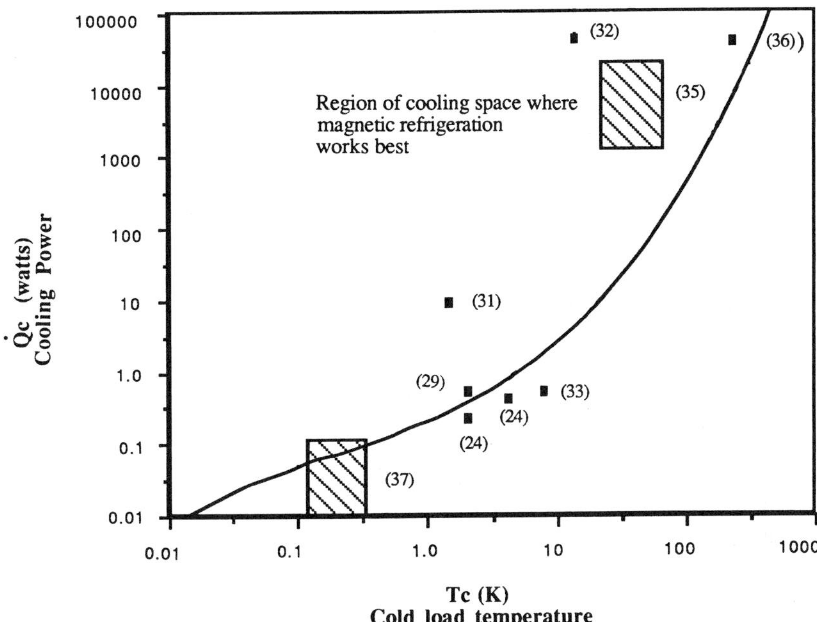

Figure 9. Cooling power - load temperature space for magnetic refrigerators. The numbers identify text references.

Magnetic refrigeration is an emerging technology in a marketplace long dominated by gas-cycle technology. As the properties of magnetic refrigerators become better known through actual validated models and performance, other incorrect preconceptions, similar to the small cooling power assumption discussed here, will be identified and clarified. Recent successes in both conductive and convective designs and new, ambitious designs continue to reveal the interesting properties of this technology.

REFERENCES

1. J.R. van Geuns, A study of a new magnetic refrigeration cycle, Phillips Research Report Supplements 6 (1966).
2. G.V. Brown, Magnetic heat pumping near room temperature, J. Appl. Phys., 47:3673 (1976).
3. G.V. Brown, Magnetic Stirling cycles - a new applicaiton for magnetic materials, IEEE Trans. Magnetics Mag., 13:1146 (1977).
4. J.A. Barclay, Magnetic refrigeration: a review of a developing technology, in: "Advances in Cryogenic Engineering," R.W. Fast, ed., Plenum Press, New York (1987).
5. J.A. Barclay, A review of magnetic heat pump technology, Proc. 25th IECEC, 2:222 (1990).
6. A.F. Lacaze, Magnetic refrigeration - an overview, Proc. of Commission A1/2 I.I.R. - I.I.F., Prague, 99 (1986).
7. Masashi Nagao, Takashi Inaguchi, Hideto Yashimura, Tadatoshi Yamada and Masatmai Iwamoto, Helium liquefaction by a Gifford-McMahon cycle cryocooler, in: "Advances in Cryogenic Engineering", R.W. Fast, ed., Plenum Press, New York (1990).

8. T. Kuriyama, R. Hakamada, H. Nakagome, Y. Toki, M. Sahashi, R. Li, O. Yosaida, K. Matsumoto, and T. Hashimoto, High efficient two-stage GM refrigerator with magnetic material in the liquid helium temperature region, in: "Advances in Cryogenic Engineering", R.W. Fast, ed., Plenum Press, New York (1990).

9. Takashi Inaguchi, Masasai Nagao and Hideto Yaosaimura, Effects of regenerators on Gifford-McMahon cycle cryocooler operating at about 4 K, in: "Advances in Cryogenic Engineering", R.W. Fast, ed., Plenum Press, New York (1990).

10. P. Kittel, Temperature stabilized adiabatic demagnetization for space applications, Cryogenics, 20:599 (1980).

11. Y. Hakuraku, H. Ogata, A static magnetic refrigerator for superfluid helium with new heat switches and a superconducting pulse coil, Jap. Journal Appl. Phys., 24:1538 (1985).

12. T. Hashimoto, Recent investigation on refrigerants for magnetic refrigerators, in: "Advances in Cryogenic Engineering", R.W. Fast, ed., Plenum Press, New York (1986).

13. T. Numazawa, T. Hashimoto, H. Nakagome, Improvement for liquefaction efficiency of the heat pipe type magnetic refrigerator, in: "Advances in Cryogenic Engineering", R.W. Fast, ed., Plenum Press, New York (1986).

14. W.P. Pratt, Jr., S.S. Rosenblum, W.A. Steyert and J.A. Barclay, A continuous demagnetization refrigerator operating near 2 K and a study of magnetic refrigerants, Cryogenics, 17:689 (1977).

15. J.A. Barclay, W.F. Stewart, W.C. Overton, R.J. Chandler and O.D. Harkleroad, Experimental results on a low-temperature magnetic refrigerator, in: "Advances in Cryogenic Engineering", R.W. Fast, ed., Plenum Press, New York (1986).

16. A.F. Lacaze, R. Beranger, G. Bon Mardion, G. Claudet, A.A. Lacaze, Double acting reciprocating magnetic refrigerator: recent improvements, in: "Advances in Cryogenic Engineering", R.W. Fast, ed., Plenum Press, New York (1984).

17. D.L. Johnson, Reciprocating magnetic refrigerator, Proc. Third NBS Cryocooler Conf., (1984).

18. J.A. Barclay, W.A. Steyert, Magnetic refrigerator development, Electric Power Research Institute, EPRI-EL 1757 (1987).

19. J.A. Barclay, O. Moze, L. Paterson, A reciprocating magnetic refrigerator for 2-4 K operation: initial results, J. Appl. Phy., 50:5870 (1979).

20. G.V. Brown, Magnetic heat pumping near room temperature, J. Appl. Phy., 47:3673 (1976).

21. C.P. Taussig, G.R. Gallagher, J.L. Smith Jr., Y. Iwasa, Magnetically active regeneration, Proc. Fourth Int. Cryocooler Conf., Easton 79, (1986).

22. A.F. Lacaze, Contribution a l'etude de la refrigeration magnetique aux temperatures de l'helium liquide, These d'Etat, INPG, Grenoble (1985).

23. Y. Hakuraku and H. Ogata, A rotary magnetic refrigerator for superfluid helium production, J. Appl. Phys., 60:3266 (1986).

24. Takenori Numazawa, Static magnetic refrigerator-cycle operation, Proc. 3rd Jap.-Sino Jt. Sem. Cry., 2nd Circular Cp24 (1984).

25. Masaaiko Takahashi, Static magnetic refrigerator II - thermal switch characteristics, Proc. 3rd Jap.-Siono. Jt. Sem. Cry., 2nd Circular Cp25 (1989).

26. V.E. Keilin, I.A. Kovalev, V.D. Kovalenko, A.V. Kortikov, I.I Mikhailov, N.V. Filin and V.A. Shaposhnikov, A static magnetic refrigerator, Proc. ICEC-12, Southhampton (1988).

27. G.F. Green, G. Green, W. Patton and J. Stevens, The magnetocaloric effect of some rare earth metals, in: "Advances in Cryogenic Engineering", R.W. Fast, ed., Plenum Press, New York (1988).

28. Koichi Matsumoto, Takatoshi Itu, and Takasu Hashimoto, An Ericsson magnetic refrigerator for low temperature, in: "Advances in Cryogenic Engineering", R.W. Fast, ed., Plenum Press, New York (1988).

29. F.C. Prenger, D.D. Hill, J. Trueblood, T. Servais, J. Laatsch, and J.A. Barclay, Preformance tests of a conductive magnetic refrigerator using a 4.5 K heat sink, in: "Advances in Cryogenic Engineering", R.W. Fast, ed., Plenum Press, New York (1990).

30. J.A. Barclay, C.K. Campenni, C.R. Cross, J.A. Hertel, D.D. Hill, S.R. Jaeger, S.F. Kral, F.C. Prenger, T.M. Stankey, J.R. Trueblood, and C.B. Zimm, Performance results of a low-temperature magnetic refrigerator, Proc. 5th Int. Cryocooler Conf., Monterey (1988).

31. S.F. Kral, P.J. Claybaker, S.R. Jaeger, and J.A. Barlcay, Final Report: Preliminary design of a high cooling power 1.8 K to 4.7 K magnetic refrigerator, Fermi National Accelerator Laboratory (1988).

32. J.A. Waynert, J.A. Barclay, P.J. Claybaker, R.W. foster, S.R. Jaeger, S. Kral, and C. Zimm, Production of slush hydrogen using magnetic refrigeration, Proc. 7th Intersociety Cry. Sym. (1989).

33. A.J. DeGregoria, P.J. Claybaker, J.R. Trueblood, R.A. Pax, T.M. Stankey, S.F. Kral, J.A. Barclay, Initial test results on an active magnetic regenerator, (to be published).

34. A.J. DeGregoria, J.A. Barclay, P.J. Claybaker, S.R. Jaeger, S.F. Kral, R.A. Pax, J.R. Rowe, and C.B. Zimm, Preliminary design of a 100W 1.8 K to 4.7 K regenerative magnetic refrigerator, in: "Advances in Cryogenic Engineering", R.W. Fast, ed., Plenum Press, New York (1990).

35. J.A. Waynert, T.J. DeGregoria, R.W. Foster, and J.A. Barclay, Magnetic heat pump for hydrogen liquefaction (to be published).

36. J.A. Waynert, et. al., Assessment of the impact of high temperature superconductors on room temperature magnetic heat pumps/refrigerators, Argonne National Laboratory Final Report 81032401 (1988).

37. Ben P.M. Helvensteijn and Ali Kasaani, Conceptual design of a 0.1W magnetic refrigerator for operation between 10 K and 2 K, in: "Advances in Cryogenic Engineering", R.W. Fast, ed., Plenum Press, New York (1990).

38. U.E. Israelsson, D.M. Strayer, H.W. Jackson, and D. Petrac, Magnetic refrigeration using flux compression in superconductors, in: "Advances in Cryogenic Engineering", R.W. Fast, ed., Plenum Press, New York (1990).

39. J.A. Barclay, The theory of an active magnetic regenerative refrigerator, NASA Report NASA-CP-2287 (1983).

40. C.R. Cross, J.A. Barclay, A.J. DeGregoria, S.R. Jaeger, and J.W. Johnson, Optimal temperature-entropy curves for magnetic refrigeration, in: "Advances in Cryogenic Engineering", R.W. Fast, ed., Plenum Press, New York (1988).

41. J.A. Barclay, A comparison of the efficiency of gas and magnetic refrigerators, Proc. 22nd National Heat Transfer Conference, Niagara Falls (1984).

1 KW CAPACITY REFRIGERATION SYSTEM AT 80 K (LINIT-R1)

E. CARATSCH

Sulzer Brothers Ltd, Winterthur, Switzerland

ABSTRACT

A newly developed and tested cryogenic system for the 80 K temperature range is presented. The plant consists of 1 compressor unit and one cold box. The system runs fully automatically, has a very low noise level and can easily be operated by not specially trained personnel. The refrigeration is produced in a nitrogen circuit in the following manner: Nitrogen, compressed by an oil injected screw compressor, enters the cryogenic heat exchangers and then passes through two gas bearing turboexpanders to achieve the required low temperature. The expanded cold nitrogen gas is sent back to the heat exchangers where it completes the refrigeration circuit by fulfilling the heat recovery duties and returning to the nitrogen compressor at the beginning of the cycle. A nitrogen gas sidestream is separated from the cooled high pressure main stream, cooled further, liquefied, undercooled and finally expanded with a throttling valve into a separator containing the cooler of the user's coolant circuit. Here the liquid nitrogen evaporates, fulfilling the plant's cooling duties, and reenters the main circuit gas flow coming from the turboexpanders.

INTRODUCTION

For many years, an increasing demand for refrigeration at the 80 Kelvin temperature level has been noted in several scientific and industrial sectors, not only for large capacities, but also for smaller requirements. The recent developments in semiconductor and superconductor applications at this temperature level in the computer-industry will increase this demand. Such refrigerating requirements are most commonly covered by means of liquid nitrogen supplied by gas distributors. However, in the case of smaller capacities, and

principally when a continuous and constant duty is required,
independent refrigerating units can represent an interesting, or
even better, solution.

Sulzer Brothers, based on its experience of over 30 years as
one of the world's leading manufacturers of cryogenic plants and
machinery, decided, in 1988, to design such units. Two
prototypes, for refrigerating capacities of 5 kW and 1 kW, have
been manufactured and successfully tested. The latter unit,
denominated LINIT-R1, will be presented in detail.

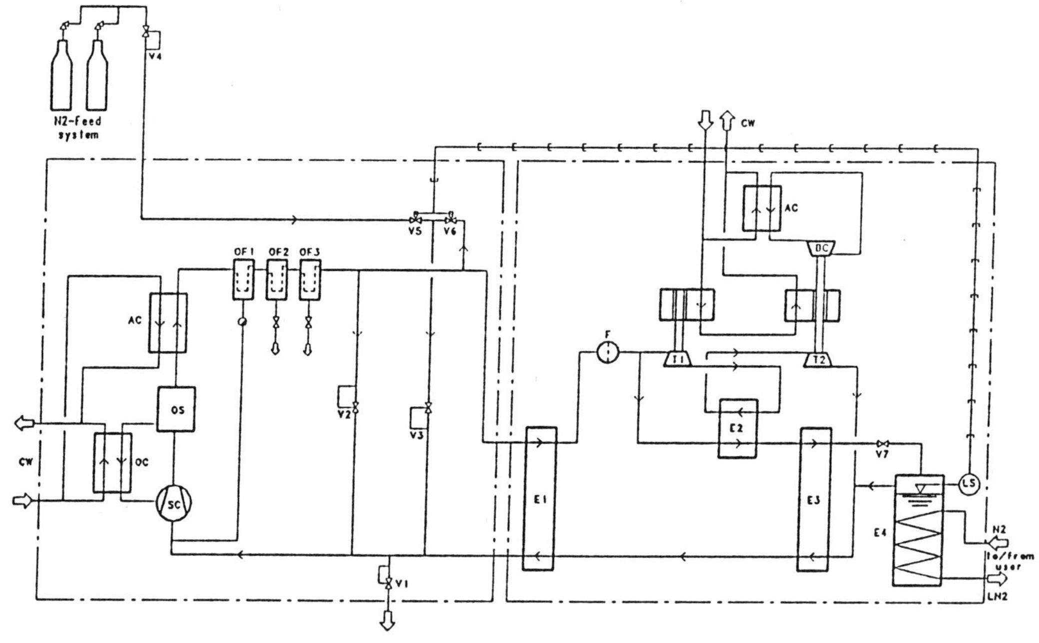

Compressor module   Cold box module

SC	Screw compressor	V1,2,3,4	Pressure control valves
OS	Oil separator	V5,6	Solenoid valves
OF1,2,3	Oil coalescing filters	V7	J-T expansion valve
OC	Oil cooler	BC	Brake compressor
AC	Aftercooler	CW	Cooling water
E1,2,3,	Heat exchangers	LS	Level switch
T1, T2	Turboexpanders	F	Filter

Fig. 1. Basic P & I diagram

PROCESS

The LINIT-R1 is based on a closed Claude cycle process, using nitrogen as the working fluid and two expanders. This solution has been chosen as the one with the best thermodynamic efficiency, after calculating and comparing various expander cycles and working fluids. The process can briefly be described as follows (see Fig. 1):

Low pressure nitrogen gas is compressed by a screw compressor. Compressor lubricating oil entrained in the nitrogen is extracted in an oil separator. The heat of compression is removed in the oil cooler and the gas aftercooler. Residual oil is removed in three oil coalescers. The compressed nitrogen enters the high pressure pass of the heat exchanger E1, in which it is precooled. The major part of the high pressure stream is expanded in turboexpander T1, then warmed somewhat in heat exchanger E2 and finally expanded again in turboexpander T2, reaching the required low working temperature. The smaller part of the high pressure stream is cooled further, condensed and subcooled in exchangers E2 and E3, and then throttled through the expansion valve V7 into a separator containing the exchanger E4 (the condenser of the user's secondary nitrogen circuit), where it evaporates, fulfilling the required plant refrigeration duty. The generated vapor combines with the exhaust of turboexpander T2. The vapor is then warmed in the low pressure passes of the heat exchangers E3 and E1, fulfilling the refrigeration circuit and heat recovery duties, and returns finally to the compressor, where the described cycle starts again.

A secondary nitrogen circuit, kept at a fixed saturation pressure in accordance with the temperature required for the refrigeration duty, serves to transfer the refrigeration capacity to the final user. This nitrogen condenses in the heat exchanger E4 and flows by gravity through a transfer line to the place of the refrigeration requirement, where it then evaporates. The formed vapour flows back through the transfer line to the heat exchanger E4, where the secondary nitrogen cycle starts again.

TECHNICAL DATA

The characteristics of the LINIT-R1 can be defined as follows:

Refrigerant	:	Nitrogen
Cycle	:	Modified Claude cycle with two expansions in series
Compressor	:	Oil lubricated screw compressor
Expanders	:	2 SULZER gas bearing turboexpanders with magnetic auxiliary bearings
Heat exchangers	:	Finned copper tubes
Cold box	:	Permanent vacuum insulated
Cooling	:	By water

The performance data for the application case of the prototype are:

Refrigeration Capacity	1	kW
transmitted to the user's		
secondary nitrogen circuit at		
Condensing temperature	80	K
Condensing pressure	0.140	MPa
	(1.40)	(bar a)
Compressor power requirement	16	kW
Electric current	208	V
	3	phases
	60	Hz
Cooling water inlet temperature	291	K
outlet temperature	302	K
inlet pressure max.	0.6	MPa
	(6.0)	(bar a)
pressure drop through unit	0.2	MPa
	(2.0)	(bar a)
Cooling water consumption		
Compressor unit	1.4	m^3/h
Cold box unit	0.1	m^3/h
Nitrogen gas requirements from feed		
system for loading the refrigerant		
circuit		
per start from warm		
conditions appr.	7	Nm^3
per start with already		
cool cold box appr.	4	Nm^3
during normal operation	no consumption	
Nitrogen gas pressure from feed system	>0.2	MPa
	(> 2.0)	(bar a)

Shown on the T-S diagram (temperature-entropy diagram) the
corresponding process looks as follows (see Fig. 2):

34.9 g/s of nitrogen gas, with a pressure of 0.102 MPa (1.02
bar a) and a temperature of 285.5 K, are compressed to a
pressure of 1.17 MPa (11.7 bar a) and cooled with water to a
temperature of 290.1 K, and then cooled in heat exchanger E1 to
122.4 K. Then 29.2 g/s are expanded, slightly reheated in heat
exchanger E2, and finally expanded in the turboexpanders to a
pressure of 0.105 MPa (1.05 bar a) and a temperature of 82.5 K.

The remaining 5.7 g/s of nitrogen gas is further cooled down, liquefied and subcooled in heat exchangers E2 and E3 to a temperature of 87.8 K and finally expanded in the Joule Thomson valve V7 to a pressure of 0.105 MPa (1.05 bar a) and a temperature of 77.6 K at which it evaporates, providing the refrigeration capacity of 1 kW. The two low pressure streams then join, are reheated in heat exchangers E3 and E1, and compressed again by the compressor.

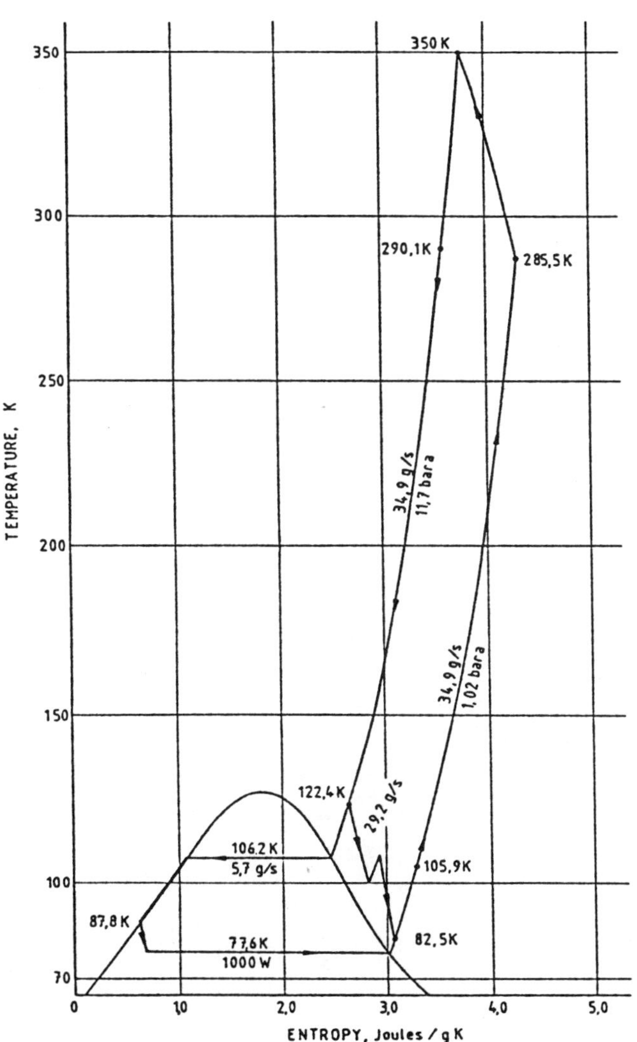

Fig. 2 Process on T-S diagram

LAYOUT

The LINIT-R1 consists of the compressor module and the cold
box module, having the following dimensions and weights (see
Fig.3):

Compressor module	width	2.050	m
	depth	1.150	m
	height	0.915	m
	weight	920	kg
Cold box module	width	0.550	m
	depth	0.570	m
	height	2.420	m
	weight	605	kg

Fig. 3 View of the LINIT-R1

The modules are connected by the ambient temperature high pressure and low pressure nitrogen lines. There is great freedom in the relative placement of these two modules: either they can be placed side by side or they can be placed apart in different rooms, floors or even buildings. The cold box is foreseen to be connected to the user of the refrigeration by a vacuum insulated transfer line for the liquid and the returning vapor of the secondary user's nitrogen circuit. These two streams will be integrated into a single line with two concentric tubes. It is recommended that the cold box and the user of the refrigeration be placed side by side, in order to minimize the thermal losses in the transfer line. The liquid nitrogen level in the refrigeration user should be positioned lower than the transfer line connection on the cold box.

The cold box module is free of noise or vibrations. For the unskilled onlooker, it is rather difficult to recognize whether the cold box is in operation or not. Also, the compressor module is free of vibrations and has a noise level of 65 +/- 3 dB(A) at 1 m distance, which corresponds, for example, to the noise of an overhead projector.

COMPRESSOR

The LINIT-R1 compressor is a Kaeser oil injected screw model AS30. It is mounted in a closed skid equipped with:

- Electric power panel
- Electric driving motor
- V-belt driven sigma profile screw
- Bulk oil removal system
- Water cooling system

OIL REMOVAL SYSTEM

The oil removal system consists of two coalescing filters in series, followed by an activated carbon filter, which are installed in the back side of the compressor unit. At the first filter an automatic oil drain is installed, the removed oil being pressed back to the suction side of the compressor. At the second and third filters a sight glass and a ball valve are installed down on the filter housing. If the sight glass is filled with oil, the oil is manually removed through the ball valve.

GAS MANAGEMENT PANEL

In the back side of the compressor is also mounted the gas management panel.

The compressor is designed to work at a pressure ratio of approx. 11 with a suction pressure of 0.102 MPa (1.02 bar a). The suction pressure fixes the temperature at the cold end.

Running or not running, the compressor suction pressure must not exceed 0.12 MPa (1.2 bar a).nor fall below 0.102 MPa (1.02 bar a). The medium operated valves V1 and V3 protect against over- and underpressurization. When the compressor stops, a pressure equalization is guaranted via the separator and the whole plant will be pressurized at 0.12 MPa (1.2 bar a).

The bypass valve V2 opens at excessive discharge pressures and protects against overpressurization. This bypass valve is able to cope with the full flow of the compressor.

COLD BOX

The vacuum insulated cold box contains the heat exchangers, the turboexpanders, the expansion valve, the separator, the filter, the piping and the measurement system. On the top of the cold box are installed the turboexpanders with their cooling water systems.

TURBOEXPANDERS

The refrigeration for the liquefaction process is provided by two turboexpanders. These are devices that extract the energy and cool the nitrogen by expanding it under near isentropic conditions.

The turboexpander rotor consists of a shaft, at one end of which is an expander impeller and at the other end a brake impeller. Compressed nitrogen is injected, through nozzles, into the expander impeller causing it to rotate. The nitrogen thus gives up energy and imparts it to the impeller. The rotational energy is transmitted via the shaft to the gas bearings and the brake impeller which circulates nitrogen in a closed circuit. The energy is further transmitted from these components to the cooling water system. This applies to the second turboexpander. The first turboexpander has no brake impeller, its energy being totally transmitted via its gas bearings to the cooling water.

The shaft is supported by dynamic gas bearings, which need no outside supply of bearing gas or any controls. The shaft generates the bearing pressure by its rotation. To support the shaft during starts and stops, it is equipped with auxiliary magnetic bearings. During the turboexpander's operation, the rotating element does not contact the static assembly. There is, consequently, no wear due to friction, thus giving the expander a virtually limitless life, and eliminating the need for maintenance.

The turboexpanders start to run as soon as the compressor starts. For supervision purposes, speed pick-ups have been installed. During cool down, the turboexpanders may run at a higher speed than at normal operation. Temperature measurements and high temperature switches guarantee a normal service.

50

EXPANSION VALVE

The expansion valve V7 produces the pressure drop and,hence,
the required temperature of the nitrogen evaporating in the
separator (Joule-Thomson effect), and also defines a correct flow
through the turboexpanders. The setting of the expansion valve
has been optimized in order to supply the required capacity.

COOL DOWN

When starting the LINIT-R1 after a long shut-down, the entire
plant is warm. The warm cold box limits the nitrogen mass flow,
and the excess delivered by the compressor will be bypassed via
the valve V2. The mass flow through the cold box grows during
cool down and, finally, the bypass valve closes. Thanks to the
small mass of the cold box and due to the good efficiency of the
turboexpanders, the service temperature of the cold box will be
reached in approx. 1 hour. During cool down, the suction
temperature of the compressor is maintained via the fill valve
V3, filling the plant with gas coming from the nitrogen feed
system via valve V5.

LOAD ADAPTATION

The cold box is dimensioned to provide 1 kW at 80 K. If the
refrigeration capacity need decreases, the liquid N2 level increases in the separator containing the user's secondary N2
circuit condenser. This will be detected by a level switch,
which changes the position of valves V5 and V6. If the
refrigeration capacity need remains low, the level in the
vessel still increases. The suction pressure decreases. Gas
from the high pressure side is bypassed via valves V6 and V3.
Consequently, the discharge pressure drops and the mass flow in
the cold box also drops. Thus, the power of the turboexpanders
decreases and the plant reaches a new balance.

If the refrigeration capacity need rises, the level in the
liquid N2 separator decreases. Flash gas lets the suction pressure rise. The fill valve V3 closes and the discharge pressure
rises too. The cold box quickly reaches its full capacity.

A refrigeration capacity of 0 % is possible. To obtain this,
the discharge pressure is reduced to about 0.32 MPa (3.2 bar
a). However, a capacity reduction to 0% means only a partial
reduction of electric power consumption by the compressor
(around a factor of 2).

It is to be noted that the compressor can suck the total
amount of flash gas from a load changing from 0 to 100% without
blowing out any gas via the overload valve V1.

PLANT STOP

When the compressor stops the plant pressure will be equalized at 0.12 MPa (1.2 bar a). The excessive nitrogen load of the plant will be blown out via the overload valve V1.

NITROGEN GAS FEED SYSTEM

The LINIT-R1 user has to provide a feed system able to supply dry nitrogen gas at a pressure of >0.2 MPa (>2 bar a), which has to be fed into the plant when it is started and blown off when it is shut down. The required capacity of the feed system depends on the mode of operation expected by the user. The feed system typically consists of a nitrogen bottle rack and a pressure reduction device (V4). If necessary, the compressor and the plant could be adapted to handle, on the low pressure side, higher pressures than 0.12 MPa (1.2 bar a), thus reducing or eliminating the nitrogen feed requirement for starting the plant.

PLANT ERECTION AND COMMISSIONING

The LINIT-R1 is very easy to erect. Both the compressor and the cold box modules are supplied completely mounted and tested from the works. They can be installed without any special foundation and need only the following connections:

- Nitrogen pipe from the feed system to the compressor module

- Nitrogen blow-off pipe from the compressor module to the atmosphere

- Nitrogen high and low pressure connection pipes between the cold box and compressor modules

- Concentric vacuum insulated liquid and vapor nitrogen transfer line between the cold box and the user of the refrigeration

- Electric power supply wiring to the compressor module

- Electric control wiring between the cold box and compressor modules

- Cooling water feeding and discharge piping to and from the compressor and the cold box modules.

The commissioning of the LINIT-R1 should be performed by a supplier's commissioning engineer, who would also properly instruct the client's operator. Depending on the availability of the necessary utilities system and on the complexity of the client's refrigeration user, the commissioning of the plant should not take more than one to two days.

OPERATION

The LINIT-R1 is designed for automatic service, at full as well as at partial load. Nevertheless, its control concept is very simple, with a minimum number of components.

The unit's operation is very easy. Provided all the valves are set in the service position, normal operation is just limited to turning the "start/stop"-switch and to opening/closing the cooling water and nitrogen feed-valves.

The main features of the LINIT-R1 can briefly be recapitulated as following:

- Very reliable operation due to a simple concept, and to the utilization of a proven screw compressor and of rotating wear-free turboexpanders

- Vibration free operation, no special foundations required

- Noiseless operation of the cold box unit and low noise operation of the compressor unit

- Flexibility in layout

- Low installation and operating costs

- Minimum maintenance requirements

- Can be operated by not specially trained personnel

BIBLIOGRAPHY

Quack, H., 1988, Expander cycles for 80 K refrigeration, in: "Proceedings of the Twelfth International Cryogenic Engineering Conference" (ICEC 12), R.G. Scurlock and C.A. Bailey, eds., Butterworth Co. Ltd., Guilford, UK.

CRYOGENIC THERMOMETRY– AN OVERVIEW

S. Scott Courts, D. Scott Holmes , Philip R. Swinehart
and Brad C. Dodrill

Lake Shore Cryotronics, Inc.
Westerville, Ohio USA

INTRODUCTION

The period from about 1965 to 1975 saw the very rapid commercial
introduction of several innovative cryogenic thermometers. The driving force was
the need for convenient, accurate temperature sensing for the burgeoning
laboratory, electronics, space and commercial markets that were gaining momentum
in that time period. This paper considers only commercially available temperature
sensors. A more comprehensive review of the dozens of existing cryogenic
temperature sensors and measurement techniques has been written by Rubin, Brandt
and Sample covering work through 1981 [1].

The first viable commercial thermometer for temperatures below 30 K was
the germanium resistance type pioneered by Cryocal™. Together with capsule
platinum thermometers, these devices provided secondary standard thermometry
from the millikelvin range to above room temperature. This combination, however,
was somewhat cumbersome and did not meet every need.

Other environmental, operational and convenience factors prompted the
rapid introduction of other thermometer types. The first category of factors
includes magnetic fields, radiation and thermal and mechanical shock. Fast thermal
response and wide range with a single thermometer fall under operational and
convenience headings, respectively. New sensors included gallium arsenide diodes,
silicon diodes, carbon glass RTs (Resistance Thermometer) and capacitance
thermometers introduced by Lake Shore Cryotronics, and the rhodium/iron
thermometer introduced by Cryogenic Calibrations, Ltd.™. Since 1975, progress
has been more evolutionary. Significant advances have been made in packaging,
sensor interchangeability and materials research, but no new sensor types have been
introduced commercially except for carbon film resistors [2] and special types of
germanium sensors used only in what were until recently Soviet block nations [3].

This paper presents general information to aid users in determining the best
sensor for their particular application and the best methods of installation and use.

Applications of Cryogenic Technology, Vol. 10
Edited by J.P. Kelley, Plenum Press, New York, 1991

THERMOMETRIC CHARACTERISTICS

Temperature sensors can be based on a variety of temperature-dependent properties. The most common sensors are resistors, diodes and capacitors. Examples of temperature characteristics of commercial sensors are included in Figures 1-2.

The first question a user should ask is whether the sensor will provide the best, or at least an adequate, signal, in the temperature range of interest. The relative temperature resolution, $\Delta T/T$, can be shown to be given by the expression

$$\frac{\Delta T}{T} = \frac{\dfrac{\Delta V}{V}}{\left| \dfrac{T}{V}\dfrac{dV}{dT} \right|}$$

where the measured quantity is assumed to be a voltage. The expression can be made to apply for capacitance or resistance measurements by replacing V by either C or R. The expression for $\Delta T/T$ neatly separates the material-specific parameters $(T/V)(dV/dT)$, known as the specific sensitivity, from the measurement system resolution, $\Delta V/V$. The specific sensitivity is a function of only the material characteristics and not the geometry of the sensor. Note that the specific sensitivity can be written as $(d \ln V/d \ln T)$, which is just the slope of the parameter vs. temperature on a log-log plot. Specific sensitivities of various sensors are plotted in Figure 3. A large specific sensitivity allows the resolution of small temperatures relative to the temperature measured, but the temperature range which can be measured would be small. As an example, resistances are easiest to measure in the 1 to 100,000 Ω range. A sensor with a constant dR/dT of 10,000 Ω/K would allow precise temperature measurements, but only over a 10 K temperature range.

The geometry of a sensor is chosen to provide a useful resistance in a desired temperature range. For example, a germanium chip can be adjusted in geometry to provide either a 500 Ω or a 1000 Ω sensor at 4.2 K. Thermocouples are an exception since they produce a given voltage independent of geometry.

The temperature resolution is also a function of the relative measurement system resolution, $\Delta V/V$ (or $\Delta C/C$ for capacitance measurements). The relative resolution of the measurement system is a function of the smallest change which can be detected, divided by the magnitude of the parameter measured. Noise, thermal emfs and the smallest signal detectable with a given instrument all limit the resolution. Note that the measurement resolution also depends on the voltage or capacitance of the sensor which depends on the sensor geometry as well as materials characteristics (except for thermocouples). Optimization of the measurement system resolution is complicated, involving tradeoffs between temperature resolution and measurement time, sensor size, equipment cost, and degree of shielding from electromagnetic or radiation fields. As an example, a germanium sensor with a resistance of 1000 Ω at 4.2 K (specific sensitivity = -2.14) excited with a 1 μA current and measured with a system with 1 μV resolution, would provide a relative temperature resolution of 4.67×10^{-4}, or an absolute

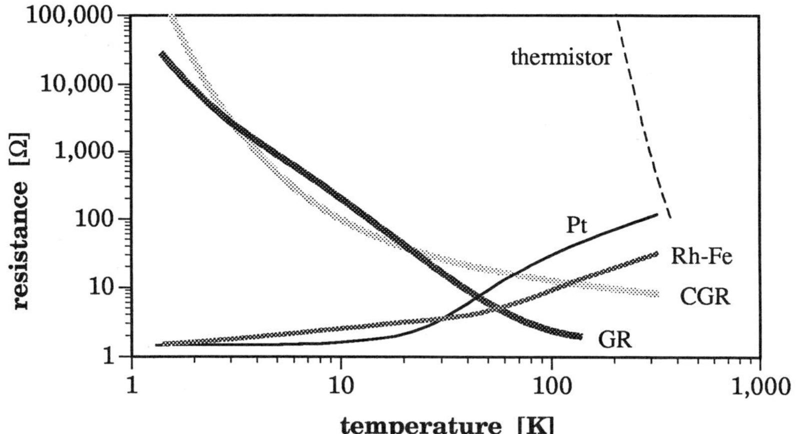

Figure 1. Characteristics of resistive temperature sensors.
CGR: CGR-1-1000 carbon-glass resistor, **GR**: GR-200A-1000 germanium resistor, **Pt**: PT-103 platinum resistor, **Rh-Fe**: RF-800-4 rhodium-iron resistor, **thermistor**: YSI 44003A thermistor.

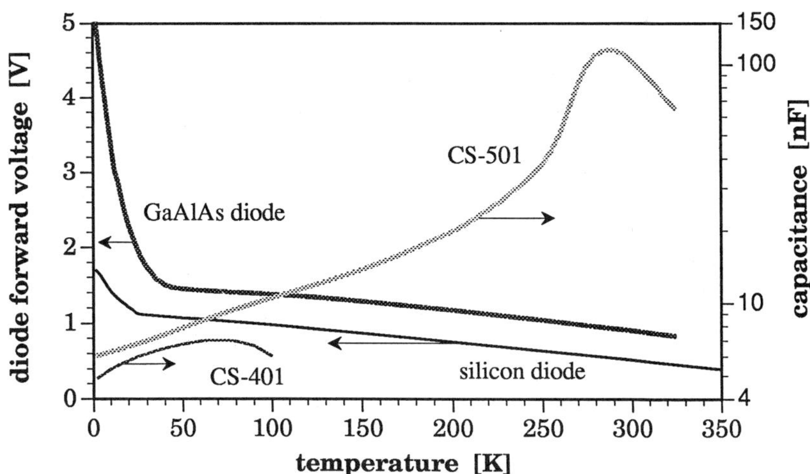

Figure 2. Characteristics of diode and capacitance temperature sensors.
silicon diode: DT-470 silicon diode @ 10 μA, **GaAlAs diode**: TG-120P gallium-aluminum-arsenide diode @ 10 μA, **CS-401**: CS-401GR-B capacitor, **CS-501**: CS-501 capacitor.

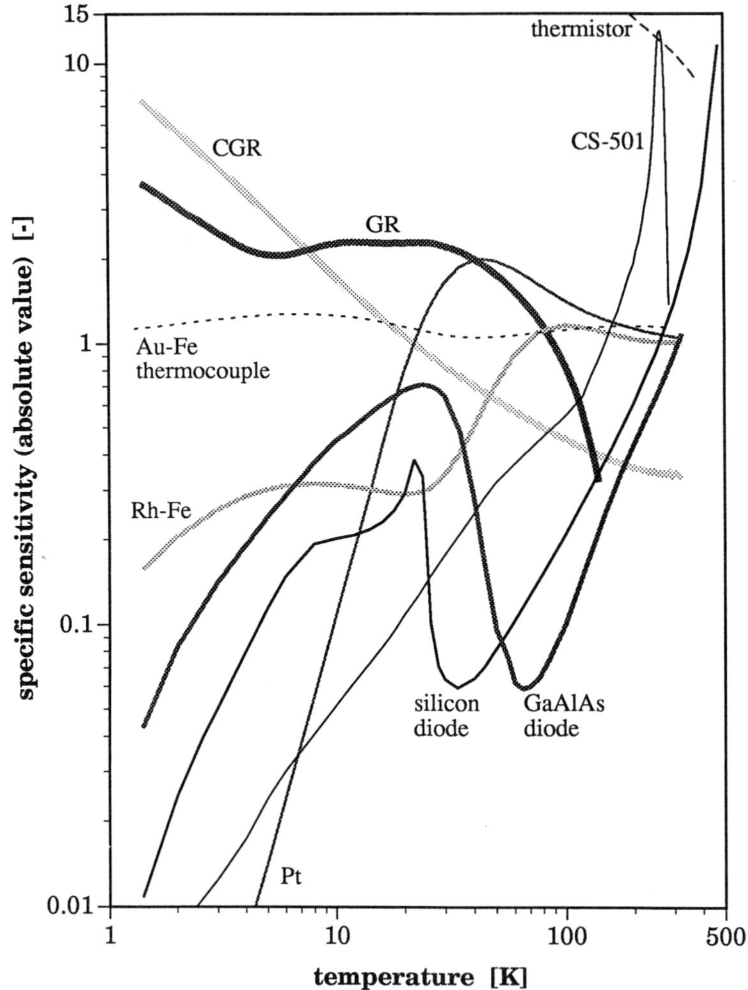

Figure 3. Absolute specific sensitivities of various temperature sensors.
Au-Fe thermocouple: KP chromel vs. Au-0.07%Fe thermocouple,
CGR: CGR-1-1000 carbon-glass resistor, **CS-501**: CS-501 capacitor,
GR: GR-200A-1000 germanium resistor, **Pt**: PT-103 platinum resistor,
Rh-Fe: RF-800-4 rhodium-iron resistor, **silicon diode**: DT-470 silicon diode @
10 µA, **GaAlAs diode**: TG-120P gallium-aluminum-arsenide diode @ 10 µA,
thermistor: YSI 44003A thermistor.

resolution of about 2 mK. Voltmeters are readily available which can measure signals of a few volts with a relative precision of one part in 10^5. Voltage measurements are typically limited to a resolution of about 0.1 μV by the resolution of the voltmeter, noise or thermal emfs from lead wire contacts. The excitation current for a germanium RT is usually varied to produce an output voltage in the 1-3 mV range to limit power dissipation, so the relative temperature resolution is only about one part in 10^4. The relative temperature resolution capability of diodes can therefore be as good as that for a germanium resistor even when the specific sensitivity is a factor of ten lower. Temperature resolutions are plotted in Figure 4 for a variety of sensors operating under specified conditions.

Given a relative measurement system resolution, improvements in resolution are obtained only by maximizing the specific sensitivity. However, a large specific sensitivity implies a narrow working temperature range because the signal quickly becomes too low or too high to measure accurately. The physical phenomenon responsible for the temperature dependence must occur in the desired temperature range. For example, if the desired working temperature range for a negative temperature coefficient RT is around 4.2 K, the device will have to remain very low in resistance from room temperature down to about 100 K in order to avoid going off scale before 4.2 K is reached. For silicon, germanium and gallium arsenide, this is accomplished by competing contributions to the resistivity which combine to keep the resistivity relatively constant until temperatures below about 100 K. The competing mechanisms are the charge carrier mobility, which increases as the temperature decreases, and the carrier population, which decreases with decreasing temperature as the carriers freeze into the impurity sites in the bandgap. Carrier freezout begins to dominate below 100 K and the resistivity begins to increase rapidly as the temperature is decreased. A hopping conduction mechanism among impurity sites in the bandgap becomes important below about 4.2 K and keeps the resistivity from rising too rapidly as the temperature decreases further. The quantity and type of dopants or defects determine the resistivity versus temperature characteristic for a given material. The resistivity vs. temperature characteristics of germanium can be varied to create a family of thermometers covering the range from below 0.05 K to 100 K.

The conduction mechanism in carbon glass involves the hopping of charge carriers across barriers that arise from the crystal structure of carbon, its dispersal in an insulating matrix, and to some extent from quantum size effects, especially in the case of carbon glass, where it is in the form of filaments. The temperature characteristic of carbon glass is monotonic from below 1 K to above room temperature, in contrast to that of germanium. However, the specific sensitivity for carbon glass in the 50 K to 300 K range is too low (0.2 as compared to 1.0 to 1.6 for platinum) to produce a high accuracy thermometer. As 4.2 K is approached, it becomes so sensitive that it becomes unusable below about 1.5 K.

Diodes, on the other hand, employ more than one conduction mechanism to maintain their status as the wide range, workhorse sensors (especially silicon diodes). A p-n junction has a well known, well-behaved temperature dependence based on the artificial energy barrier created by the dopants. The temperature characteristic is fairly linear and monotonic from above room temperature to about 30 K, where the resistivity of the bulk material in which the junction is fabricated

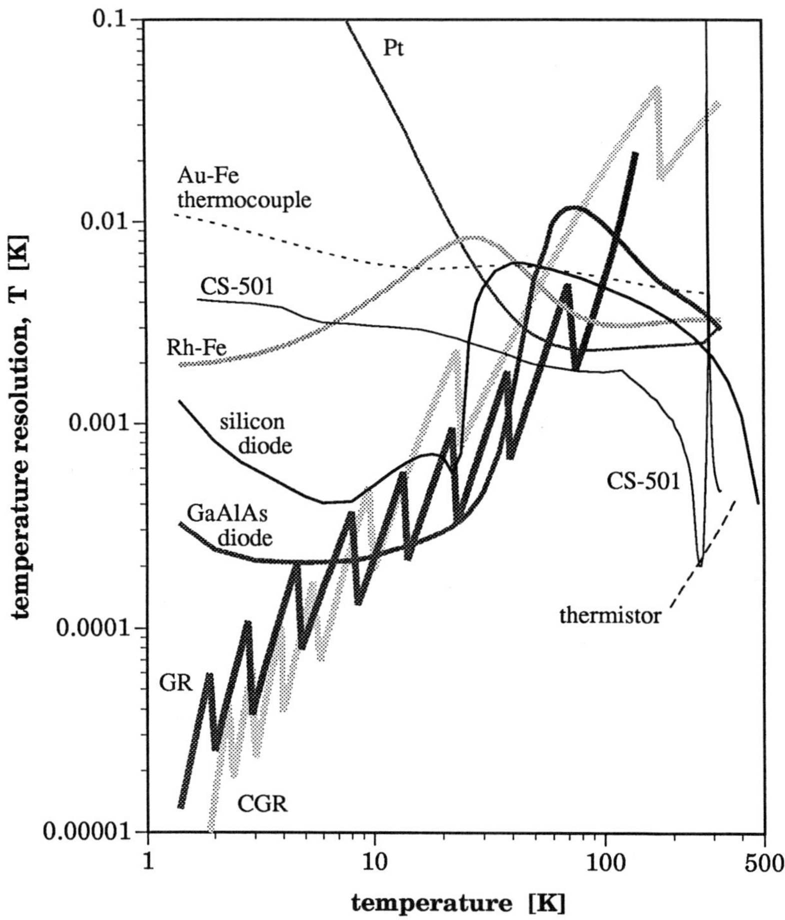

Figure 4. Temperature resolutions of various temperature sensors under specified operating conditions. **Au-Fe thermocouple**: KP chromel vs. Au-0.07%Fe thermocouple, **CGR**: CGR-1-1000 carbon-glass resistor @ 1-3 mV or I=0.1 μA min, **CS-501**: CS-501 capacitor, **GaAlAs diode**: TG-120P gallium-aluminum-arsenide diode @ 10 μA, **GR**: GR-200A-1000 germanium resistor @ 1-3 mV or I=0.1 μA min, **Pt**: PT-103 platinum resistor @ 100 μA, **Rh-Fe**: RF-800-4 rhodium-iron resistor @ 300 μA, **silicon diode**: DT-470 silicon diode @ 10 μA, **thermistor**: YSI 44003A thermistor @ 1 μW. Relative measurement system resolutions: 0.1 μV or ΔV = V/100,000; 0.1 pF or ΔC = C/100,000; whichever is LARGER.

begins to dominate. The sensitivity increases by a factor of 10 at 4.2 K. It is this combination of near linearity and sizable sensitivity in the high temperature range with an increased sensitivity at 4.2 K that gives the silicon diode its unique status in thermometry. Note from Figure 3, however, that below 100 K its specific sensitivity does not compare with that of carbon glass. Another important feature of the silicon diode is that it is one of the few cryogenic sensors that can be made with sufficient consistency to allow interchangeability with respect to a standard curve [4,5]. However, every diode lot and even each wafer made in the same furnace run will have slightly different characteristics, requiring families of standard curves if temperature measurement errors are required to be less than about ± 1% of the measured temperature.

Gallium arsenide and gallium aluminum arsenide are direct bandgap materials that, compared to silicon, suffer in specific sensitivity at high temperatures but are superior at low temperatures. They are also less robust than silicon diodes and cannot presently be made interchangeable. They share the general diode attributes, but are less strongly influenced by magnetic fields.

Platinum RTs have a very long history, are one of the three most common electronic thermometers (along with thermocouples and thermistors) and are interchangeable to varying degrees. The most stable units are wire wound and suspended strain free in a hermetic capsule. These are very large and thermally slow for most cryogenic uses, however. The industrial ceramic-encapsulated and thin-film units are usable down to 14 K if pushed, although a 30 K limit is recommended. They do not conform to a DIN-type standard below about 77 K due to the influences of packaging and finite conductor size on conduction in the platinum.

Rhodium-iron RTs are lower in specific sensitivity than platinum RTs near room temperature, but have an added magnetic impurity that causes the sensitivity to remain usable well below 4 K. They can be used to 800 K if some temperature shift is permissible. This is the only sensor other than a thermocouple that can be baked at high temperatures and then used to 1 K, for instance in an ultra-high vacuum (UHV) environment. Temperature shifts of about 1% of temperature have been observed in rhodium-iron sensors baked for one hour at 800 K between calibrations. Another Coles alloy, Pt-Co, has shown some promise for applications in moderate magnetic fields [6].

Thermocouples are advantageous where very low thermal mass or a differential measurement is required. They are difficult to use with high absolute accuracy, however, because thermoelectric effects occur over the entire length of the wire and at every junction. As discussed by Walstrom [7], spatial variations in thermocouple wire composition or strain, or the presence of a magnetic field in a region with a temperature gradient result in temperature measurement errors. On the other hand, thermocouples such as Au-Fe can be used over a very wide temperature range, since the specific sensitivity remains near 1.0 from 1 K to 300 K, as seen in Figure 3.

Capacitance sensors are ceramic chip capacitors made from ferroelectric materials to obtain a significant temperature dependence. They are substantially immune to magnetic fields, which is the only reason they are used. Unfortunately, a phase change that occurs below room temperature causes them to drift following thermal cycling. Interruption of their excitation can also initiate drift. Therefore, they are used only as control sensors with temperature being obtained from another sensor at zero magnetic field. Use of a capacitance temperature sensor involves 1) waiting for signal stabilization following cooldown, 2) calibration in zero magnetic field against another temperature standard, and 3) use at low temperatures in a magnetic field.

TEMPERATURE SCALES AND PRACTICAL ACCURACIES

Temperature scales are established using thermodynamic quantities such as fixed points of nature (e.g., triple points) and by formulating methods for carrying the scale between these points. These methods include specified interpolation formulas and the physical means of making the measurements, such as platinum RTs. Realization of the scale to fractions of a millikelvin is a very specialized undertaking, but must be done by national laboratories in order to allow laboratories and companies supplying calibrations to users in the field to do so without undue degradation of the thermometer accuracy. Harris-Lowe and Turkington [8] presented examples of difficulties commonly encountered in calibration against secondary standards.

A new scale has recently been adopted, the International Temperature Scale of 1990 (ITS-90) [9]. It differs from the previous scales (ETP-76 and IPTS-68) by a maximum of 0.02 K below room temperature and increases in deviation up to about 0.20 K at 900 K. These differences are significant for some applications. All sensor calibrations should now be based on ITS-90. Users should consider recalibration of old sensors to the new temperature scale.

Inaccuracies build up from the primary to the secondary standards and from the secondary to the working standards maintained by the commercial calibration facility. Additional error is propagated in the transfer from the working standards to the field thermometer, and further errors are introduced by installation and measurement instrumentation. A comparison is performed every six months between Lake Shore's secondary standard platinum and germanium thermometers and the working versions. Similarly, the calibration instruments are sent back to the manufacturers periodically for recalibration. The resulting transfer accuracy to the user's thermometer is dependent on the sensitivity and stability of the thermometer type. For 1000 Ω germanium, it is typically 4 mK in the helium range and 40 mK at 100 K. Carbon glass, which is more sensitive at high temperatures, but not as stable, attains a 65 mK transfer at 100 K and 250 mK at 300 K. Platinum RTs typically transfer with a 10 mK loss at 100 K and 20 mK at 300 K. Typical values for silicon diodes are 12 mK below 20 K and 25 mK at higher temperatures. Further losses in accuracy depend on how the sensor is installed and instrumented and on how much stress it endures. For a scientist or engineer doing expensive work that depends heavily upon accurate thermometry, recalibration and rotation (intercomparison) of thermometers is relatively inexpensive insurance.

Variation of the excitation current of a sensor can also contribute errors to temperature measurement. The non-linear forward diode characteristics yield smaller changes in voltage with current changes than does the linear resistor. In some cases, the current cannot be regulated well enough to prevent serious errors even with a diode, but can be measured accurately. For a resistance sensor, there may be no degradation of accuracy, but the current-voltage relationship must be known for a diode. If the constant n in the diode equation exponent, qV/nkT, is known, corrections can be made for $\pm 10\%$ changes in current to better than ± 50 mK from 40 K up. Here q is the electronic charge, k is the Boltzmann constant and n depends on the mix of conduction mechanisms. An error is also produced if the diode excitation current has an ac component [10].

THERMOMETER STABILITY

Studies have been performed world-wide to establish the stability of all types of thermometers and fixed points [11]. It is well known that germanium and platinum resistors can suddenly shift a significant amount in resistance after many cycles, but be as stable with repeated cycling after the shift as before. Not as much attention has been paid to diodes because they are not as stable as secondary standard RTs, but they enjoy such wide-spread use that they merit similar testing.

There are two types of tests commonly performed on diodes to establish their stability. The first is thermal shock cycles from room temperature into liquid nitrogen and then into liquid helium. When first testing a new type of diode or a new production lot, several hundred or thousand cycles may be performed. When confidence is established, subsequent production operations may include only a few cycles. One type of diode exhibited a good temperature characteristic, but was found to have an unacceptable decline in voltage after about 50 cycles into liquid helium. The DT-470-SD diode currently in use demonstrates better short term cycling stability: ± 30 mK over hundreds of thermal cycles.

The second type of stability test is provided by recalibration of the sensor over the entire range. Data from a commercial calibration facility is presented in Table 1. The diode sensors are contained in ceramic/sapphire flat packages which were clipped to a copper calibration block with a thin film of thermally conducting grease between the sensor package and the copper block. All calibrations were performed with the sensors in vacuum. They were stored on the shelf in room ambient conditions between runs. The temperature points chosen from the full range calibrations for presentation include the common liquid points and several which appeared to have the most drift. The 2 K point is subject to more error than the higher temperature points because the thermal resistance of the link between the sensor and the mounting block increases with decreasing temperature and because reestablishment of an identical thermal link is impossible upon remounting. Some diodes must be thermally aged in order to reach the level of stability illustrated.

Construction techniques are critical in maximizing the stability of thermometers. Data on the differences between strain-free and encapsulated sensors is available elsewhere (e.g. [12], pp. 28-35). Epoxy encapsulation also can have a large effect on the short term stability of diodes, as discussed by Krause and Swinehart [4]. Differential thermal contractions between epoxy encapsulation and a diode or other sensor can produce calibration shifts due to the strain on the sensor. The strain-induced shifts may or may not be reproducible.

Table 1. Cumulative, mean recalibration temperature shifts [mK] for a group of fifteen DT-470 silicon diode sensors. The temperature shifts relative to the initial calibration are tabulated for several temperatures.

	measurement temperature				
Calibration	4.2 K	13 K	30 K	77.35 K	305 K
19 May 1986 day 0	0	0	0	0	0
4 Nov 1986 day169	-25.31	-17.47	-20.51	-19.24	-10.76
12 Aug 1987 day 450	-21.52	-22.53	-16.46	-13.67	0.00
8 Feb 1988 day 630	-22.78	-29.11	-18.35	-23.04	-2.66
28 Sep 1989 day 862	-29.75	-37.97	-18.10	-21.26	-0.95

THERMOMETER INSTALLATION

The three major problems that must be solved in assuring that the thermometer reads the temperature of the workpiece are: (1) conducting away internally dissipated heat (self-heating), (2) preventing external heat sources from affecting the sensor and (3) matching the thermal time response of the sensor and its thermal link to that of the workpiece. As can be seen from the power dissipation data in Figure 5, diodes will be more difficult to install from a self-heating point-of-view than germanium and carbon glass resistors. Grease, epoxy or solder are frequently used to maximize the area and thermal conductivity of thermal links, although care must be taken because mechanical stress or overheating can sometimes cause an apparent temperature shift or mechanical failure. External heating sources are usually taken care of in careful cryostat design by shielding, baffling and heat-sinking of wiring at as many intermediate temperatures between the sensor and room temperature as possible. Small diameter wires with high thermal and electrical resistance are used, such as 36 AWG phosphor bronze or manganin. The leads consist of single strand wires insulated with a varnish to allow intimate contact with the heat sinks and sealing into vacuum-tight feedthroughs. Strain-free packages often rely on thermal conduction down the lead wires, so the lead wires must be in good thermal contact with the object to be measured. In extreme cases, thin-film sensors must be applied directly to the workpiece to provide sufficiently rapid thermal response.

The magnitude of the self-heating for silicon diodes has been investigated by the authors by comparison between different mounting configurations. In one case, four DT-470-SD diodes were calibrated and then inserted into holes in copper bobbins (CU adapters) without being attached to the copper. Four turns of the lead

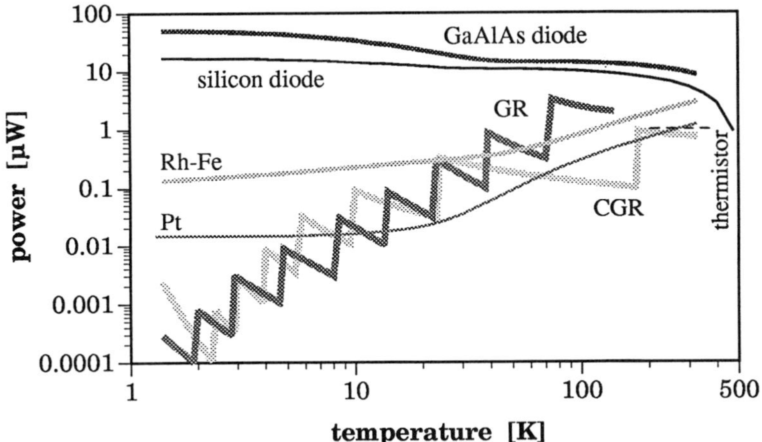

Figure 5. Power dissipation in various temperature sensors under specified operating conditions. **CGR**: CGR-1-1000 carbon-glass resistor @ 1-3 mV or I_{min} = 0.1 μA, **GR**: GR-200A-1000 germanium resistor @ 1-3 mV or I_{min} = 0.1 μA, **Pt**: PT-103 platinum resistor @ 100 μA, **Rh-Fe**: RF-800-4 rhodium-iron resistor @ 300 μA, **silicon diode**: DT-470 @ 10 μA, **GaAlAs diode**: TG-120P @ 10 μA, **thermistor**: YSI 44003A @ 1 μW.

wire were made around the bobbin and potted in epoxy as usual. The sensors were recalibrated and the deviation between the second and first calibrations was calculated. The lead heat-sinking was not sufficient to remove heat from the diodes, and they read much warmer below 10 K. Direct connection to the copper is necessary to prevent this. On the other hand, germanium and carbon glass RT's, which are suspended inside their packages on 50 micrometer gold wires, transfer most of their heat through the leads. This is successful because of their low power dissipation. An example in which self-heating is a problem is with the calibration of some germanium resistors intended for use at 0.05 K. Below 1.5 K they are calibrated with an AC bridge which has a lower minimum power output than the DC system used from 1.5 K to 6 K. There is a difference between the two calibrations of up to 50 mK at 1.5 K.

CALIBRATION CURVE FITTING

Errors of a largely mathematical nature can occur in fitting interpolation routines to discrete data points. References on appropriate numerical techniques abound [13-16]. A few points are worth mentioning here, however. A data set can be broken into smaller ranges to reduce fitting errors, but a trade-off exists between reduced fitting errors and the increased difficulty of determining the appropriate fit to use. Reduction of the fitting error substantially below the calibration uncertainty is not beneficial. Several data points should always be taken outside the fitting range of interest because polynomial fits cause the curves to wobble or bulge near the ends. Relationships best fit with non-linear parameters such as $\ln T$ or $1/T$ should have calibration data taken with uniform spacing on the non-linear parameter axis (i.e. take calibration points uniformly spaced in $\ln T$ or in $1/T$). Also, if the data is not very smooth, and the points are not spaced correctly for the degree of non-linearity present, the fit will not be smooth. In this case, a simple least squares fit is the best that can be performed.

THERMOMETRY IN MAGNETIC FIELDS

The introduction of a magnetic field vastly complicates low temperature thermometry. Some practical solutions are given by Rubin, et al. [17]. As noted previously, capacitance sensors are very useful as control sensors, but are not stable enough for thermometry, and require high frequency (1 kHz to 1 MHz) excitation with coaxial cabling. Carbon glass RT's are very useful from 1.5 K to 77 K because the offsets are only a few percent of temperature and can be corrected for in a reasonable way [18]. Platinum RT's can be used the same way in moderate magnetic fields [19]. Silicon diodes have serious temperature errors at low temperatures and high fields and are not recommended. Gallium arsenide and GaAlAs diodes have only a few percent change in temperature indication up to 5 tesla, allowing the advantages of diode thermometry to be utilized in moderate magnetic fields.

THERMOMETRY IN THE PRESENCE OF NUCLEAR RADIATION

Much of the data on the radiation dependence of thermometers is classified, published in difficult-to-obtain reports, or performed in environments not typical of cryogenic situations. For instance, thermal annealing and ionic conduction will occur in materials irradiated at room temperature and above, but will not occur at very low temperatures. The effects of irradiation are thus dependent on the sensor temperature during irradiation. Data on sensors irradiated at cryogenic temperatures is especially scarce. Room temperature irradiations of various temperature sensors have been performed by the authors. The temperature reading shift data plotted in Figure 6 indicates that carbon glass and platinum RT's may be an adequate combination for many purposes, and even silicon diodes may be used for low doses.

CHECKING SENSOR CALIBRATIONS IN LIQUID BATHS

The convenient liquid baths are helium, nitrogen, and water or oil. Most of the other cryogenic liquids, such as neon, are too expensive. Oxygen and hydrogen can be dangerous, as can simple hydrocarbons such as methane. Carbon dioxide is a possibility, but it must be pure. Water, too, must be pure, either at the ice point or the boiling point. These baths can be more difficult to use reliably than the unwary might suppose. There is sufficient literature published to coach the new experimentalist at these temperatures (see appendix in ref. [12]).

One fact that is not widely realized is that diode thermometers are high impedance devices, and small ionic conductivities can shunt them. Near room

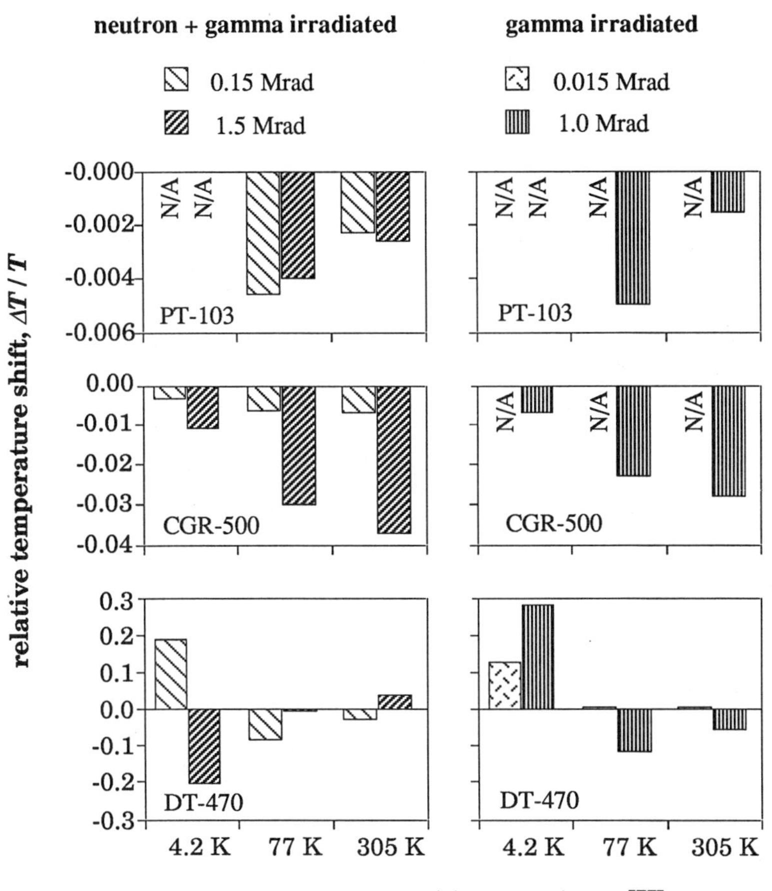

Figure 6. Effects of radiation on temperature sensors. The relative temperature measurement shift is indicated at 4.2, 77 and 305 K. **PT-103**: platinum resistor, **CGR-500**: carbon-glass resistor, **DT-470**: silicon diode.

temperature and above, diodes should not be immersed directly in the liquids. In an ice bath, the diode can be placed in a thin rubber sheath such as a surgical glove, and in a hot oil bath it should be placed in a metal probe.

Liquid helium is generally a well-behaved temperature check-point, especially if only 50 mK accuracies are required. Problems with dilution by other gases are minimal because all other elements are solids in liquid helium and because access of foreign materials is restricted by the precautions taken to restrict heat input. If the dewar pressure can be measured accurately, accuracies of a few millikelvins can be expected. Pressure upsets due to probe insertion must be allowed to die out, and if the temperature is to be reduced below 2.17 K by pumping on the bath, familiarity must be gained with the unusual properties of superfluid helium.

Liquid nitrogen (LN) is another case. Blundell and Ricketson [20] have discussed some of the vagaries of this liquid. In general, special precautions must be taken to obtain accuracies of better than $\pm$ 0.25 K, or short term repeatability of better than $\pm$ 0.05 K. The temperature of a LN bath is about 75 K at an altitude of 1.6 km. Most baths are open to the air, so liquid oxygen is condensed into them. The boiling point of liquid air at standard pressure is about 81 K. There are head pressures, stratification and stirring variables to worry about if great accuracies are required. A sensor cannot be attached to heavy copper lead wires and immersed in a shallow nitrogen dewar, such as a thermos bottle, and produce a reliable check point. The best course of action for quick checks of sensor calibrations is to use several calibrated sensors attached to a copper block so intercomparisons can be performed.

REFERENCES

[1] L. G. Rubin, B. L. Brandt and H. H. Sample, Cryogenic thermometry: a review of recent progress, II, Cryogenics 22:491-503 (1982).

[2] W. B. Bloem, A cryogenic fast response thermometer, Cryogenics 24:159-165 (1984); commercial distribution by Rivac Technology bv, Valkenswaard, Holland.

[3] L. M. Besley and A. Szmyrka-Grzebyk, Stability studies on Kiev cryogenic germanium resistance thermometers, Rev. Sci. Instrum. 61:1303-1307 (1990).

[4] J. K. Krause and P. R. Swinehart, Reliable wide range cryogenic diode thermometers, in: "Advances in Cryogenic Engineering, Vol. 31," Plenum Press, New York (1985) 1247-1254.

[5] B. C. Dodrill, J. K. Krause, P. R. Swinehart and V. Wang, Performance characteristics of silicon diode cryogenic temperature sensors, to be published in: "Applications of Cryogenic Technology - Vol. 10," Plenum Press, New York (1990).

[6] F. Pavese and P. Cresto, Search for thermometers with low magnetoresistive effects: platinum-cobalt alloy, Cryogenics 24:464-470 (1984).

[7] P. L. Walstrom, Spatial dependence of thermoelectric voltages and reversible heats, <u>Am. J. Phys.</u> 56:890-894 (1988).

[8] R. F. Harris-Lowe and R. R. Turkington, Comparison of calibrated temperature sensors: 4-300 K, <u>Cryogenics</u> 24:531-535 (1984).

[9] H. Preston-Thomas, The international temperaure scale of 1990 (ITS-90), <u>Metrologia</u> 27:3-10 (1990); Erratum, <u>Metrologia</u> 27:107 (1990).

[10] J. K. Krause and B. C. Dodrill, Measurement system induced errors in diode thermometry, <u>Rev. Sci. Instrum.</u> 57:661-665 (1986).

[11] L. M. Besley and H. H. Plumb, Stability of germanium resistance thermometers at 20 K, <u>Rev. Sci. Instrum.</u> 49:68-73 (1978).

[12] E. G. Eckert and R. J. Goldstein, "Measurements in Heat Transfer," 2nd Ed., Hemisphere Publishing Corporation, New York (1976).

[13] R. L. Burden, J. D. Faires and A. C. Reynolds, "Numerical Analysis," Prindle, Weber & Schmidt, Boston (1978).

[14] W. H. Press, B. P. Flannery, S. A. Teukolsky and W. T. Vetterling, "Numerical Recipes," Cambridge University Press, Cambridge (1986).

[15] C. F. Gerald, "Applied Numerical Analysis, 2nd Ed.," Addison-Wesley, Reading (1978).

[16] H. J. Hoge, Useful procedure in least squares, and tests of some equations for thermistors, <u>Rev. Sci. Instrum.</u> 59:975-979 (1988).

[17] L. G. Rubin, B. L. Brandt and H. H. Sample, Some practical solutions to measurement problems encountered at low temperatures and high magnetic fields, <u>in</u>: Advances in Cryogenic Engineering, Vol. 31, Plenum Press, New York (1986) 1221-1230.

[18] H. H. Sample, B. L. Brandt and L. G. Rubin, Low-temperature thermometry in high magnetic fields. V. Carbon-glass resistors, <u>Rev. Sci. Instrum.</u> 53:1129-1136 (1982).

[19] T. Haruyama and R. Yoshizaki, Thin-film platinum resistance thermometer for use at low temperatures and in high magnetic fields, <u>Cryogenics</u> 26:536-538 (1986).

[20] D. J. Blundell and B. W. Ricketson, The temperature of liquid nitrogen in cryostat dewars, <u>Cryogenics,</u> 19:33-36 (1979).

MONITORING RAPIDLY CHANGING TEMPERATURES OF THE OSCILLATING WORKING

FLUID IN A REGENERATIVE REFRIGERATOR[*]

Wayne Rawlins[**], Ray Radebaugh[***], and K. D. Timmerhaus[**]

[**]University of Colorado
Department of Chemical Engineering
Boulder, Colorado

[***]Chemical Engineering Science Division
National Institute of Standards and Technology
Boulder, Colorado

ABSTRACT

Characterization of an orifice pulse tube refrigerator requires measurements of the instantaneous gas temperatures at various locations in the refrigerator. This presents several challenges. The temperature probe has to fit inside a 3 mm diameter tube with minimum disturbance to the flow. Void volumes, introduced by placement of a temperature probe into the system, have to be kept at a minimum. The temperature sensing device has be robust to survive pressure waves and mass flows oscillating at frequencies of up to 30 Hz. It also must have a fast response time to monitor the rapidly changing temperatures in the system. The temperature resolution has to be on the order of 10 mK. This paper discusses rapid temperature measurements with both thin-foil thermocouples and fine-wire resistance thermometers. A 4 μm diameter tungsten wire was found to satisfy these diverse requirements.

INTRODUCTION

There has been considerable interest in pulse tube refrigerators recently because of their potential for high reliability.[1] The

---

[*]Research sponsored by NASA/Ames Research Center. Contribution of NIST, not subject to copyright.

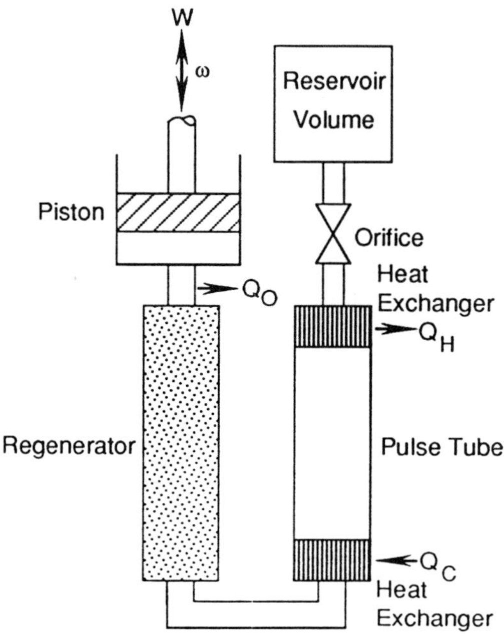

Fig. 1. Schematic of the orifice pulse tube refrigerator.

most useful type is the orifice pulse tube refrigerator, shown in
Figure 1. It operates in a thermodynamic cycle similar to that of
a Stirling refrigerator, except there is no moving displacer. The
major loss term in the cycle is associated with the regenerator
ineffectiveness. An apparatus has been designed to evaluate the
regenerator performance in an orifice pulse tube refrigerator[2] (see
Figure 2). The working fluid is helium gas. A reciprocating
compressor generates the oscillating pressure wave in the pulse tube
refrigerator. The pressure in the system oscillates around an
average, non-zero value while the mass flow periodically reverses,
oscillating about an average value of zero. Since the pulse tube
refrigerator operates with higher mass flow rates than a Stirling
refrigerator for the same refrigeration power, improved regenerator
performance is required.[1] Evaluation of the regenerator effectiveness, ϵ, requires a knowledge of the instantaneous gas temperature
in the system.[3] Regenerator effectiveness in an ideal heat exchanger
is unity. Since the effectiveness for a well designed regenerator
can be very close to unity, it is convenient to define regenerator
ineffectiveness as $\lambda = 1-\epsilon$. A simplified relation for the ineffectiveness of a regenerator is given by

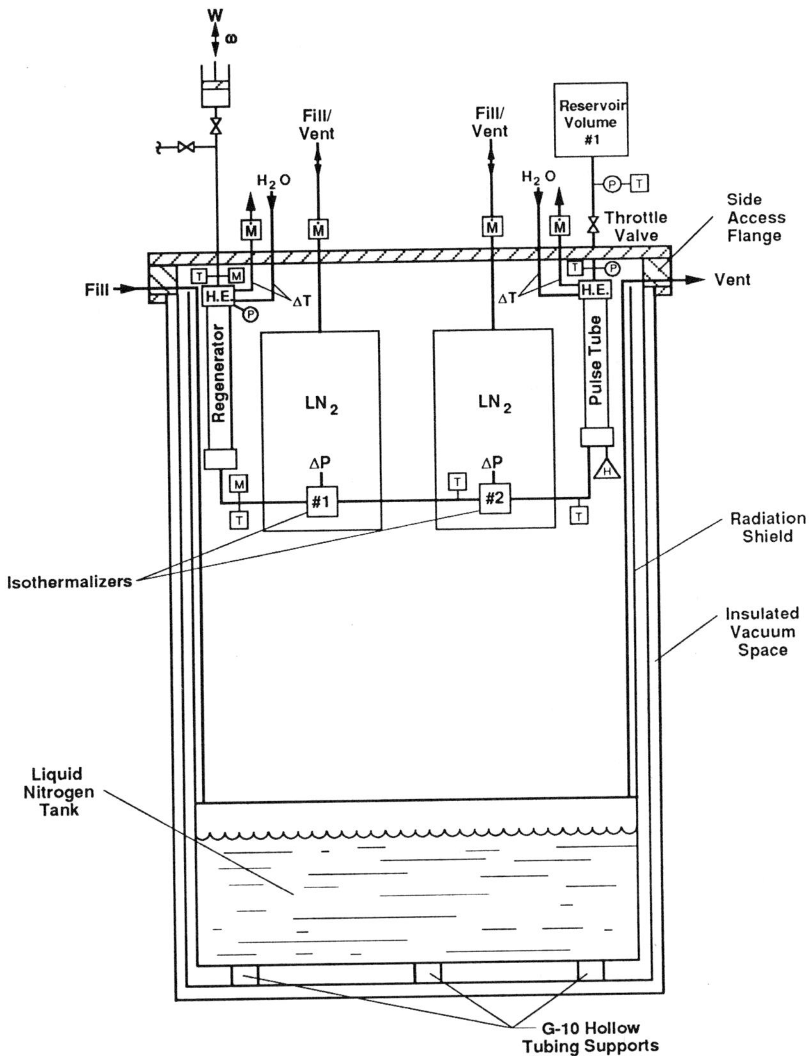

Fig. 2. Schematic of the apparatus designed to evaluate regenerator performance.

$$\lambda = \frac{\oint \dot{m} T \, dt}{m \Delta T},$$
(1)

where $\dot{m}$ is the mass flow rate, m is the mass of gas which flows in one half cycle, T is the gas temperature, ΔT is the temperature difference between the ends of the regenerator, and t is the time.

Both $\dot{m}$ and T can be evaluated anywhere within the regenerator but must be measured at the same location. Generally, it is convenient to make these measurements at the hot and cold ends of the regenerator.

The system being evaluated operates at frequencies of up to 30 Hz. Void volumes introduced into the system must be much smaller than the pulse tube volume. Temperature measurements are made inside 3 mm stainless-steel tubing. The thermometer support and electrical connections, located outside the 3 mm tubing, must also fit within the limited space of the surrounding cryostat. Pressure inside the tubing, where measurements are taken, can be as high as 5.0 MPa, while the surrounding cryostat is maintained at high vacuum; therefore, vacuum integrity is a necessity. Ideally, the temperature probe should neither restrict the gas flow through the tubing nor disturb the flow profile, while measuring the gas temperature as near to midstream as possible. Temperature measurements are to be at room and liquid nitrogen temperatures.

These design parameters severely restrict acceptable thermometer choices. Monitoring the temperature wave with as little lag as possible at 30 Hz requires a response time of less than 0.3 ms. The mass flow rate, with its periodic reversal in flow, is continually changing and, hence, the value of the heat transfer coefficient is continually changing. To obtain the desired response, the sensor must be small, have a small thermal mass, and have as high a thermal conductivity as possible. On the other hand, locating the sensor at midstream in the oscillating flow places large stresses on it, requiring that the sensor be as robust as possible. These factors present serious conflicts between what is needed and what can be practically achieved.

To evaluate potential thermometers, two parameters were considered to be of primary importance. These were the theoretical thermal penetration depth and the thermal time constant. The thermal penetration depth in a thermometer subjected to an oscillating flow is obtained from the relation

$$D_t = \sqrt{\frac{k}{\pi \rho C_p \nu}}, \tag{2}$$

where D_t is the thermal penetration depth, k, C_p, and ρ are the thermal conductivity, heat capacity, and density of the thermometer material, respectively, and ν is the frequency of oscillation. Equation (2) is valid for a semi-infinite plane within the thermometer. The thermometer time constant, or response, is calculated from

$$\tau = \frac{\rho V C_p}{hA},\qquad\qquad\qquad (3)$$

where V and A are the volume and surface area of the thermometer, respectively, and h is the convective heat transfer coefficient between the thermometer surface and the fluid. In meaningful temperature measurements, the thermal penetration depth should be much greater than the thermometer thickness and the thermal response time must be less than 0.3 ms.

TEST PROCEDURES

Thermocouples and resistance thermometers are two temperature measurement devices which offer the potential for rapid response and miniaturization. Miniature vacuum fittings, which use a metal gasket, were modified to allow the temperature probes to be inserted in the 3 mm tubing. These fittings allow the pressure integrity of the system to be maintained, and provide for convenient insertion and removal of the temperature probes within the limited available space. The temperature probes can be fabricated in several different ways, and their assembly is discussed in the following sections.

A cylindrical tube, with a temperature probe and a pressure transducer inserted at one end of the tube and connected to a compressor at the other end, was used to test the probes. The tube, or test volume, has a diameter of 38 mm, a length of 621 mm, and a volume of 670 cm^3. Nearly adiabatic behavior during compression and expansion of the gas in the test volume was assumed, since the length and diameter of the tube were large compared to the gas thermal penetration depth. This arrangement permitted the pressure amplitude to be measured near the location of the temperature probe. Since the probe is located near the end of the test volume, gas velocities are low, and the associated convective heat transfer coefficient between the gas in the test volume and the thermometer approaches its lowest value. This configuration represents the worst-case condition that occurs in the pulse tube refrigerator. Once the pressure ratio of the test volume at the thermometer is known, and since the gas in the cylindrical tube is assumed to behave adiabatically, a theoretical value for the gas temperature variation can be calculated and compared to the experimentally determined value. Durability of the probe, on the other hand, is tested by subjecting it to both oscillating pressure and oscillating mass flow in a separate test apparatus.

TEST THERMOMETERS

Thermocouple probes

All thermocouple probes tested were fabricated from 5 μm thick chromel/constantan foil. The theoretical thermal penetration depth of this foil at 30 Hz and 300 K is approximately 250 μm. The theoretical time constant for the foil in helium gas at a pressure of 2.2 MPa, a temperature of 300 K, and assuming free convection, is 42 ms. Even though this time constant is several orders of magnitude larger than desired, the larger dimensions made probe fabrication simpler, since the probe could initially be constructed of the more durable 5 μm foil and then etched in place to the desired thickness.

All thermocouple probes of this type were constructed by imbedding the foil and attached wires in a cast epoxy rod, which was then epoxied into a modified vacuum gland. Several thermocouple foil configurations were tried before one was developed that would not break after being subjected to the simulated operating conditions of the pulse tube. Figure 3 provides a schematic of the successful configuration. Tests on this version of the thermocouple probes showed that at low frequencies the measured temperature lagged the pressure wave by about 16 ms, but as the frequency of oscillation was increased, the lag of the temperature wave to the pressure wave decreased to approximately 4 ms (see Figures 4 and 5). This effect appears to be due to the thermal conduction between the thermocouple foil and the probe support. The magnitude of the temperature oscillation also decreased with increasing frequency due to the

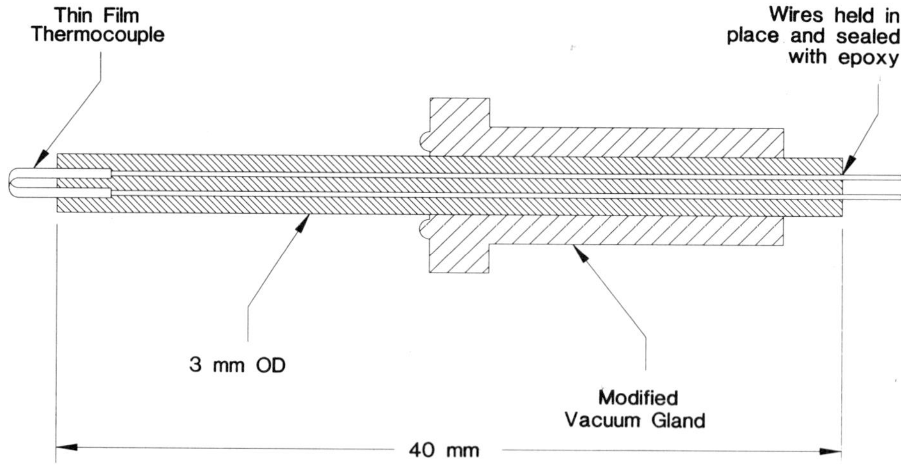

Fig. 3.   Schematic of the probe design for the 5 μm chromel/constantan foil thermocouple.

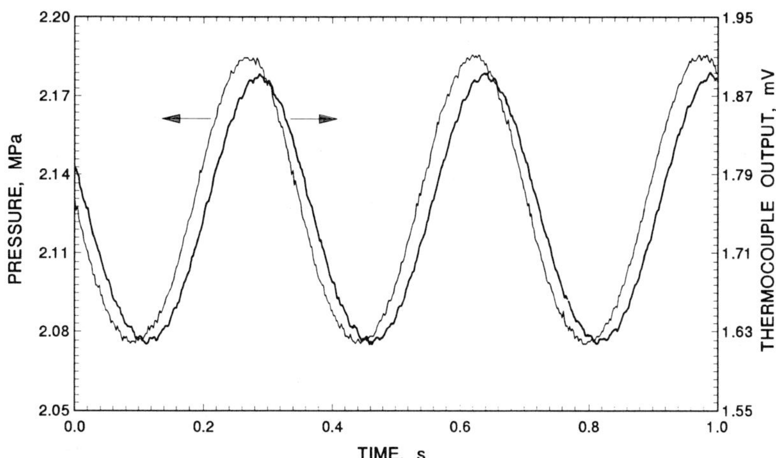

Fig. 4. Temperature oscillations of the 5 μm chromel/constantan
thermocouple vs. pressure in the test volume using helium
gas at 2.2 MPa, and a frequency of 3 Hz.

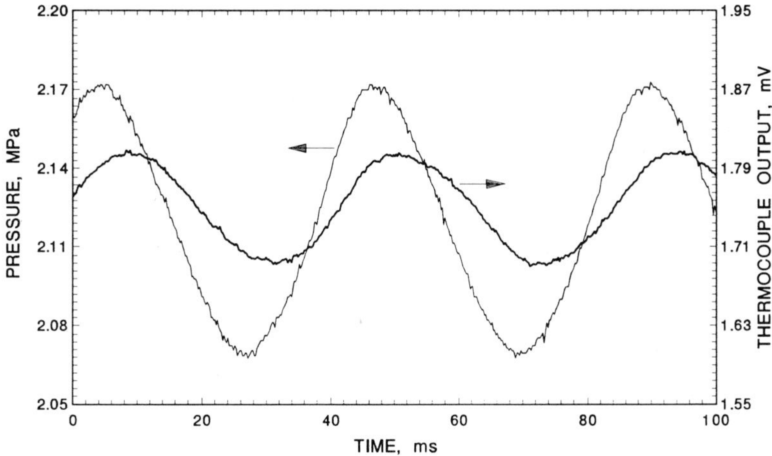

Fig. 5. Temperature oscillations of the 5 μm chromel/constantan
thermocouple vs. pressure in the test volume using helium
gas at 2.2 MPa, and a frequency of 23 Hz.

decrease in the thermal penetration depth. For some application this configuration may be satisfactory, but it is unacceptable for the intended application. Etching the exposed foil to a smaller dimension to improve the response time would have decreased the conduction problem. However, the lead/lag problem was of such a magnitude that it would have required a reduction in foil dimensions to the point where fragility of the probe would again have become a problem.

Platinum resistance thermometer

Platinum wire is commercially available in a wide variety of wire diameters and has a large data base supporting its thermal characteristics as a resistance thermometer. The thermal penetration depth of platinum at 30 Hz and 300 K is greater than 500 μm, which is more than adequate.

The probe body for the platinum resistance thermometer was constructed from epoxy resin, and supports for the wire were tapered needles (see Figure 6). Fine platinum wire for the probe was available in a form known as Wollaston wire. Wollaston wire consists of a platinum wire core encased in silver. The outside silver is 50 to 100 μm in diameter, making the wire relatively easy to handle. The platinum wire center is available in many different small diameters. Platinum wires of 1, 2, and 5 μm diameter were selected for the temperature probes. Theoretical time constants of 11, 44, and 276 μs, respectively, were calculated for the three wire diameters, assuming free convection, a helium gas pressure of 2.2 MPa, and a temperature of 300 K. After the Wollaston wire was

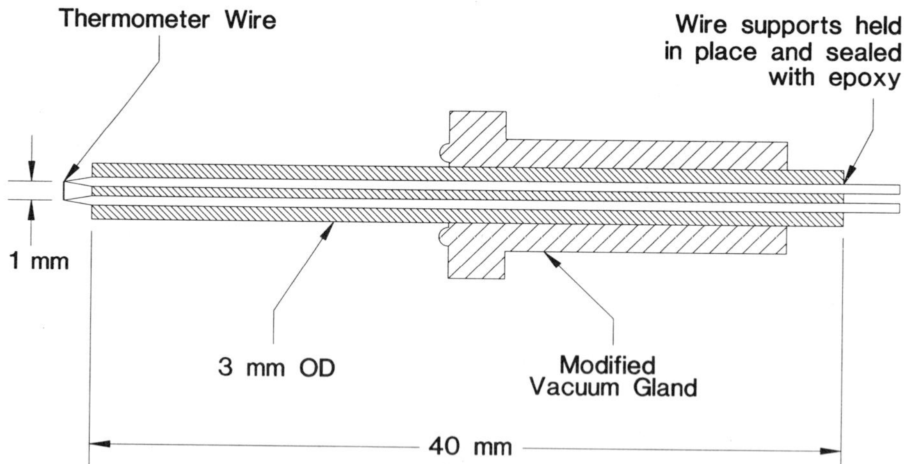

Fig. 6.   Schematic of the probe design for the resistance thermometer probes.

soldered to the ends of the probe supports, the silver was etched
away from the mid-point of the probe by a drop of nitric acid, which
does not attack platinum.

Temperature probes using any one of the three platinum wire
diameters provided satisfactory operation in unidirectional flow,
but they were destroyed in a matter of seconds in oscillatory flow.
It was decided not to try larger diameter wires because of the
slower response times and the lower resistance values. Increasing
the wire diameter also would have decreased the length-to-diameter
ratio of the wire, thereby increasing heat losses by conduction to
the probe supports. These losses could have become just as significant as those experienced with the thermocouple probes. Finally, a
51 μm diameter quartz wire covered with a thin film of platinum was
used in the temperature probe. This wire showed good durability but
exhibited an excessive time lag.

Tungsten resistance thermometer

Problems with the platinum wire led to the use of tungsten
wire. Tungsten wire is much more durable than platinum, has a good
temperature coefficient of resistivity, and is available as a fine
wire. Tungsten, however, has several problems which make it a more
difficult material to handle. It is not available in an easy-to-work
form such as Wollaston wire. Since tungsten oxidizes easily and has
a much rougher surface texture, it is prone to surface contamination. These properties can result in a calibration shift with
time. Tungsten wire is available commercially with a thin film of
plated platinum. This minimizes the oxidation and contamination
problems, but not the handling problem. Fortunately, a commercial
probe was available that, with a few minor alterations, could be
made to fit within a vacuum gland. This involved enlarging the
gland opening to accommodate the larger diameter of the probe.
Furthermore, a sleeve had to be glued around the portion of the
probe which extended beyond the vacuum gland to increase this
diameter and to minimize the excess void volume in the vacuum
fitting (see Figure 7). After being epoxied in place, the probe was
shipped back to the vendor, who then soldered the tungsten wire in
place using a sophisticated micromanipulator. This added greatly to
the cost of the probe but eliminated the handling problems. The
wire diameter used for the probes tested here was 4 μm.

The theoretical thermal penetration depth of tungsten at 30 Hz
and 300 K is greater than 800 μm. The theoretical time constant of
a 4 μm tungsten wire in helium gas at a pressure of 2.2 MPa, a
temperature of 300 K, and assuming free convection, is 160 μs. The
4 μm tungsten wire should therefore give the required response time
at 300 K; the response at 70 K is even faster. The response time is
not expected to change significantly with pressure. This configuration proved to be very rugged; it was tested for durability in

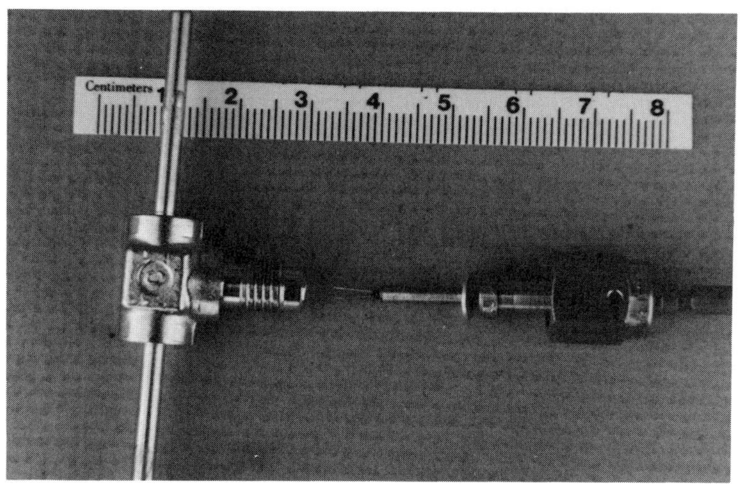

Fig. 7.   Photograph of the modified tungsten resistance probe and the modified vacuum fitting.

extended operation and during on/off cycling. The resistance of the 4 μm tungsten wire at room temperature is 5.94 Ω and it can be operated at 3 mA negligible self-heating. The probe is operated at two temperature levels: room temperature and liquid nitrogen temperature. Calibration curves for the probe are presented in Figure 8. Over the range of temperatures from 77 to 300 K the curve is linear, with a slope of approximately 0.0225 Ω/K, and therefore has good resolution in the region of interest. The intrinsic sensitivity, defined as

$$\beta = \left(\frac{T}{R}\right)\frac{dR}{dT},\tag{4}$$

where T is the temperature and R is the resistance of the tungsten thermometer, is 1.032 at 277 K. This value remains nearly constant for temperatures down to 77 K.

To check the response of the probe, a self-heating test was performed. The test was performed by placing the thermometer in a Wheatstone bridge and applying a step function change in the bridge's input voltage. The higher input voltage caused the thermometer to self-heat. The thermal time constant, as shown in Figure 9, was found to be approximately 260 μs for both heating and cooling. The time constant was determined by using a digital oscilloscope and evaluating the time at which the sensor voltage achieved 63.2% of its final steady state value after applying the voltage step input to the bridge. The probe was then evaluated in the test volume for its temperature response due to oscillating pressure. The response of the probe at 5.4 Hz and 30 Hz, both

80

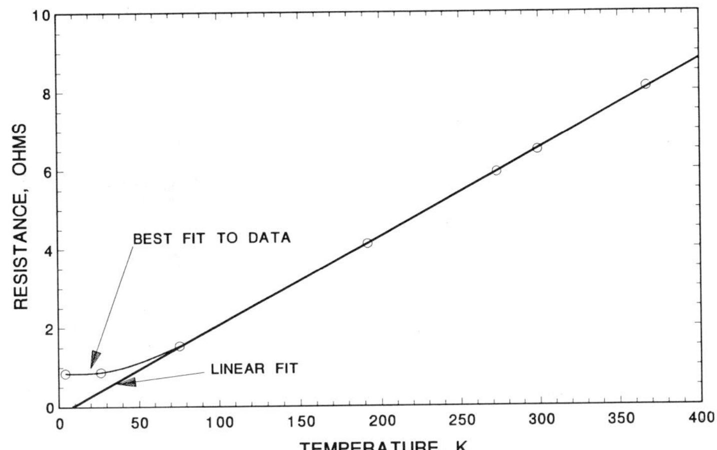

Fig. 8. Calibration curve for the tungsten resistance thermometer. The straight line is the best linear fit through the
points at liquid nitrogen temperatures and above, with a
correlation coefficient of 0.9999.

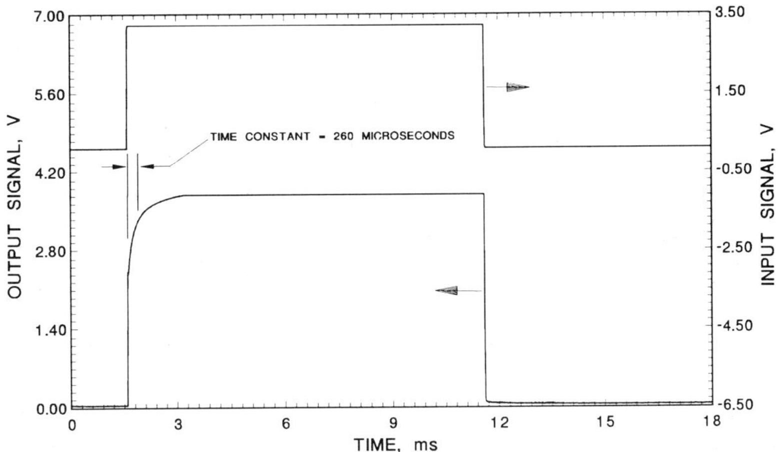

Fig. 9. Frequency response test of the 4 μm tungsten resistance
thermometer.

subjected to a pressure of 2.2 MPa, can be seen in Figures 10 and 11, respectively. The piezoresistive pressure transducer can easily respond to frequencies of over 1000 Hz with negligible phase shift. In comparing these figures it is seen that the amplitude of the temperature remains approximately constant for both frequencies. The figures also show that the pressure and temperature oscillations, as measured by the probe, are in phase at both frequencies, which is in agreement with the theoretical behavior for an adiabatic volume. The change in the voltage of the resistance thermometer indicates that the system had a temperature ratio (the maximum measured temperature divided by the minimum measured temperature) of 1.0152 ± 0.0008 at 5.4 Hz and 1.0144 ± 0.0001 at 30 Hz. The theoretical temperature ratios, calculated from the system pressure ratios and assuming adiabatic conditions, were 1.0211 ± 0.0001 at 5.4 Hz and 1.01900 ± 0.0001 at 30 Hz. These small discrepancies between experimentally and theoretically determined temperature ratios are probably due to the system not being truly adiabatic. The differences between experimentally and theoretically determined temperature ratios also decrease with increasing frequency. This trend follows expectations, because with an increase in frequency there is less time for heat transfer in the helium gas; thus, system operation more closely approaches adiabatic conditions.

CONCLUSION

Several different temperature probes have been constructed and

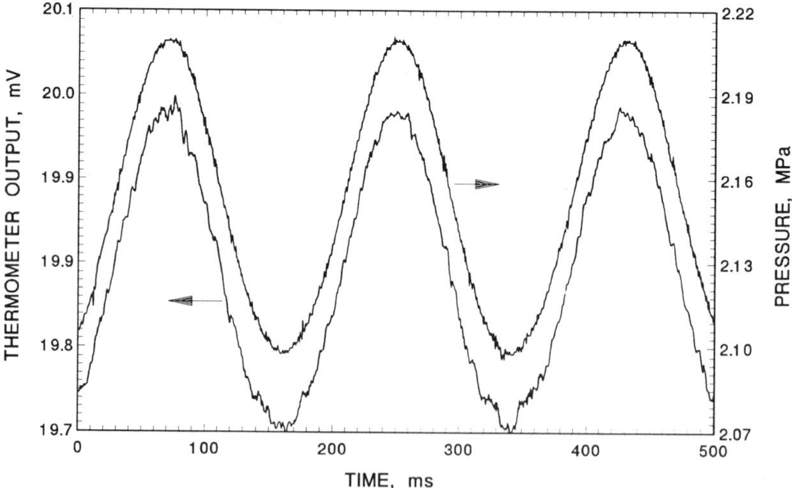

Fig. 10. Temperature oscillations of the 4 μm tungsten resistance thermometer, run at 3 mA, vs. pressure in the test volume using helium gas at 2.2 MPa, and a frequency of 5.4 Hz.

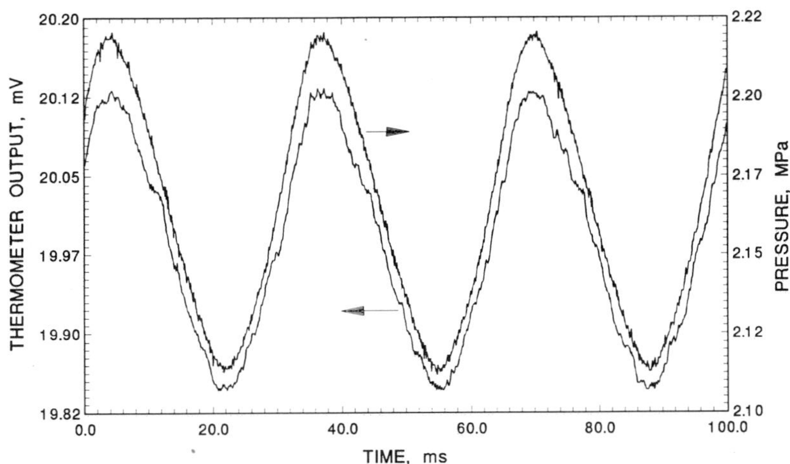

Fig. 11. Temperature oscillations of the 4 μm tungsten resistance
thermometer, run at 3 mA, vs. pressure in the test volume
using helium gas at 2.2 MPa, and a frequency of 30 Hz.

tested in an attempt to meet the desired design criteria of fast
response, durability, and small size. Thermocouple probes potentially promised a relatively fast response time but exhibited large
conduction paths to the probe supports and proved unacceptable.
Platinum resistance probes offered a good response time, were well
characterized, and were relatively easy to work with, but lacked
durability. The 4 μm tungsten wire, though more difficult to work
with, had all the desired characteristics. The tungsten wire was
small, and, when supported on fine wire supports, was relatively
unobtrusive even when located in midstream of a gas flow. It showed
good thermal characteristics, had a good response time, and
exhibited excellent durability.

REFERENCES

1. R. Radebaugh, Pulse tube refrigeration, a new type of
    cryocooler, <u>Japanese Journal of Applied Physics</u>, Vol.
    26 (1987), p.2076
2. W. Rawlins and R. Radebaugh, An Apparatus for the measurement
    of regenerator performance in pulse tube refrigerators, in:
    "Advances in Cryogenic Engineering," Vol. 35, R. Fast, ed.,
    Plenum Press, New York (1990), p. 1213.
3. R. Radebaugh, D. Linenberger, and R. O. Voth, Methods for the
    measurement of regenerator ineffectiveness, in:
    "Refrigeration for Cryogenic Sensors and Electron Systems",
    NBS special publication 607 (1981), p. 70.

PERFORMANCE CHARACTERISTICS OF SILICON DIODE

CRYOGENIC TEMPERATURE SENSORS

B. C. Dodrill, J. K. Krause
P. R. Swinehart, and V. Wang

Lake Shore Cryotronics, Inc.
64 E. Walnut St.
Westerville, OH 43081

ABSTRACT

Performance characteristics with current
technology for silicon diode temperature sensors are
outlined and data presented which illustrates the
commonly encountered problems associated with their
use. The packaging which resulted in a wide range,
hermetically packaged diode sensor is presented. Data
on the matching characteristics and reproducibility of
the sensor are discussed. In addition, a means of
enhancing accuracy over the temperature range from 50K
to 335K to better than 0.15K with the aid of a two
point calibration is illustrated.

INTRODUCTION

Diode temperature sensors have been in use for
over thirty years and have become firmly established as
rugged and reliable temperature sensors. The
popularity of diode sensors is due to the advantages
they offer over alternate forms of thermometry; wide
range, high sensitivity, high signal level, ease of
use, and inexpensive operation. Several papers cover
these aspects in detail.[1,2,3,4,5,6,7,8,9,10,11,12,13]

Applications of Cryogenic Technology, Vol. 10
Edited by J.P. Kelley, Plenum Press, New York, 1991

Diode sensors historically have been available in a wide variety of packages and package configurations. The practice of potting the diode chip in epoxy or potting a glass encapsulated commercial diode was the consequence of the demand for small, low thermal mass sensors. These configurations also permitted the small manufacturer to make small quantities of sensors by hand, which was necessary in the developing years of the cryogenic sensor market. In this "traditional" diode package, the sensing element itself was usually in intimate contact with the epoxy or glass. Thermal expansion mismatches between the epoxy or glass package and the diode sensing element could result in instabilities of the temperature sensor with time. In addition, the poor thermal properties of the materials used in the package construction are detrimental to the overall performance and reliability of the temperature sensor.

This paper first presents self-heating data taken at Lake Shore Cryotronics, Inc. on an epoxy encapsulated diode package to illustrate what aspects are critical in the design of a temperature sensor and how they influence performance. A thermal design is then presented which addresses these issues and supporting data illustrating the device performance are given. In addition, current design and understanding of the silicon diode sensing element allows the user to select a Lake Shore silicon diode sensor and, with a two point calibration, achieve accuracies which approach 0.1K over the temperature range from 50K to 335K.

SELF-HEATING

Self-heating in a sensor refers to the situation where the power dissipated in the sensor causes the sensing element to locally warm to a higher temperature than its package and/or surrounding environment. Self-heating is especially critical in cryogenic temperature sensors because thermal conductivities near liquid helium temperatures may be smaller by orders of magnitude than they are near room temperature. Self-heating in a sensor is not necessarily bad as long as the effects due to the self-heating are reproducible and create minimal temperature uncertainties. Unfortunately, they usually are not, and heat flow from and to the sensor must be carefully considered.

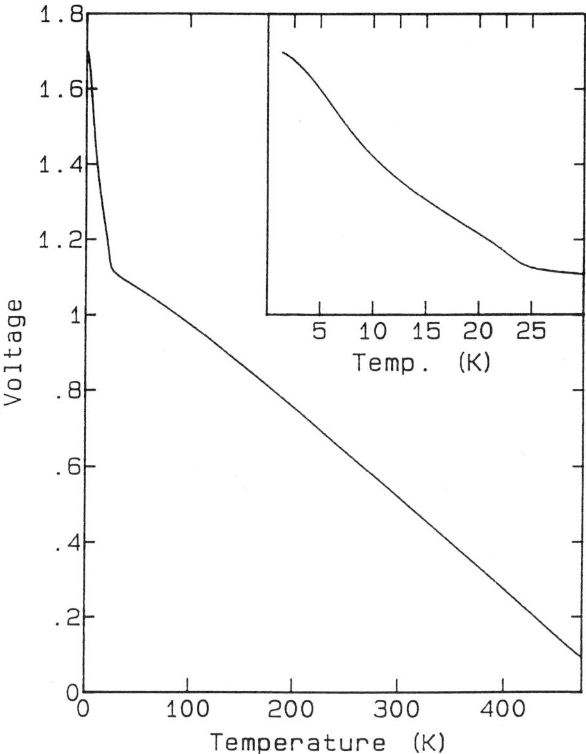

Figure 1. Voltage characteristic of the silicon diode
temperature sensors discussed in the text.

Self-heating of the junction in a silicon diode
sensor is unavoidable due to the relatively high power
levels at which the sensor is operated. A typical
silicon diode sensor, at 4.2K with a bias current of 10
microamperes, dissipates approximately 20 microwatts
(Figure 1), which is nearly three orders of magnitude
larger than the typical power dissipation associated
with reading a 1000 ohm germanium resistance
temperature sensor at the same temperature. Consequently, the potential for self-heating errors in diode
thermometry makes the packaging and mounting of the
diode extremely critical to its overall performance and
reliability as a temperature sensor.

Self-heating tests for resistance sensors are
straight forward. At a given temperature, the current
is set and the resultant voltage is measured. The
calculated resistance (V/I) should exhibit no current
dependence. Any deviation in measured resistance at

constant temperature with increasing current is the
result of self-heating and, hence, the associated
temperature error for a given current level is readily
measurable. However, the forward voltage for a diode
has a non-linear relationship with current, and
self-heating is not as easily measured.

SELF-HEATING TEST RESULTS

Two different tests have been developed for
examining the self-heating effects in diode temperature
sensors. The first test involves mounting the diode
sensors onto an isothermal copper block as they would
normally be mounted in use. Calibrations against
germanium resistance thermometers are carried out below
4.2K with the block first submerged and directly in
contact with a pumped helium bath (tops of sensors
exposed to liquid). The sensors are then recalibrated
with the copper block mounted in a vacuum. A thermal
link is used between the copper block and the bath to
provide the necessary cooling. This test exposes
mounted sensors to two drastically different thermal
environments. Note that the sensor is never dismounted
during the course of these tests. An ideal package
design would result in no difference in temperature
measurement between the two calibrations described.

Averaged results for the vacuum/liquid calibration
comparison for a group of 5 miniature epoxy
encapsulated silicon diode sensors are shown in Figure
2. The temperature range is between 1.5K and 4.2K and
current excitations are 1, 10 and 100μA where the
deviation from the corresponding vacuum calibration is
shown on the vertical axis. The sensors indicate a
warmer temperature when mounted in vacuum. As
expected, self-heating induced errors increase as the
excitation current is increased. This current-
dependent trend is fairly typical of all diode packages
examined, except that the sapphire-based hermetic
package suffers about a factor of ten less offset than
the less efficient designs.

The results in Figure 2 indicate that the solution
to the power dissipation problem would be to reduce the
operating current to 1μA, but this raises the static
impedance of the diode into the megohm range, which
lowers the signal-to-noise ratio, resulting in tighter
constraints on the measurement electronics as well as

the care required in shielding (installation) of the
sensor and its leads.[14] Consequently, 10μA is a
compromise associated with minimizing errors due to
self-heating and errors due to noise pick-up. Some
manufacturers have historically specified 100μA as the
operating current in order to avoid oscillations and
hysteresis in their temperature characteristics, but
this current level adds dramatically to self-heating
problems and is therefore not recommended.

     The second test for self-heating effects requires
testing for effects related to the mounting of the
sensor. If a sensor can not be dismounted and then
remounted reproducibly from a thermal viewpoint, self-

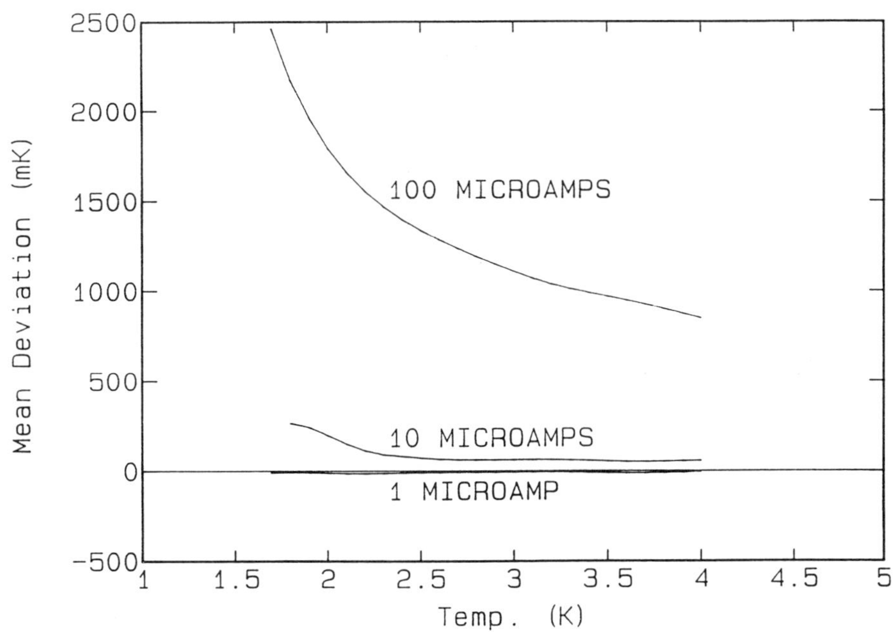

Figure 2.   Self-heating comparison for a group of
                   five epoxy encapsulated silicon diode
                   sensors for temperatures below 4.2K and
                   currents of 1, 10 and 100μA. Tempera-
                   ture error is defined as vacuum
                   calibration voltage minus liquid
                   calibration voltage divided by vacuum
                   voltage sensitivity (dV/dT).

heating can generate temperature measurement errors.
The best test for mounting related self-heating (the
first test only examined power related effects) is to
place an unmounted sensor directly in a liquid helium
bath and then mount the sensor in an isothermal copper
block situated in vacuum at the same temperature as the
bath. This test exposes the sensor to two totally
different thermal environments and two totally
different mounting situations. Again, an ideal sensor
should yield the same reading in both cases.

This test was conducted for 140 miniature epoxy
encapsulated sensors identical to those used in the
power tests above. The average voltage offset between
liquid and vacuum corresponded to a temperature error
of 534mK with an operating current of 10 μA. Note that
this is considerably greater than the offset shown in

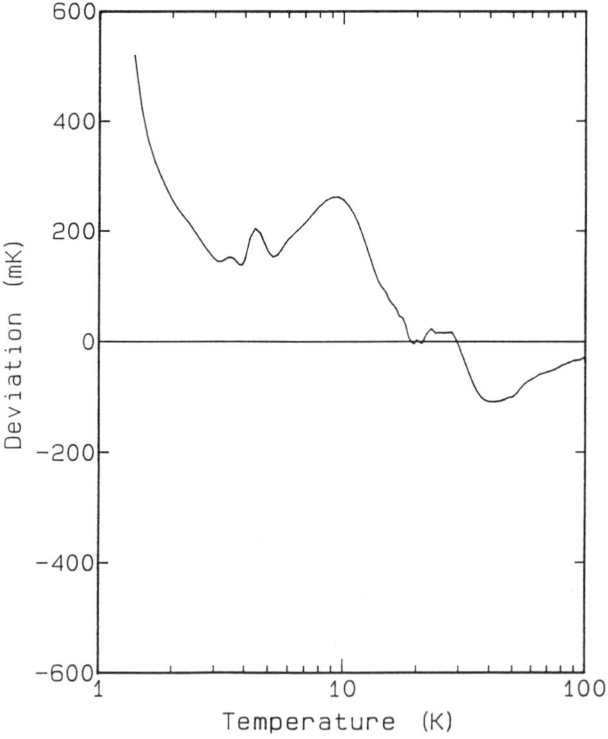

Figure 3. Epoxy Sensor - Baseline: Calibration in
vacuum, sensor greased to calibration block.
Second calibration in vacuum, after
dismounting and remounting.

Figure 2 and demonstrates the importance of providing a
means to reliably mount a sensor.

Figure 3 shows an example of the wider range
deviations which can occur. A small epoxy encapsulated
sensors was calibrated, dismounted, remounted, and then
recalibrated. The deviation between the two
calibrations below 20K can be attributed solely to
self-heating and inability to reproduce the same
thermal characteristics upon remounting the device.
Typical vacuum/liquid offsets for epoxy/glass
encapsulated diodes examined range from a tenth of a
kelvin to as high as one kelvin.

THERMAL DESIGN CONSIDERATIONS

The major thermal problem with traditional epoxy
and glass sensor packages is that the thermal contact
between the sensing element and the outer environment
is through a variety of uncontrolled thermal paths
which depend on the physical potting of the device
(Figure 4a). Therefore, the diode temperature sensor's
response curve is dependent upon the exact mounting
configuration and the conditions of the thermal
environment. Since this environment will undoubtedly
change from calibration to use, package design can
become the most critical part of diode sensor design.
One means around the mounting related problems is to
calibrate the sensor in situ, but in most cases this is
impossible or at least very impractical to do.

The solution to self-heating of the sensing chip
is to have a sensor package with a controlled, low
thermal resistance path between the sensing element and
its outside environment; a thermal path which can be
reliably and easily reproduced when remounting the
temperature sensor. Figure 4b is a schematic view of
such a sensor package.[15] Briefly, the construction
details are as follows. The substrate material is
single crystal sapphire, for high electrical isolation,
yet good thermal conductivity. The base bottom is
fully metalized with molybdenum/manganese and plated
with nickel and gold so that the sensor can be readily
soldered to a mounting surface. The die attach pad is
fired into the sapphire and is surrounded with an
alumina body. The leads are gold-plated Kovar and are
thermally sunk to the substrate to minimize temperature
errors resulting from heat conduction along the leads.

An alumina lid is sealed on under vacuum to form a
moisture free hermetic seal. Since the inside of the
cavity is evacuated, the sensing element is thermally
isolated from the environment except through the
sapphire substrate.

THERMAL DESIGN TEST RESULTS

To determine the magnitude of self-heating effects
for this package design, 140 hermetically sealed diode
sensors were subjected to the vacuum/liquid comparison
at 4.2K described previously. The average measurement
difference was 9.7mK, indicating well controlled self-
heating effects.

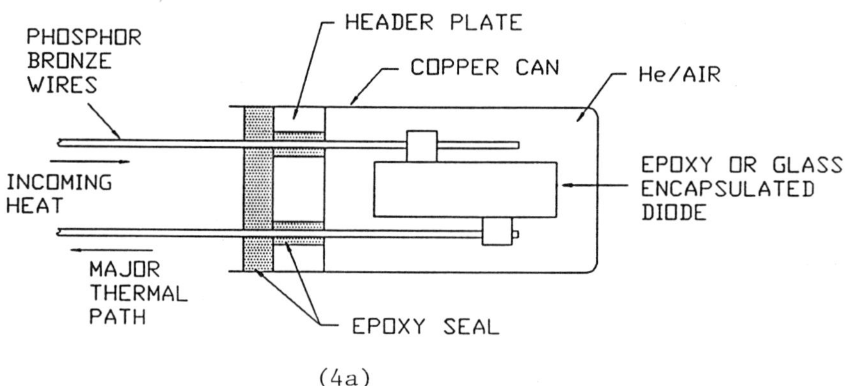

(4a)

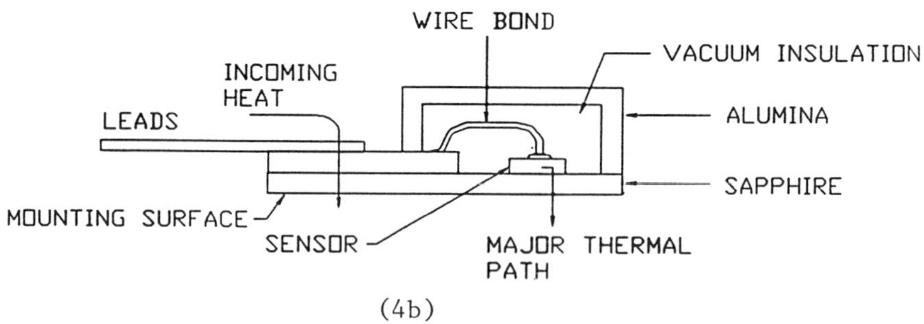

(4b)

Figure 4a.   Schematic view of a typical non-hermetic
               diode package.
Figure 4b   Schematic view of hermetic diode
               package.

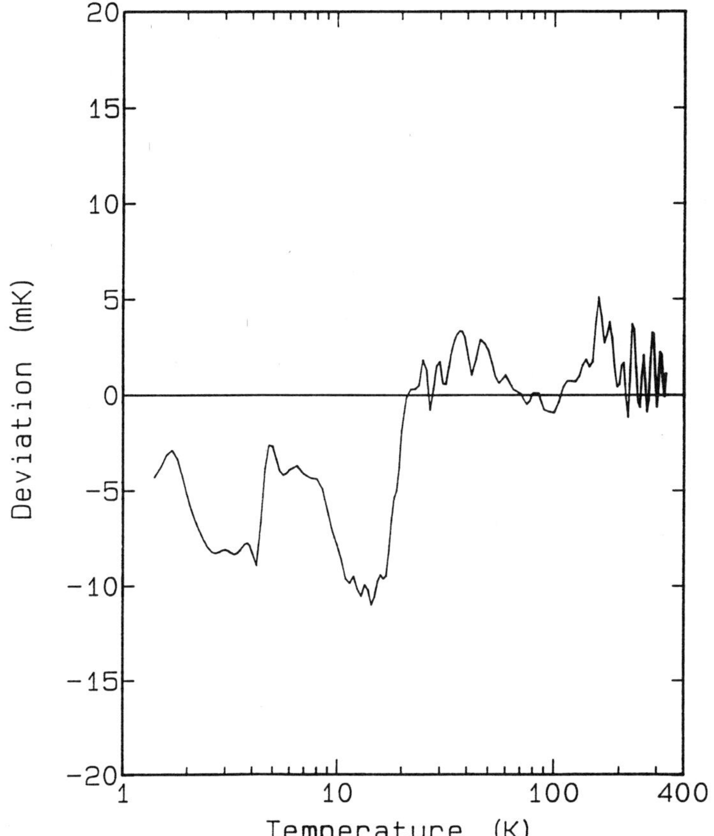

Figure 5.   Hermetic diode sensor - Baseline:
                 Calibration in vacuum, spring clipped to
                 copper calibration block. Second
                 calibration under same conditions after
                 dismounting and remounting. Note scale
                 change from Fig. 3.

Figure 5 illustrates a typical recalibration
comparison for these devices when they are calibrated,
dismounted, remounted, and recalibrated. The
deviations are well within experimental uncertainties
and indicate no self-heating effect. These results
should be compared to those presented in Figure 3.

Obviously, when using commercial packages designed
for room temperature applications and modified for

cryogenic applications, care must be taken to understand the conditions under which the sensor was calibrated as well as having an understanding of the repeatability of the thermal path for the sensor. The hermetic package for diode sensors reduces this concern to the level of ±0.25% of T at 4.2K, i.e., ±10mk or less. This compares very favorably with some sensor/package combinations which we have measured which repeat to levels of no better than ±10% below 10K upon remounting and recalibration, yet recalibrate within about 0.1% if the mounting is left undisturbed.

INSTALLATION AND USE

    The importance of properly mounting temperature sensors has been discussed in detail elsewhere and, hence, will not be discussed extensively here.[16] The important points to keep in mind when mounting any temperature sensor are: (a) maintain good thermal contact between the sensor and the sample being measured and (b) minimize the amount of heat flowing through the sensor due to radiation or conduction loads. Even a relatively large package, such as a copper bobbin, will not prevent the sensor from reading differently than the temperature of the object to which it is attached if there is a large flow of heat through the package.

    The thermal installation of the leads which run to the sensor is also of primary importance. An excessive heat flow through the connecting leads to any temperature sensor can create a situation where the active element (in this case the diode chip) is at a different temperature than the sample to which the sensor is mounted. This is then reflected as a real temperature offset between the sensor and the true sample temperature. These errors can be minimized by the proper selection and installation of the connecting leads. This too, has been covered elsewhere[17,18], but the following discussion is warranted since the hermetic diode sensor package serves as its own heat sink to compensate for any heat which may be coming down the leads to the device.

    As a measure of how well the devices perform as heat sinks, sensors were operated with an additional power load (typically, a few milliwatts) input to the leads near the sensor. The diodes were monitored at

Table I. Heat sink capacity of hermetic diode package
in various mounting configurations and
associated temperature rise at 4.2K for
Copper(Cu) and Constantan(C) 36 gauge leads,
each one meter in length.

| MOUNTING METHOD | HEAT SINK CAPACITY(mK/μW) | | | Temperature Rise(mK) | | |
	20K	10K	4.2K		4-77K	4-300K
NO HEAT SINKING OF LEADS						
A	1	3	6	Cu C	5400 80	12600 400
B	0.02	0.04	0.15	Cu C	135 2	315 10
C	0.005	0.02	0.15	Cu C	135 2	315 10
LEADS SOLDERED TO SAPPHIRE HEAT SINK						
C	0.002	0.008	0.03	Cu C	27 0.4	63 2

A. Sensor physically clamped in place with grease
B. Soldered to copper adapter & inserted into hole with grease
C. Soldered to copper plate

For 36 gauge wire, the following data was assumed for
temperature differences as shown:
4K - 77K - Copper (900μW/m); Constantan (13μW/m)
4K - 300K - Copper (2100μW/m); Constantan (68μW/m)

their normal 10μA excitation current and the apparent
change in temperature was recorded. Several different
mounting configurations were chosen for this test to
illustrate the importance of heat sinking of sensor
leads. The results are tabulated in Table I. The
temperature rise was determined by calculating the heat
flow down a 1 meter long, 36 gauge wire thermally
anchored at the two temperature extremes with no additional thermal anchoring. The above table demonstrates
why copper wire is seldom used for sensor applications
and thermal lagging of the sensor wires is very
important if temperatures near that of liquid helium
are to be measured.

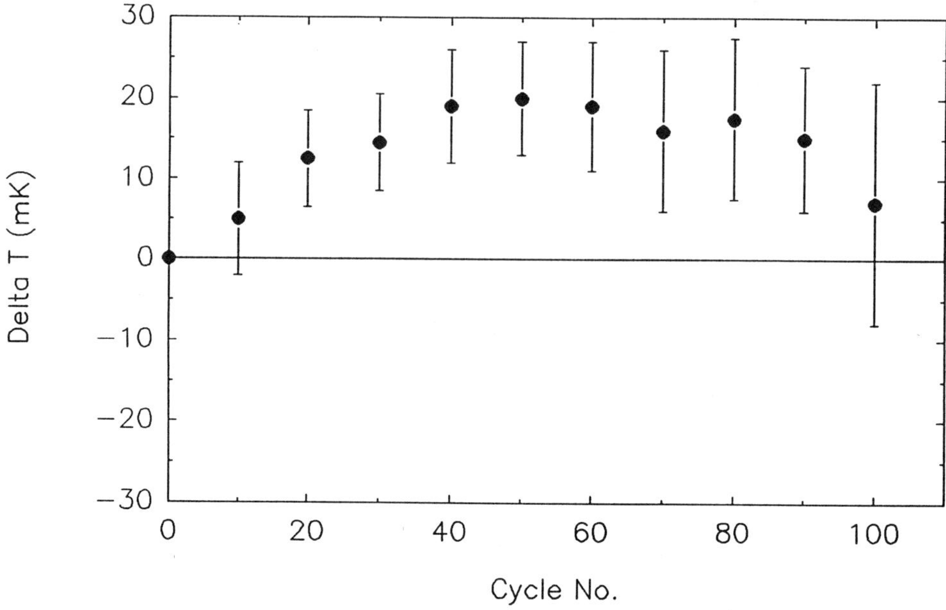

Figure 6. Short term repeatability for a group of
eleven hermetic diode sensors cycled 100
times between room temperature and liquid
Helium temperature. The data was corrected
for temperature variation of the bath using a
germanium resistance thermometer. A reading
was taken every tenth cycle and compared to
the first helium reading.

SHORT-TERM REPEATABILITY

To determine short term repeatability for the
sensors, two different tests were conducted. First, a
group of eleven hermetic diode sensors was cycled 100
times from room temperature into liquid helium. Data
was measured for each sensor every tenth cycle. Figure
6 is a graphical representation of the average repeatability of the sensors at 4.2K. Deviations were
calculated with respect to the first 4.2K dip value for
each sensor and the results averaged. The vertical bars
represent plus and minus one standard deviation. As was

mentioned earlier, this test only determines short-term
repeatability at a single temperature and is,
therefore, not a measure of full range reproducibility.

LONG-TERM STABILITY

In order to determine the exact reproducibility or
stability characteristics for the hermetic diode
sensors, several groups were recalibrated periodically
over an extended period of time. The devices were
dismounted and stored at room temperature between
calibrations. No other stress conditions were applied
to them.

Figure 7 is a recalibration comparison for a group
consisting of twenty sensors. The sensors were
calibrated eight times over a twenty month period. The
final calibrations were compared to the original calibrations to determine mean deviations as a function of
temperature. The mean deviation for the group of
twenty diodes is shown in Figure 7. The worst case
deviations between calibrations at all temperatures
between 2 and 330K were less than 50mK. In comparing
recalibrations of sensors, the accuracy of the
calibration itself must be considered along with the
reproducibility of the sensor. Combining the stability
specification of ±30mK with the calibration accuracy
specification of ±20mK and an allowance for customer
instrumentation errors yields an estimated worst case
long term uncertainty of about 100mK. The results for
this group of twenty sensors recalibrated with the same
instrumentation were a factor of two better than this
specification at all temperatures for all sensors.

THERMAL TIME CONSTANTS

Thermal time constants are often of importance to
sensor users. Thermal time constants were measured at
liquid helium, liquid nitrogen, and room temperatures
for the hermetic diode sensor. Due to the package's
low mass (about 35 milligrams), the response times are
relatively fast. They are typically less than 10
milliseconds at 4.2K, less than 100 milliseconds at
77.35K, and on the order of a second at room
temperature.

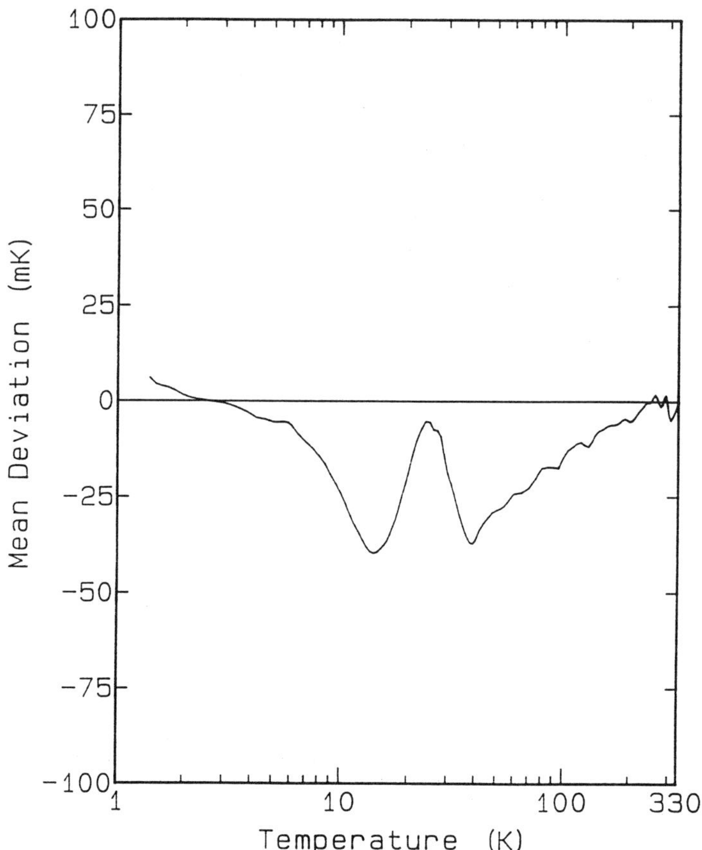

Figure 7.   Mean deviation as a function of
temperature for twenty sensors upon
recalibration after twenty months.

CURVE CONFORMANCE/INTERCHANGEABILITY

An important feature for any temperature sensor is
the uniformity between sensors and the ability to
provide standardized specifications for those sensors.
The ability to establish a standardized calibration
curve means that individual sensor calibrations for
many applications need not be performed.

Unlike the situation that exists for industrial
grade platinum resistance sensors (where standardized
specifications have been established), there have been
a wide variety of diode voltage/temperature curves, all
with slightly different thermometric characteristics,

available from several different manufacturers. For
platinum resistance sensors, the temperature-resistance
curve is determined primarily by the platinum purity.
The physics of silicon diode temperature sensors are
much more complicated; the voltage-temperature
characteristics depend on junction area, starting
material, doping elements and their densities, junction
depths, impurity concentrations, packaging and
conduction mechanisms and other parameters.

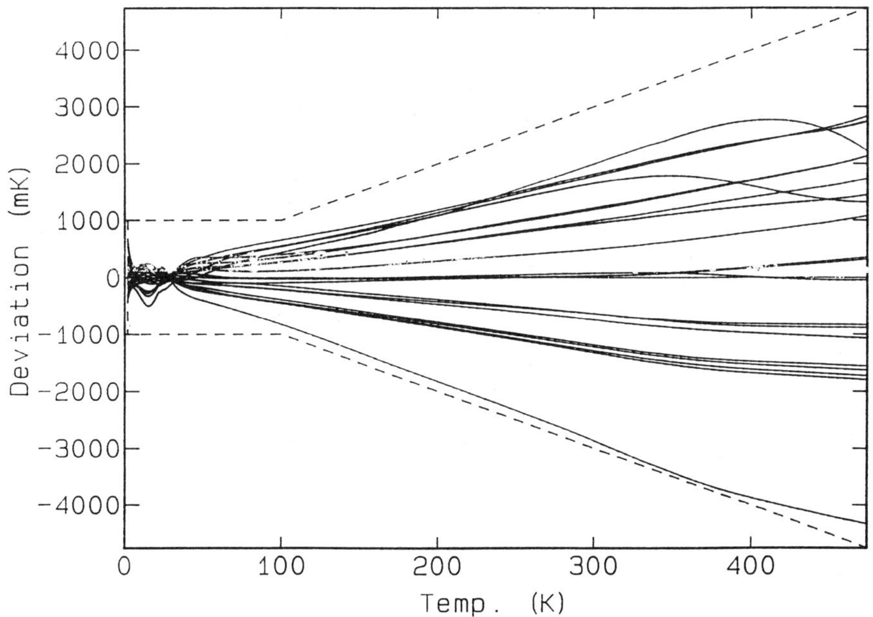

Figure 8. Deviation in temperature of several hermetic
          diode sensors from the standard curve over
          the temperature range from 1.4K to 300K.

INTERCHANGEABILITY

     A standardized calibration curve is generated for
diode sensor type by calibrating a large group of

sensors and determining the mean curve from the
calibration data. The curve shown in Figure 1 (Lake
Shore Curve 10) serves as the standard curve for these
diode sensors. Deviations of individual sensor
calibrations from Curve 10 are shown in Figure 8. They
fall within ±1K or ±1% of the absolute temperature,
whichever is greater. The uniformity of these sensors
in their deviation from the standard curve is of
paramount importance. This uniformity allows the
grouping of these sensors into tolerance bands similar
to the approach used for platinum resistance
thermometers.

In practice, testing of an individual sensor for
conformance to the standard curve can be achieved by
checking at three temperatures; liquid helium, liquid
nitrogen, and room temperature, provided it has been
established statistically from many calibrations that
the diodes do not exceed the specified tolerances
between these temperatures. As can be seen in Figure
8, the calibration curve variations below 40K can allow
a sensor to be very close to the standard curve at
4.2K, but out of tolerance at lower and higher
temperatures. For the diode sensors in question here,
tolerances have been established for selection at the
three temperatures that provide a high confidence level
for tolerance band conformance over the 2K to 475K
temperature range.

The closer the tolerance, the lower the proportion
of sensors meeting the specifications. The closest
tolerance to which reasonably large numbers of sensors
can be provided is ±0.25K from 2K to 100K, ±0.50K from
100K to 305K, and ±1.00K from 305K to 475K. Higher
accuracy matching can be obtained, but with higher cost
and lower availability.

The highest accuracies require the sensors to be
individually calibrated. It is most cost effective to
calibrate a sensor that does not match the standard
curve well (matching tolerances have no effect on
sensor stability or other measures of sensor quality).
Typically, a full range calibration will require about
60 calibration points, which is a relatively expensive
and time consuming process. An intermediate
calibration procedure involving a limited number of
data points (one to three) is the SOFTCAL™ procedure
described below.

SOFTCAL™ "CALIBRATION" FOR T>30K

Due to the uniform nature of the temperature characteristics of the hermetic diode sensors above 30K and the manner in which they converge at approximately 28K, it is possible to generate a precision calibration curve from only one or two calibration points with the standard curve (see Figure 1) serving as a reference curve.

Using a two point calibration measurement and a linear correction to the standard curve, the "calibration" will be given as follows:

$$V(T) = mV_{st}(T) + [V(T_1) - mV_{st}(T_1)]$$

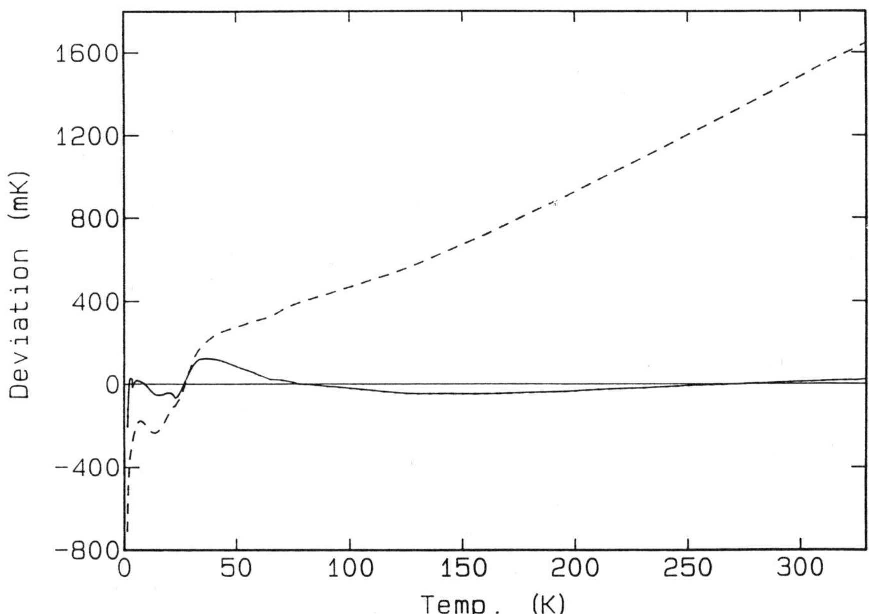

Figure 9. Effect of a SOFTCAL™ "calibration". The dotted line illustrates the accuracy of a poorly conforming sensor when compared to the standard curve. The solid line illustrates the improved accuracy of the same sensor with a three-point SOFTCAL™ "calibration".

with the slope m defined as

$$m = (\ V(T_2) - V(T_1)\ )/(\ V_{st}(T_2) - V_{st}(T_1)\ )$$

where

T_1, T_2	= known test temperatures with known voltages $V(T_1)$ and $V(T_2)$
$V(T)$	= calibration curve of sensor
$V_{st}(T)$	= Standard curve for hermetic diode sensors.

Note that m and the quantity in brackets are the slope and intercept of the correction. In use, V(T) is measured and the corresponding value $V_{st}(T)$ is determined from the equation. The standard calibration

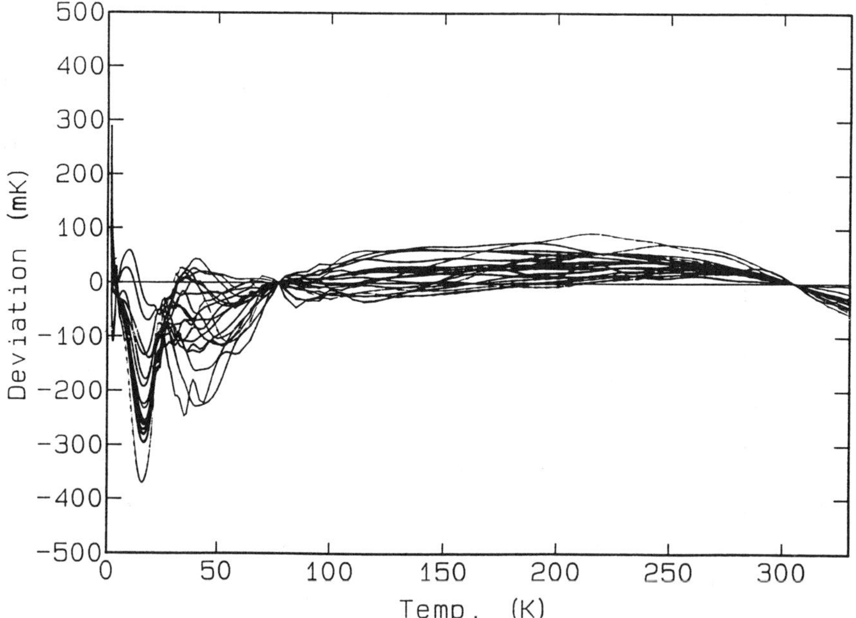

Figure 10. Comparison of a three-point SOFTCAL™ "calibration" with corresponding full-range calibrations of several hermetic diode sensors. SOFTCAL™ calibration points: 4.2K, 77.35K and 305K.

curve is then used to determine the temperature.

ONE-POINT SOFTCAL™ "CALIBRATION"

A one-point calibration will give T_2 and $V(T_2)$, but a pinning point (i.e., where the actual calibration is assumed to match the standardized curve value) is also required. The point selected is $V_s(T_1)=V(T_1)=1.111$ volts, which corresponds approximately to $T_1=28K$. This point was chosen since the hermetic diode sensors tend to converge to within ±0.25K at this point.

The selection of the calibration point is critical in determining the accuracy of the calibration. The closer the point is to 28K, the less accurate the calibration corrections become. For operation in the range T>30K, the calibration point should be chosen in the range 200<T<350K. For example, the ice point (0°C) is a convenient point. The resulting accuracy obtained using this technique is ±0.1K for a temperature span of ±40K around the calibration point, ±0.5K elsewhere in the 30 to 373K range, and ±1K above 373K. The error to which the calibration point is known must also be added to these accuracy specifications.

TWO-POINT SOFTCAL™ "CALIBRATION"

A two-point calibration requires that voltages be measured at two temperatures (T_1 and T_2) as well as defining the pinning point at 28K. To achieve the best wide range results, one calibration point should be near the ice point with the other near liquid nitrogen temperature. The resultant SOFTCAL™ "calibration" will yield an accuracy of ±0.1K between the two calibration points and ±0.2K accuracy elsewhere over the 30 to 373K range. Above 373K, the accuracy degrades to ±1K. Figure 9 illustrates such a SOFTCAL™ "calibration" for a sensor with a temperature error approaching 0.6% of temperature above 100K. The resultant two point correction above 28K improved the "calibration" accuracy to better than 100mK over the temperature range between 50K and 325K. It should be noted, however, that liquid nitrogen and ice baths do not automatically give well defined temperature points. Both can vary by as much as ±0.5K if proper techniques are not used.[19]

SOFTCAL^{TM} BELOW 28K

The same procedure can be used for a correction to
the curve below 28K. Unfortunately, a correction at
4.2K will improve the accuracy to only approximately
±0.3K at 20K. Figure 10 shows the resulting accuracy
of a three point SOFTCAL^{TM} correction for a number of
calibrated sensors. The deviation indicated is the
difference between the SOFTCAL^{TM}-corrected curve and
the actual calibrations for a group of twenty sensors.

SUMMARY/CONCLUSIONS

An improved diode sensor package has been
developed which minimizes the self-heating and
remountability problems commonly encountered with trad-
itionally packaged diode thermometers. Problems arise
from self-heating as a result of either poor or
inadequate thermal design of the sensor package or from
improper installation of the sensor and its leads.
Unfortunately, these problems are neither easily
measured nor found by the user.

Due to the thermal design of the hermetic sapphire
package, the sensing element responds with good
independence from specific mounting configurations or
the thermal environment. Self-heating or mounting
related temperature errors are, on the average, reduced
to the level of 10-30mK at 4.2K from levels of 0.1-1K
due to previously available packages.

Efforts toward the fabrication of a uniform
sensing element have yielded a diode chip with
thermometric characteristics that can be matched to a
standardized calibration curve to tighter tolerances
than previously available. Furthermore, since the
uniformity of the sensors' response is such that they
conform to the standard curve over the full range of
use rather than at only a few points, the specific
level of curve conformance can be routinely determined
by checking the sensor at the discrete temperatures of
liquid helium, liquid nitrogen, and room temperature.

A method of providing improved accuracy from only a few
calibration points, called SOFTCAL^{TM}, has been
developed. Three calibration points can provide
accuracies of 0.1K to 0.3K (plus the errors due to the
inaccuracies of the calibration points) in the range

from liquid helium temperatures to room temperature.
This capability can be provided built into a readout
instrument. The user can calibrate the sensor by
simply establishing the sensor at the calibration
temperature and pushing a button.

Acknowledgements

We would like to thank Dr. J. M. Swartz for
critical comments and fruitful discussions during the
course of this investigation. We also thank L. Smith
for the graphics presentation and J. Salopek for the
manuscript preparation.

REFERENCES

1. A. G. McNamara, Semiconductor Diodes and Tran-
sistors as Electrical Thermometers, _Rev. Scien.
Instrum._, 33:330 (1962).

2. J. Unsworth and A. C. Rose-Innes, Silicon pn-
junctions as Low Temperature Thermometers,
Cryogenics, 6:239 (1966).

3. J. M. Swartz and D. L. Swartz, Cryogenic
Temperature Sensors, _Instrum. Tech._, 33 (1974).

4. D. L. Swartz and J. M. Swartz, Diode and
Resistance Cryogenic Thermometry: A Comparison,
Cryogenics, 14:68 (1974).

5. R. V. Aldridge, On The Behavior of Forward
Biased Silicon Diodes at Low Temperatures, _Solid
State Electron._, 17:607 (1974).

6. R. W. Treharne and J. A. Riley, A Linear
Response Diode Temperature Sensor, _Instrum.
Tech._, 59 (1978).

7. V. Chopra and G. Dharmadurai, Effect of
Current on the Low Temperature Characteristics of
Silicon Diodes, _Cryogenics_, 20:659 (1980).

8. J. Sondericker, Production and Use of High
Grade Silicon Diode Temperature Sensors, _in_: "Advances
in Cryogenic Engineering", Vol. 27, R. W. Fast, ed.,
Plenum Press, New York (1981), p. 1163.

9. M. G. Rao, Semiconductor Junctions as Cryogenic Temperature Sensors, _in_: "Temperature, Its
Measurement and Control in Science and Industry",
American Institute of Physics, New York (1982),
p. 1205.

10. M. G. Rao, R. G. Scurlock and Y. Y. Wu,
Miniature Silicon Diode Thermometers For Cryogenics, _Cryogenics_, 23:635 (1983).

11. M. G. Rao and R. G. Scurlock, Recent Advances
in Si and Ge Diode Thermometers. _in_: "Proceedings of
the Tenth International Cryogenic Engineering
Conference", H. Collen, P. Berglund and M. Krusius,
Ed., Butterworth, Guildford (1986), p. 617.

12. J. K. Krause and P. R. Swinehart, Reliable
Wide Range Diode Thermometry, _in_: "Advances in Cryo-
genic Engineering", Vol. 31, R. W. Fast, Ed.,
Plenum Press, New York (1985), p. 1247.

13. R. M. Igra, M. G. Rao and R. G. Scurlock,
Thermometric Characteristics of the Improved
Miniature Si Diodes. _in_: "Proceedings of the Eleventh
International Cryogenic Engineering Conference",
G. Klipping and T. Klipping, Ed., Butterworth,
Guildford (1986), p. 617.

14. J. K. Krause and B. C. Dodrill, Measurement
System Induced Errors in Diode Thermometry, _Rev.
Scien. Instrum._, 4:57 (1986). This paper gives
a thorough discussion of potential errors associated
with induced currents picked up through the leads
to the diode sensor.

15. The hermetic diode sensor is commercially
available as the DT-470 series diode sensor from
Lake Shore Cryotronics, Inc.

16. J. K. Krause and P. R. Swinehart, Demystifying
Cryogenic Temperature Sensors, _Photonics Spectra_,
(1985).

17. A. C. Rose-Innes, "Low Temperature Laboratory
Techniques", The English University Press, Great
Britain, (1973).

18. G. K. White, "Experimental Techniques in Low
Temperature Physics", Clarendon, Oxford, (1979).

19. D. J. Blundell and B. W. Ricketson, The
Temperature of Liquid Nitrogen in Cryostat Dewars,
<u>Cryogenics</u>, 19:33 (1979).

THERMALLY-COUPLED CRYOGENIC PRESSURE SENSING

David L. Clark

Martin Marietta Astronautics Group
Denver, Colorado

ABSTRACT

Improved methods for measuring low pressures in cryogenic
systems are described. Pressure transducers and associated
sensing lines are thermally coupled to the cryogen to eliminate
thermo-acoustic-oscillations (TAO) caused by thermal gradients
between the flow line and the transducer. The applications include
flow and density measurement, and vapor bubble detection. Fully
immersed and externally mounted transducers are used in these
methods to reduce heat leak and measurement uncertainty.
Implementation considerations are discussed, including transducer
selection and temperature compensation.

INTRODUCTION

Many operational benefits can be obtained using thermally-
coupled pressure transducers for cryogenic applications. Locations
where the sensing line comes into contact with the cryogen are best
suited for thermally-coupled techniques since the external heat
leak can be minimized and TAO will be suppressed. By mounting
the transducer in the same thermal environment as the cryogen,
many measurements which were previously impossible can now be
made. Some of these advanced applications include density and
phase measurement as well as improved flow measurement. The
advent of commercially available transducers engineered
specifically for low-temperature use enables the cryogenic system
designer to employ these concepts with greater confidence and
reliability.

The lighter cryogens, such as hydrogen and helium, are very

susceptible to external heat sources, due to their low heat of vaporization. More exotic cryogenic fluids, such as slush hydrogen and superfluid helium, are even more sensitive. System designers must use every means possible to reduce the heat leak, particularly in flow lines where the sensing lines must contact the liquid. The heat transfer through the sensing lines can generate TAO and amplify the heat leak to a value much greater than that from conduction alone. Pressure measurements under the influence of TAO will show large fluctuations, sometimes greater than the actual pressure being measured. By thermally-coupling the transducer and associated sensing lines to the cryogen, external heat leak is dramatically reduced and TAO are suppressed.

APPLICATIONS

The basic application of the thermally-coupled transducer involves the measurement of pressures along cryogenic flow lines. The concept mounts the transducer and the sensing lines to the inner flow line, inside the vacuum jacket and under Multi-Layer-Insulation (MLI) as shown in Fig. 1. A mounting saddle is used to improve the conduction between the transducer and the inner line. The sensing lines are also thermally anchored to the inner line with some allowance for expansion. When the installation is complete, the only heat leak is through the electrical leads with some added loss from the bulge of the MLI around the transducer. Heat will also be generated by the transducer itself from the excitation current. This installation is particularly useful for flow measurements using fixed-geometry restrictions in light cryogens, since the differential pressure generated is extremely low

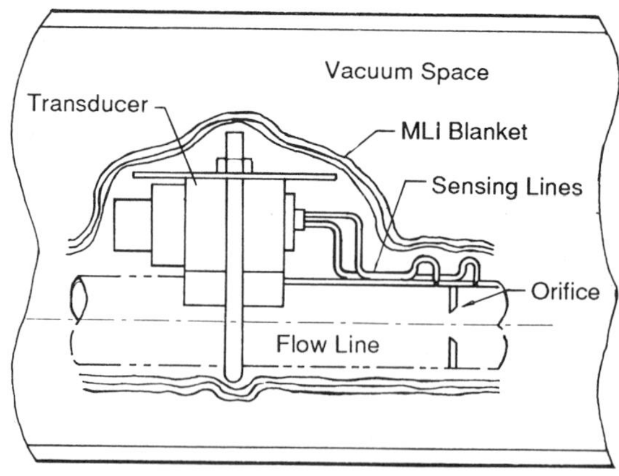

Fig. 1 Thermally-Coupled Flow Meter

and any externally induced pressure fluctuations can drastically degrade the measurement.

An extension of this method enables the measurement of many parameters that were previously impossible or extremely difficult to measure. When the externally induced fluctuations are eliminated, the actual flow perturbations can be detected. This leads to applications for thermally-coupled transducers which can record two phase flow, flow distributions, and entrained vapor bubbles.

Another interesting application uses a differential transducer fully immersed in the fluid with a fixed length reference tube extending vertically from the low pressure port. This leg will maintain a pressure at the low side port and any change in the fluid density in the region of the reference tube will register as a differential pressure. This method was developed for measuring the increased density of solid particles entrained in a triple-point liquid such as slush hydrogen. This sensor also responds to vapor bubbles in the fluid and can be used to detect boiling.

OPERATIONAL CONSIDERATIONS

The most important aspect of thermally-coupled applications is the selection of the transducer. Previous testing involved the use of piezoresistive transducers[1] and variable reluctance transducers[2]. The variable reluctance transducers have been used for superfluid helium testing with some success. The NASA SHOOT experiment is using variable reluctance type transducers. NASA has calibrated and flight qualified selected devices. Some of the problems experienced with these type transducers in cryogenic use are a tendency to develop open sensing coils when the sensor is chilled down too quickly and very poor predictability for temperature effects. Since these transducers are not rated by the manufacturer for cryogenic operation, these problems should be expected. The piezoresistive transducers performed well at the Applied Superconductivity Center but experienced failures upon cold shock. Recently, commercially available piezoresistive transducers have been developed which are rated for cryogenic temperatures and rapid cool-down rates[3]. These devices are predictable in both the zero-shift and gain effects due to the temperature change.

Good thermal contact with the fluid of interest is important for proper use of the thermally-coupled concepts. When the transducer is mounted to the vacuum insulated process line, the transducer should have a saddle to increase the thermal contact, and the use of a conductive grease is recommended. For most applications, the transducer can be assumed to be at the liquid temperature if it is well insulated inside the MLI in a hard vacuum.

However, the flow line must be chilled down and have enough time to cool the transducer. For more accurate results, the actual temperature of the sensing diaphragm must be known. For the piezoresistive transducer, this can be accomplised by two means. The transducer can be ordered with an integral temperature sensor or measurements of the bridge resistance of the piezoresistive transducer can be used to infer the temperature.

Temperature compensation requires a complete understanding of the characteristics of the device at the operating temperature. For the piezoresistive transducer, the zero shift is less than 3 percent of the maximum output which indicates that a sensor temperature measurement error of 10% would only result in a zero-compensation error of less than 0.3%. The gain-shift for the piezoresistive transducer is roughly 1 mV per Kelvin in the low temperature range which translates to 0.28% error per Kelvin at liquid nitrogen temperatures. Clearly, the gain-shift requires a more accurate temperature measurement and any installation should consider this problem. However, these temperature effects are predictable and consistent between transducers.

Another design consideration is the accessibility of the sensor. Since the transducer is installed inside the vacuum space, provisions should be made to access the device for replacement or recalibration. Typically, this involves the addition of a vacuum flange and leak-free, reusable fittings on the sensing lines.

RESULTS

Testing performed at the Engineering Propulsion Laboratory, Denver, CO, has verified applications of thermally-coupled transducers in Internal Research and Development (IR&D) projects. The applications were originally developed in response to measurement challenges and are now widely used where critical cryogenic pressure sensing is required.

The first applications used variable reluctance transducers fully immersed in the cryogen to eliminate heat leaks. These devices were calibrated at cryogenic temperatures using a helium pressurant and a quartz manometer reference. These sensors performed well when they were cooled down overnight to their final operating temperature. The zero-shift and gain-shift were unpredictable and varied anywhere from 20% to 80%. One transducer experienced a zero-shift of 75% of its operating range.

The thermally-coupled flow measurement technique was implemented on a hydrogen transfer line where a differential transducer was thermally-coupled to the line for sensing pressure drop across a flow nozzle. This installation used a previously cold-

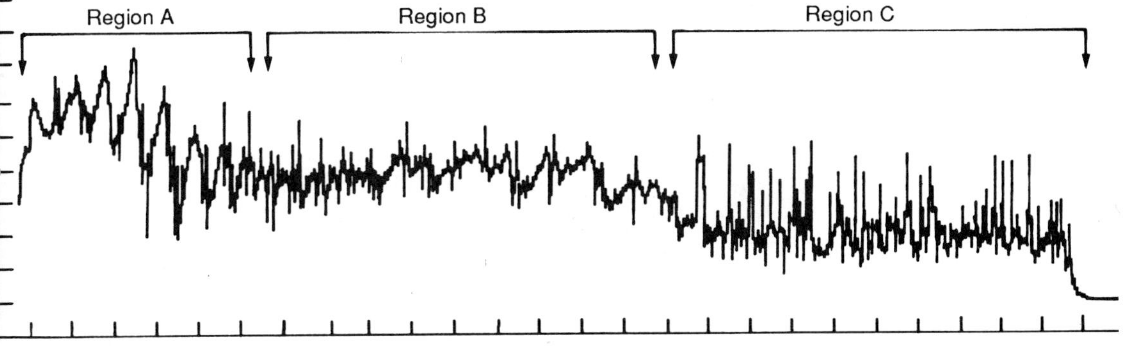

Fig. 2 Thermally-Coupled Flowmeter Data

calibrated variable reluctance transducer. As shown in Fig. 2, this test clearly demonstrated the improvements from the technique as the sensor showed the transition from slug flow in Region A to full flow in Region B to the breakdown of flow as the pressurant was ingested in Region C.

A later test used the thermally-coupled transducer to detect the ingestion of pressurant as the capillary retention of a propellant acquisition device broke down. These results are shown in Fig. 3 and Fig. 4. The trace for Fig. 3 shows the thermally-coupled transducer as the capillary acquisition device ingests pressurant. The flow shows a small rise and then the bubbles pass through the nozzle, registering as negative spikes.

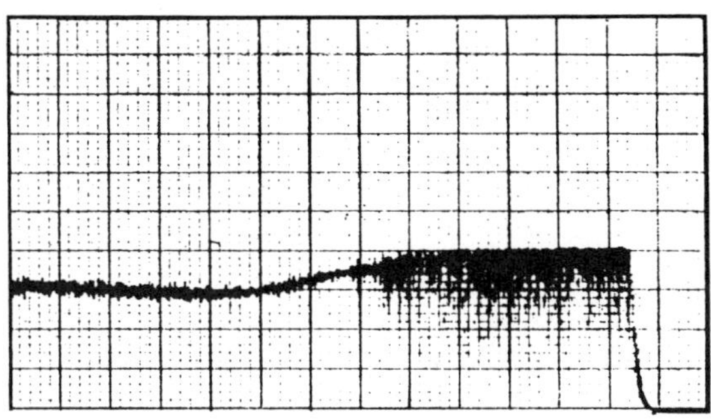

Fig.3 Thermally-Coupled Bubble Detector

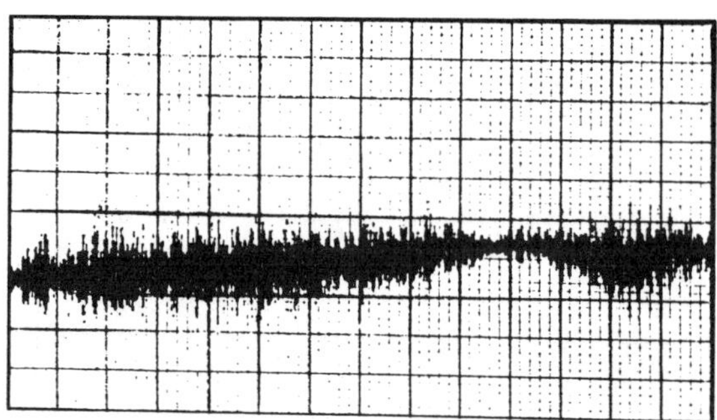

Fig. 4 Externally Mounted Transducer

Fig.4 shows the same portion of the outflow with an externally mounted transducer measuring the differential pressure. The large fluctuations are due to boiling in the sensing lines and tend to mask the actual flow perturbations. The actual point of flow breakdown is almost impossible to determine and the pressure to use for flow calculations is not obvious.

With the demonstrated success of the concept, a more ambitious application is being fabricated for Dr. John Anderson's hydrogen research in Denver. This system will be used as the liquid hydrogen supply manifold for developing transfer and chill-down techniques. Since the quality of the hydrogen is extremely important, every method was used to reduce the heat leak. The manifold consists of three branches plus a bypass line as shown in Fig. 5. On each of the three distribution legs a venturi flowmeter is installed, and a piezoresistive transducer is thermally-coupled to the associated line. Each line and transducer is wrapped with MLI, and the entire manifold is installed inside a vacuum cannister. An aluminum saddle is installed under each transducer to insure good thermal contact. Since all connections and plumbing are vacuum jacketed, the entire installation will have very low heat leak. Any entrained vapor bubbles will register immediately if the conditions reach saturation.

CONCLUSIONS

The concept of thermally-coupled cryogenic pressure measurements has demonstrated the potential to reduce heat leak, minimize measurement-induced TAO, and enable measurements that are impossible with standard techniques. The availability of cryogenic rated transducers opens new possibilities for cryogenic system designers. Any application which requires low heat leak or low differential pressure measurements can benefit from these techniques.

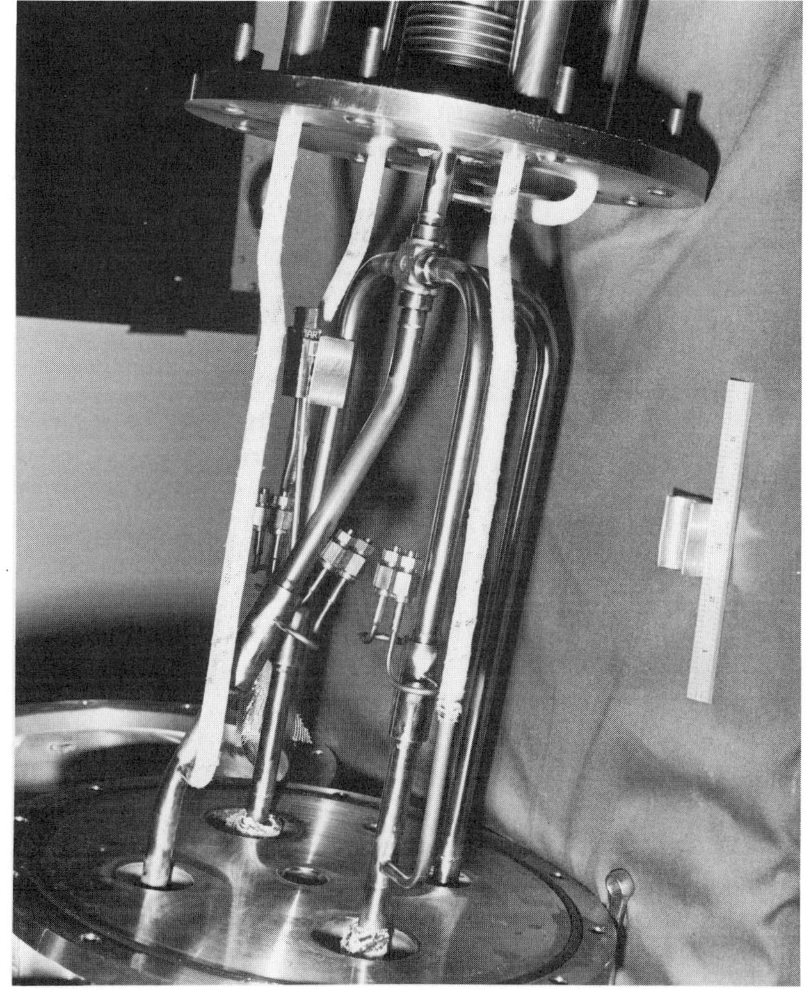

Fig.5 Liquid Hydrogen Distribution Manifold

REFERENCES

1. P. L. Walstrom and J. R. Maddox, Use of Siemens KPY Pressure Transducers at Liquid Helium Temperatures, <u>Cryogenics</u>, 27:439 (1987).
2. A. Kashani, A. L. Spivak, R. A. Wilcox and C. E. Woodhouse, Performance of Validyne Pressure Transducers in Liquid Helium, <u>Proc. 12th Int. Cryo. Eng. Conf.</u>, Southhampton, U.K. (1988)
3. C. Boyd, D. Juanarena and M. G. Rao, Cryogenic Pressure Sensor Calibration Facility, <u>in</u>; Advances In Cryogenic Engineering, Vol. 35, R.W.Fast,ed., Plenum Publishing Corp., New York (1990), pp 1573-1581

CRYOGENIC COOLING OF BIOLOGICAL SAMPLES

FOR ELECTRON AND OPTICAL MICROSCOPY

G. J. Fuld, R. G. Hansen, and N. Loes

R. G. Hansen and Associates
Santa Barbara, CA

ABSTRACT

Cryogenic liquid transfer systems have proven to
be the method of choice for most cooling applications
for spectrophotometric studies. The application of
such methods to biological systems requires a few
modifications. The cryogenic systems can operate in
any position but most liquid samples must be placed in
cuvettes in the vertical position from the top. Temperature control from 77K to room temperature can be accomplished using a liquid nitrogen supply, a 20 ohm
strip heater, a silicon diode transducer and an
automatic temperature controller.

Various designs are presented with limited test
data, where standard cuvettes are mounted on OHFC copper in a vacuum shroud necessary for the low temperature studies. Some suggested approaches to expanding
the versatility of the systems are given. A unique application is the use of a vacuum cooled sample plate
and injecting the test sample into the chamber through
a rubber septum giving instantaneous cooling of the
sample.

INTRODUCTION

The equipment for cryogenic cooling of biological
samples during laboratory instrumentation procedures is
certainly not well defined in the literature. In 1971,
B. Meyer[1] stated,

"...commercial equipment made by a reputable
manufacturer has the advantage that one knows what
he will obtain, what the equipment can do, when it
will be delivered, what it will cost, and, most
important, that it will work. Presently,
commercial sophistication is such that, for a
specified task, a commercial component will almost
always outperform a laboratory-built tool."

ELECTRON MICROSCOPY

The available literature on cryobiologic instrumentation cooling is heavily weighted towards scanning electron microscope (SEM) and transmission
electron microscope (TEM) electron microscopy.
Published data on spectroscopy cooling is much more
sparse, although our corporate involvement tends to
indicate a high level of interest in this area. A
summary of experiences with electron microscopy will
prove beneficial. The combination of high vacuum and
high energy electron beams necessitates various
pretreatments designed to render the sample resistant
to such drastic treatments.[2] Franks[2] suggests that
freeze fixation is often the preferable method, but
recrystallization is a serious problem. Rapid cooling
to below the devitrification temperature of ice, 143K,
is essential.

The methods employed for stage cooling of SEM and
TEM units vary considerably.[3]   A SEM stage tends to be
quite large and capable of handling a bulky specimen,
while a TEM stage needs less cooling but needs much
better thermal and mechanical stability and precision.
There are three methods of cooling a stage; direct
contact of the liquid cryogen below the stage (direct
conduction), connecting a braided copper cryogen
chilled line to the stage, and use of high pressure gas
and a Joule-Thomson effect cooler beneath the stage.
Probably the most efficient cooling method is the
direct conduction method. This is the basis of a
number of SEM cooling systems manufactured by our
firm.[4]   Robards and Sleytr[3] feel that the nitrogen
cooled stages used in the past are adequate for most
SEM's. Helium cooling is preferred for high resolution
TEM's and scanning transmission electron microscopes
(STEMs). If only the direct temperatures of the
boiling point of liquid cryogens is needed, i.e., 4.2K
or 77K, an auxiliary stage heater is not needed.

    If fine control is desired at any temperature in
these ranges, a small strip or donut heater of about
20-25 watts power must be installed. The heater is
used in the feedback loop with a temperature sensor
such as a gold doped iron-Chromel thermocouple or,
preferably, a silicon diode detector. Using one of the
readily available cryogenic temperature controllers can
provide fine control to ±0.1K.[5]

    When using stage cooling, preferably by conductive
cooling, it is necessary that the sample chamber itself
have a dry, inert atmosphere. The obvious selection is
a vacuum of at least 1.33 x 10^{-3} N/m^2 (10^{-5} Torr) or
better, easily achieved with a good roughing pump with
the added effect of cryopumping from the cold finger.
If potential evaporation of moisture is a problem, very
dry nitrogen or helium gas can be introduced into the
chamber. This can be achieved by flushing the chamber
with cold gas, but this will still cause heat transfer
losses well in excess of that from the vacuum atmosphere, which is preferred. Certainly, if temperatures
below 10K are desired, excess radiative heat losses can
be reduced by building a radiation shield around the
sample holder. The radiation shield should be of thin
polished aluminum, possibly gold plated, completely
surrounding the sample area. The only openings should
be to allow beam access to the sample itself.

    As an example, let us examine the design of a
liquid transfer cooling system for a SEM stage, using
either liquid helium or nitrogen. A block diagram is
shown for a typical setup in Figure 1. For the liquid
cryogen supply, a self pressurizing storage dewar,
Model RGH/CF-50-SP-LL, equipped with a pressure and
vacuum safety valve, pressure gauge and a meter level
gauge, is ideal. It can be refilled while feeding
liquid through a transfer line. A fully vacuum
shrouded, stainless steel transfer line to the cryostat
unit is a necessity, especially when transferring
liquid helium. A typical line is the RGH Model 3063,
which has a 1.83 m. (6-foot) flexible section, and a
right angle bend to a 1.27 cm. (1/2") insertion tube to
the dewar. The delivery end is a 1.27 cm. (1/2")
stainless tube that screws into the cryostat. Varying
the turns on the large brass nut will shorten or
lengthen the transfer bayonet. Varying the bayonet
length operates a brass needle valve which allows
approximate gas flow setting.

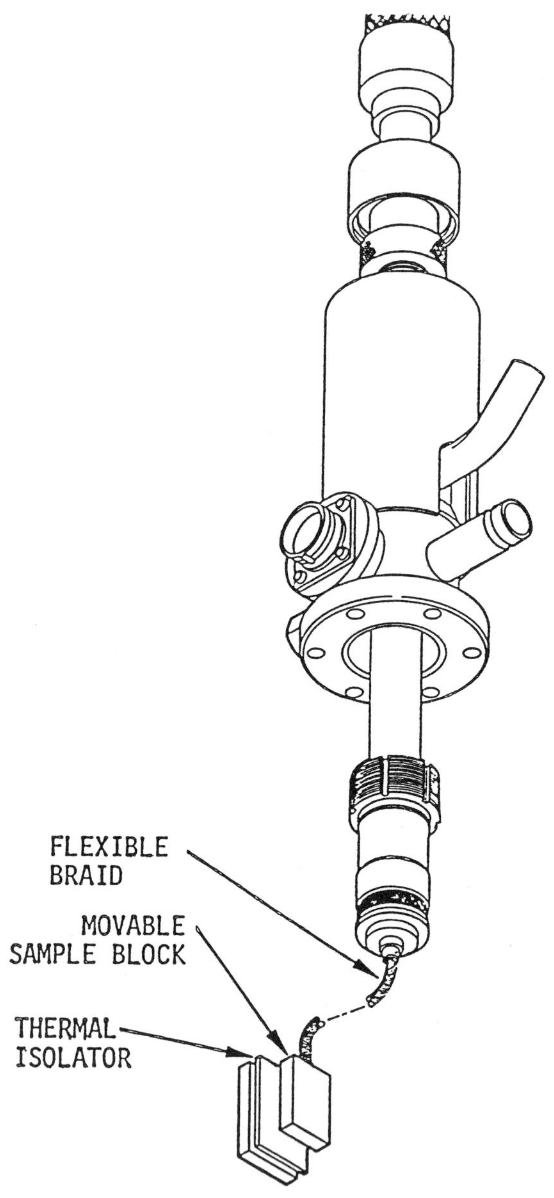

FIGURE 1
HIGH-TRAN SEM COOLER

The cryostat itself, shown in Figure 1, is the
R. G. Hansen & Associates' "High-Tran" unit. It is
fabricated of 304 stainless steel, with the cold tip
and the radiation shield station fabricated of OFHC
copper. The unit is entirely electron beam welded,
internally where possible. It is versatile, and has
either an o-ring seal attached to the vacuum shroud or
a 6.99 cm (2-3/4") conflat at the end of the instrumentation skirt for direct attachment to a vacuum
system. This conflat would attach to a mating conflat
at the rear wall of the vertical wall plate on the SEM.
The cold finger of OFHC copper projects into the sample
chamber, and a braided copper strap of 0.318 cm.(1/8")
x 1.27 cm. (1/2") is attached. Although it would be
desirable to attach the cold finger directly to the
sample stage, the necessary motion in the X and Y
direction precludes this solution. A specially
machined replacement for the top of the cold stage is
needed so that it has either a sapphire plate or a G-10
phenolic wafer below the plate, still allowing the
tongue and groove movement of the stage. If the
thermal insulator is not utilized, a very serious
(probably even critical) heat leak would occur into the
body of the microscope. It is clearly desirable to
have a radiation shield in place around the sample
holder. It is a necessity for helium temperatures.
Although there are undoubtedly many homemade electron
microscope cooling units in use, basically, only two
commercial companies are offering this to the researcher.[6]

It is useful to note several recent applications
that apply to cooling of electron microscopes. A
specimen holder for biological samples is described by
Bastacky, Goodman and Hayes.[7] It is, basically, a
copper foot with a variable holder, magnification calibration screen and a grid. Another application describes a computer controlled temperature system for
cryomicroscopy.[8] With today's modern cryogenic controllers, available with RS232 or IEEE-488 interfaces,
it is relatively simple to interface computers with
cooling systems. The importance of rapid low temperature sample preparation is emphasized in a recent paper
by Sargent.[9] He describes a cryogenic transfer system,
the Hexland CT 1000A, which is commercially available
for accomplishing this task. By freezing in nitrogen
slush, doing sample preparation in the same chamber
area and transferring the specimen, finally, to the
stage at 93K, most artifacts are avoided.

OPTICAL MICROSCOPES

Although the previous discussion has been mostly confined to various forms of electron microscopy, cryomicroscopy with optical microscopes is still of emerging importance. A very thorough review of this area of study, supplementing an earlier review, was made by Diller[10] in 1988. It is interesting to note that the first cryocooled microscope, made in 1897 by Molish, can be viewed in the museum of the University of Vienna. Diller's ground rules for design are simple.

"The design of a cryomicroscope follows a few straight forward principles relating primarily to considerations of performance and fabrication. Generally it is desired to be able to affect a preplanned thermal protocol on a specimen mounted on the cryostage and to be able to evaluate the response to the defined thermal stress by direct and continuous observation throughout the protocol at an optical resolution sufficiently high to identify important phenomena within the tissue and its environment."[10]

Since the current work is concerned with techniques rather than applications, only novel techniques will be cited. Steponkus et al.[11] have described their system and its performance in detail. They use high resolution microscopy, examining specimens with a combination of video and computer image enhancement. Korber and his associates[12] in West Germany have used similar sophisticated technology for biological studies. On the freezing stage of their system, compensatory heating is used. A small portion of the system adjacent to the sample is heated, while heat is withdrawn through a copper block flushed with liquid nitrogen or cold gaseous nitrogen. A solid state controller is used to control the temperature via a closed loop feedback system. Through the use of sophisticated scanning, image analysis, and video equipment, excellent data acquisition is obtained.

POURFILL DEWARS

An inexpensive, simple cryogenic cooling unit can be made using the pourfill dewar, such as the RGH pourfill dewar (PFD) 12.5 system[13]. By design any pourfill dewar operates in the vertical position, while a liquid transfer system can operate in any orientation. A spectrometer application using a pourfill dewar

utilizes a horizontal light beam thus allowing windows
to be vertical. For microscopes, the viewing windows
must be horizontal to allow a vertical light path. A
modification of the RGH PFD unit to mount the windows
on their side, is shown in Figure 2.

The cold finger of the modified PFD is made at
right angles to the vertical dewar, so that a truncated
vacuum shroud is available. Various sample holders are
available, but a simple OSM-1 sample holder with a
round 2.54 cm (1") screw fitting is most often used.
The optical windows would be optically flat Suprasil
quartz. In operation, the sample is placed onto the
copper cold finger by a 1/4-28 NF thread in the mount.
After the vacuum shroud is attached, the system is
pumped down with a mechanical pump, which is left
running. Liquid nitrogen is then added (capacity about
0.2 liters) and cooling should be complete in 5 to 10
minutes. The hold time is about 2 hours. All
components of the PFD dewar are of 304 SS except the
cold finger, which is of OFHC copper.

If automatic temperature control from the range of
77K to about 150K is desired, the cold finger is
instrumented with a small 20 watt strip heater and a
silicon diode sensor for temperature control. Using a
Scientific Instruments Model 9620 dual sensor controller, sensitive control to +/-0.1K is readily ob-
tainable. A vacuum pump-out with a removable actuator
(Cryolab, Inc.) is mounted on the system, along with a
19 pin hermetic electrical feedthrough for the attachment of the heater and sensor leads. It is important
that this wiring be heat sunk on the stem of the pourfill so that drift errors are prevented in the control
operations.

Using the same set-up as above, one of the blank
windows can be replaced by a modified blank having a
metal port with a small hole that is covered by a rubber "vaccine" septum. A microscope slide can be inserted in the chamber, attached to the sample mount,
and precooled under vacuum. The sample can then be
introduced through the septum with a hypodermic
syringe, and the sample frozen "instantly" on contact.
The configuration of this unit, P/N 2467, as a
replacement for a blank window is shown in Figure 3.

SPECTROSCOPY

Another simple adaptation to hold a quartz cu-
vette, which operates with both a pourfill dewar as

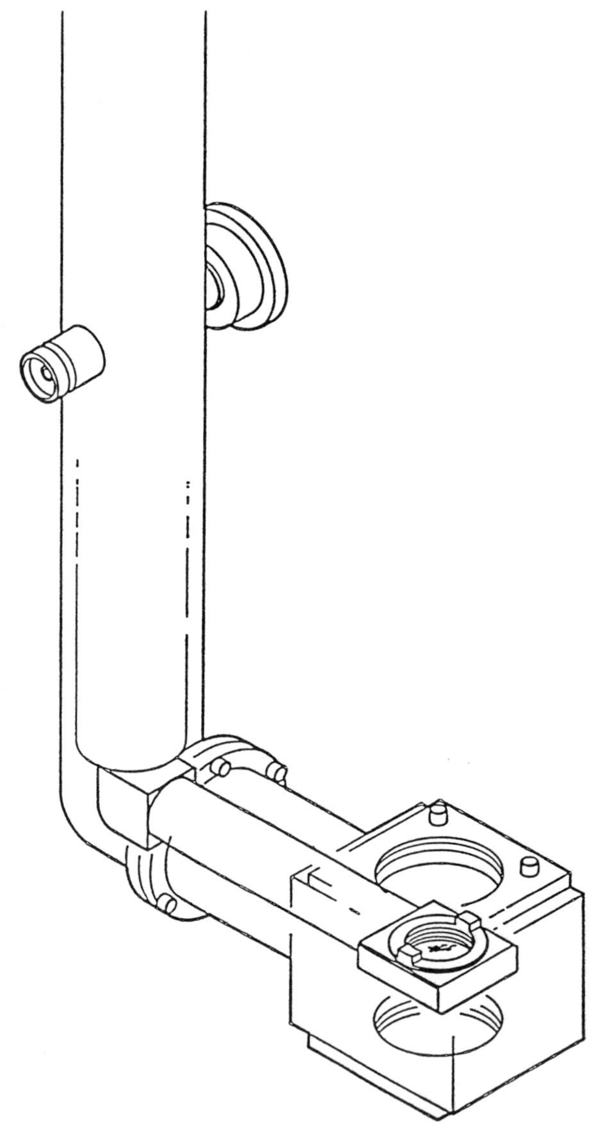

FIGURE 2
POURFILL OPTICAL MICROSCOPE DEVICE

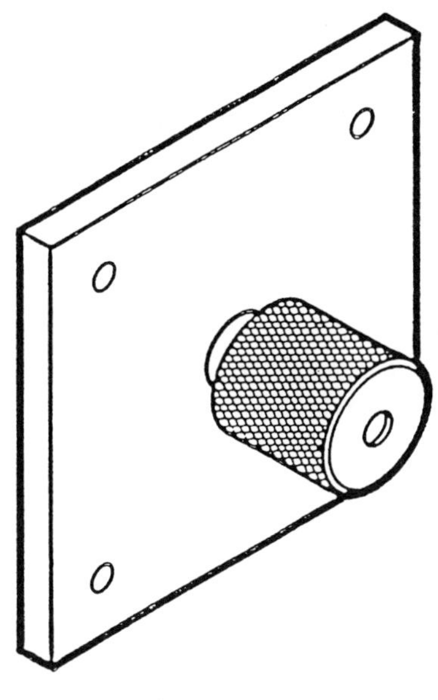

FIGURE 3
INJECTION PORT

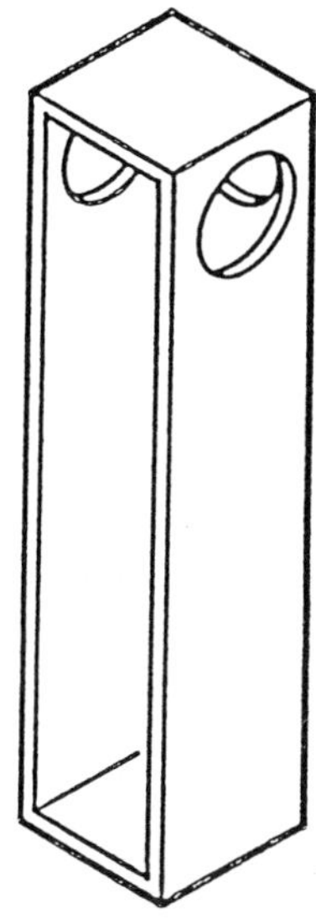

FIGURE 4
QUARTZ COUVETTE CELL

shown in Figure 2 and on a liquid transfer system as
shown in Figure 1, is illustrated in Figure 4.

All types of spectrophotometers have been supplied
with cryogenically cooled sample holders. These
include FTIR, UV, Raman, IR and many others. In the
biological area, the low temperature requirements are a
bit more specialized, especially in regard to speed of
cooling and ice recrystalization and devitrification,
which occurs at 143K. The literature has numerous references to low temperature systems and studies, but
most use homemade systems. For instance, Badulescu et
al.[14] made a system consisting of a pair of AgCl windows with a spacing of 0.1 mm. The cell was placed in
a cylinder and surrounded by circular tubing cooled
with liquid nitrogen. The whole assembly was then
placed in a metal container, with a pair of NaCl
windows, with vacuum applied to the container. A
copper-constantan thermocouple was used for monitoring.
The system was efficient as excellent data was obtained. Another recent application of spectroscopy in
the EPR area was by Harder et al.[15] which used a fully
glass cell, chamber and working area, merely froze the
sample tube in liquid nitrogen. Using Stryofoam cups,
the cell itself is insulated. Although this setup appears crude compared to today's sophistication, it was
effective for these experiments.

We have manufactured numerous liquid transfer
systems as well as closed cycle cooling systems for
spectrophotometers. The liquid transfer systems have
almost exclusively been of the "High Tran" type,
similar to that shown in Figure 1. A cube shaped
vacuum shroud head with five openings has proved to be
the most versatile. Usually two windows of optical
material are installed, along with three blank fillers.
Window composition can be Suprasil quartz, BaFl, NaCl,
KRS-5, CsI or KBr. Nominal window apertures can be
0.75" (1.9 cm.) or 1.25" (3.17 cm.). The most efficient manner is to mount the window in a metal retainer
with epoxy. The window itself then seats against a
Viton o-ring on the shroud. The standard gap in these
systems from window to window is about 2.2" (5.59 cm.).
It should be pointed out again that these liquid
transfer systems operate equally well with either
liquid nitrogen or liquid helium.

Liquid helium price structure has prompted a
number of researchers to consider closed cycle systems
although they have a moderate cost disadvantage. A

closed cycle Gifford-McMahon unit, such as the CTI-22A,
can readily obtain 10K with an expected 10,000 hours or
more before minor maintenance. To operate in the 10K
region, these units require a radiation shield, and the
instrumentation skirt, vacuum shroud, shield, heater
and silicon diode temperature sensor have to be added
to the basic system. However, the maximum cooling
power at 10K is usually 0.5 watts, somewhat less than a
liquid flow system.

SUMMARY

In summary, we have tried to examine some of the
cryogenic cooling systems in use in microscopy and
spectrometry. Although the general consensus of opinion is that most researchers are using homemade equipment, commercially available equipment will do the job
more efficiently and less expensively for even the most
exotic of applications.

REFERENCES

1.   B. Meyer, "Low Temperature Spectroscopy"
      American Elsevier Publishing, New York
      (1971).

2.   F. Franks, "Biophysics and Biochemistry at Low
      Temperatures," Cambridge University Press,
      Cambridge (1985)

3.   A. W. Robards, and U. B. Sleytr, Low Temperature
      Methods in Biological Electron Microscopy in
      "Practical Methods in Electron Microscopy," Vol.
      10, Am M. Glauert, ed., Elsevier Press, Amsterdam
      (1985)

4.   R. G. Hansen & Associates, Private Communication
      and "High-Tran Liquid Transfer System," data
      sheets, 1988.

5.   Units supplied by Scientific Instruments, West
      Palm Beach, FL, or LakeShore Cryotronics,
      Westerville, OH, are generally applicable.

6.   SEM stage cryogenic devices are offered by Oxford
      Instrument Co., Oxford, England, and R. G. Hansen
      & Associates, Santa Barbara, CA.

7. J. Bastacky, C. Goodman, and T. L. Hayes, A Specimen Holder for Low-Temperature Scanning Electron Microscopy, _J. Electron Microscopy Technique_, 14:83-84 (1990).

8. S. J. Shah, K. R. Diller, and S. J. Aggarwal, A Personal Computer-Based Temperature Control System for Cryomicroscopy, _Cryobiology_, 24:163-167 (1987).

9. J. A. Sargent, Temperature Scanning Electron Microscopy: Advantages and Applications, _Scanning Microscopy_, 2:835-849 (1988).

10. K. R. Diller, Cryomicroscopy _in_ "Low Temperature Biotechnology Emerging Applications and Engineering Contributions," J. J. McGrath and K. R. Diller, eds. ASME, _BED Vol 10, HTD Vol 98_, New York (1989).

11. P. L. Steponkus, M. F. Dowert, J. R. Ferguson, and R. L. Levin, Cryomicroscopy of isolated plant protoplasts, _Cryobiology_, 21:209-233 (1984).

12. Ch. Korber, S. Englich, P. Schwindke, M. W. Scheiwe, G. Rau, A. Hubel, and E. G. Cravalho, Low temperature light microscopy and its application to study of freezing in aqueous solutions and biological cell applications, _J. Microscopy_, 141:263-276 (1985).

13. R. G. Hansen & Associates, "Liquid Cryogen Pourfill Dewar," Bulletin PFD 189.

14. S. Badulescu, R. Bicca De Alencastro, H. Le-Thanh, G. Richer, C. Sandorfy, P-P. Vaudreuil, and D. Vocelle, The Protonation of a Retinyl Schiff Base: A Study by FTIR Spectroscopy at Low Temperature in Solution, _Photochemistry and Photobiology_, 49:313-318(1989).

15. S. R. Harder, B. A. Feinberg, and S. W. Ragsdale, A Spectroelectrochemical Cell Designed for Low Temperature Electron Paramagnetic Resonance Titration of Oxygen-Sensitive Proteins, _Analytical Biochemistry_, 181:283-287 (1989).

TECHNOLOGY COMMERCIALIZATION: OPPORTUNITIES AND CHALLENGES

K. L. Crandell

ARCH Development Corporation
1101 E. 58th St., Chicago, IL   60637

ABSTRACT

Commercialization of technology from university and national laboratory settings has changed from a passive activity in the early 1970's to an active process in the 1990's, involving exclusive licensing and new venture formation. Researchers are playing an increasingly important role in the development of promising technology and markets as consultants to licensee firms and stakeholder participants in new ventures. The challenges facing the researcher and business development professional involved in the licensing and venture processes will be discussed.

DISCUSSION

Technology transfer at national laboratories and universities has changed from a passive activity in the 1970's to an active process in the 1990's. The national laboratory mission began with a focus on weapons technology and then centered on nuclear reactors. In the 1960's, the mission changed to producing good science with some demonstration plants for energy technologies. Technology transfer for the most part was viewed as a by product of program completion.

Now good science is still important, but the public expects an added return. This change was driven by a growing public perception that the United States' international economic competitiveness is slipping and that the enormous government investment in research could be brought to bear to the advantage of the United States. The government spent some $60 billion in 1989 on research, much of it at the 400 federal laboratories spread across the country. Six billion of this was spent at the twelve government-owned, contractor-operated national laboratories like Argonne National Laboratory, which is operated by the University of Chicago.

One measure of the level of technology transfer from the federal laboratory system is the annual revenue generated from the technology licensed from the system. Currently, the government receives less than $4.0 million per year in licensing revenue with the majority derived from one license from the National Institute of Health (NIH). Using this measure, public sector technology transfer has not been a substantial success.

Beginning in 1980, a number of pieces of legislation were enacted by Congress which culminated in the Technology Transfer Acts of 1986 and 1989. This legislation enabled not-for-profit and small business contractors to take title to the intellectual property (patents, copyrights) resulting from government funded research. It also allowed researchers to share in the proceeds generated from their inventions.

In practice, there were several key aspects to this legislation. First, contractors could grant exclusive licenses to firms. This is important because it increases the possibility that the investment needed to develop a product from an invention will generate a return commensurate with the risk of the project. Prior to this legislation, the government granted primarily non-exclusive licenses. If a firm had an interest in an exclusive license, the government would advertise the company's interest in an exclusive license in the Federal Register for a period of time to determine if any other party was interested. If more than one firm expressed interest, non-exclusive licenses were granted. Many firms were not interested in potentially receiving non-exclusive licenses after telegraphing their research interests to the industries in which they compete.

Second, researchers/inventors could receive a significant portion of the value created by the invention they created (i.e., Incentives are O.K.). A corollary to this was that it was " O.K." for inventors to be actively involved in the commercialization of their inventions. This is important because patents and publications alone do not make a product. In many cases there is a considerable amount of " know how" needed to successfully refine a laboratory innovation into a product to meet the needs of a specific application. Often the areas where innovation occurs are quite new and there is a limited pool of professionals skilled in the art. In some fields the inventors may represent a majority of the available experienced talent pool.

Let me take half a step back and discuss for a few minutes the nature of the technology transfer process. Like most activities, there is a spectrum with varying levels of intensity. Since one of these, publication of research articles, is routine and at the relatively low intensity end of the spectrum, I will focus on the intensive licensing and new enterprise creation.

For a technology to be suitable for licensing, it must represent at least an incremental improvement in a product or technology for which there is a market. Second, a firm must exist that has an established infrastructure with which to integrate the innovation into its operations. Third, those that have the rights to the technology and the potential licensee firm must reach an agreement.

In theory, the university or laboratory based licensing process has five steps. In the first, the researcher invents something that is novel, not obvious and useful. In the second, the inventor submits the invention in report form to the university or laboratory administration which recognizes its promise. Third, a patent application is prepared and filed. Finally, a group of potential licensees are identified of which one or two are selected and a license agreement is completed.

In practice, this is rarely the case for a series of reasons. First, researchers do not always realize when they have invented something with significant market potential. Second, administrations do not always have either the staff or the expertise to evaluate the technical and market potential of the inventions. In the past, market reviews have been the most neglected. Many administrators simply do not have the experience, skill, time or inclination to develop business contacts and collect relevant market information. Third, the administrations rarely have the budgets to always have the staff or expertise to actively market the invention to firms that might be interested. In a way this is understandable, since technology transfer is quite new to many administrations and researchers.

A rough estimate of the number of inventions a technology transfer operation can expect is one invention per \$3 million in research funding per year. Out of a hundred inventions disclosures received by a technology transfer operation, 20 to 30 might be candidates for patenting and 10 to 15 are typically suitable for licensing. On average, each invention disclosure will take between sixteen and twenty-four hours of aggressive analysis by a skilled individual (technical degree and several years of industry experience) to be able to make a preliminary assessment of the potential and much more if the innovation is promising.

The outcome of this preliminary assessment may be that the invention has significant, little or uncertain potential. In cases where the invention is a concept with little confirming data, a thorough analysis usually raises more questions than it answers. This last point is important because it illustrates a little understood facet of the technology commercialization process. The facet is that innovation may be the result of a flash of insight but additional data is what makes transfer of the innovation possible (and is also what makes good technology and strong patents).

Reducing the uncertainty surrounding the potential of an invention may be hampered because data collection on key performance figures is not funded or will take months to complete. Under these conditions it is naive to expect a meaningful technical or market analysis and an informed patent decision in a short period of time. Balancing this is the need to publish, the fact that creating and sharing knowledge with other members of the academic community is one of the fundamental reasons for research. Some institutions have addressed this quandary by accepting larger proportions of false positive (i.e., money is no issue) or false negatives (i.e., disappointed inventors and lost opportunities). In practice, there can be a healthy tension between early disclosure and protecting rights that brings increased efficiency to the analysis and patent decision processes.

Once a licensee is identified, the terms and conditions of the license must be decided. These will likely include performance milestones to insure that the licensee firm goes to the trouble to develop the technology. The license will likely involve a specific field of use in which the licensee firm can commercially develop the technology. This allows multiple licenses to be granted covering the same invention. The license will also likely include provisions for exclusivity and royalties. Royalties range from less than one percent to ten percent of net revenue in some medical device-related inventions. Equity may also be given in lieu of cash in the case of smaller licensee firms. The magnitude of the proceeds from the license are in part based on the economics of a specific industry, the amount of development left to do, and the customs of the groups involved. The best way to figure out what an invention is worth is to shop it to a number of potential licensees and ask other licensing administrators what were the terms of similar past licenses.

The most intensive type of technology transfer is spinning off a new company based on an innovation. Companies can be formed around spin-off technologies for several reasons. They may result from an inventor's wish to pursue research that he or she can no longer continue inside the institution and may evolve into a small contract R & D or Small Business Innovative Research (SBIR) company. A firm may be founded because an innovation is recognized as potentially important by a firm or partnership willing to invest capital. The point here is that there are may reasons for spinning-off companies, many of which turn in to perfectly good enterprises that produce worthwhile goods and services and make inventors and researchers very happy. They do no always warrant investing large amounts of capital or wind up serving large markets.

A subset of these spin-offs are the venture capital backed enterprises that are financed with the expectation that, if the development and market risks can be overcome, significant economic value will be created. The innovations on which the firms are based will likely be revolutionary with the potential to reorder the markets that will be impacted. Fewer than one invention disclosure in a hundred has the potential to reach this level.

Let us say we believe our innovation is worthy of outside investment. One of the first questions that confronts an entrepreneur is what kind of investor is best. The choices include independent venture capital, corporate venture capital, state venture capital and wealthy private investor. In early stage investing, the entrepreneur would like seasoned active investors that will resemble a partner helping to build the business. In point of fact, these are quite hard to find because as venture capitalists become more successful they tend to raise and manage larger pools of capital. Since partners can only do so many deals in any given year, larger funds with the same fixed life time require larger sums per investment to insure all capital is invested over the life of the fund. The result is that successful venture capitalists tend to have " shelf pressure" and are looking for deals in which they can put millions of dollars to work quickly. In addition, the skill set required to be successful in the later stage investing tends to be more transaction oriented.

A quick test to determine if a venture capital firm is likely to do an early stage deal is to take the total dollar amount of their fund and divide by the number of partners; if the number is greater than $10 million per partner, they are unlikely to be investing in seed deals. The other end of the spectrum are the venture capitalists or wealthy private investors whose fund is so small that they simply do not have the staying power for the later funding rounds that a venture may require. This increases the likelihood that the firm valuation in the later rounds may be lower than otherwise might be the case. The best way to find a good venture capitalist is to talk to entrepreneurs, venture capital attorneys, accountants and incubator managers for their insights on the venture capitalists in one's geographic area. One way to gauge the level of operational involvement that a venture capitalist undertakes is to ask him to discuss his role in his other deals.

Corporate venture capitalists are probably not a potential source of seed capital for a new venture, although they may invest along-side other qualified investors. When corporate venture capitalists invest, generally they have some strategic interest in technology or markets related to their company's business. In many cases, corporate venture capitalists are staff members who must sell their investment decisions to management at the division level in addition to the Board of Directors. One way to demonstrate that a specific division of a corporation has interest in your venture is to solicit a research and development contract from the division ahead of time. One of the lessons here is to evaluate the businesses at the division level before approaching the corporate venture capitalist and tailor the presentation to fit with the strategic goals of the division. The second lesson is to line-up the local venture capitalist who will lead the deal before you approach corporations.

Another question to be answered is how much money should be raised. The short answer is enough to achieve milestones that will clearly demonstrate the progress the firm has made and reduce the venture's risk in eyes of future investors. The reason for this is that asking investors for the entire amount that you perceive you will need, especially if this figure is in the million-plus range (which it is likely to be in the materials or biotechnology areas), will not be seen as prudent by the investor and may be seen by the investor as a sign of naivety on the part of the entrepreneur.

The type of milestones that are appropriate are beta site tests completed, product sales revenue, management team assembled, scale-up completed and strategic partnership formed. This might cover eighteen months of operations with the milestones as a form of near term deliverable for the investors. In general, when raising sums of money in the million dollar range, it may also be appropriate to spread out or "stage" the investment further by taking the second half of the investment after some intermediate goals are accomplished.

Another issue that arises is determining a fair valuation for the venture. In business schools around the country it is taught that the value of an equity security is the net present value of the expected future cash flows discounted by a rate that reflects the risk of the cash flows. At this early stage, it is very difficult to assess accurately the expected future cash flows or the appropriate discount rate for a company with no operating history or product sales. Several other less rigorous approaches have been used to arrive at valuations. Some entrepreneurs would like to raise several million dollars and do not want to give up more than 50% of the equity. In this case, the valuation calculations tend to arrive at figures approaching $2 million before investment. Some venture capitalists will not put in less than $500K and want 40 to 60% of the equity, placing the valuation around $250K before investment.

Seasoned venture capitalists draw on their experience in investing in other early stage companies to arrive at a valuation with which they feel comfortable. Entrepreneurs can use a similar approach by developing a knowledge base of valuations of other early stage companies via their contacts and estimate a value for their firm by assessing the relative attractiveness of their venture. One estimate by Grant Skeens, manager of the State of Illinois' venture capital program (which has a large number of investments in early stage companies in the Midwest), indicated that most seed stage deals are in the $250K to $750K before investment valuation range. One way to resolve honest differences in opinion on valuation is to set out milestones for the corporation, that if accomplished would increase the size of the entrepreneurs' equity stake.

When considering equity investors and dilution, an important point to keep in mind is the difference between operational control and board control. In all likelihood, even if the company is successful, the founders will own a minority stake (that is also quite valuable). However, if the founders have been successful at an operational level, it would be unwise for the investors to exercise control at that level. The point being that if management is successful, the board would be unlikely to change it. A final point to keep in mind is that real valuation is predicated on an investment reaching closing. If after all is said and done there is no investment, then talk about offered terms and conditions is worthless.

From the venture capitalist perspective, establishing as quickly as possible the size of the potential market, the degree to which the technology is proprietary, management capability and whether the venture capitalist can work effectively with the entrepreneur are very important. To the degree that the business plan and supporting information provided by the entrepreneur address these issues, the process will be expedited with increased potential for a successful outcome.

I would like to spend the last part of the talk on a topic that has been in the news quite recently: conflict of interest. When first confronted with the complexities of ventures and licensing, administrators' and department chairmen's reactions tend to separate into two types. One group feels that incentives for inventors involved in commercializing their inventions is unethical, possibly evil and should be fought. The second group believes that commercialization and everything associated with it are good. A case can be made that neither position is optimal for the long haul because neither includes provision for management's continued roll in managing the risks inherent in these new activities.

Reuven Cohen, Associate Counsel of Brigham and Women's Hospital in Boston has written a particularly insightful article entitled "Monitoring University Hospital-Industry Relationship" which I will try to summarize here. As you may be aware, the first set of NIH guidelines covering conflict of interest were withdrawn. These guidelines, if instituted, would have included two provisions with the potential to adversely impact university-industry relationships. These provisions were: complete financial disclosure by the scientist receiving NIH funds and that academic scientists participating in NIH sponsored research could have no financial relationship with a company that stood to benefit from the research. While this particular set of guidelines was withdrawn, the concerns that brought them about are very real.

Cohen believes that in dealing with industry, representatives of the academic world would do well to preserve the openness that has been a hallmark of the academic world. The NIH guidelines, however, would have required everything to be disclosed to everyone, which is clearly draconian. In order to avoid another wave of poorly thought-out policies, administrations would do well to institute disclosure, oversight and review policies in cases involving academic industry relationships. The effort this focused approach takes will serve the interests of the scientist, institution and industry in the long run.

CONCLUSIONS

From the inventor's perspective, what all this means is that he or she will have to be more actively involved in the technology transfer process than they might have expected. For the business development professional or university research administrator considering this area, the same holds true. Even at the handful of institutions that have outstanding technology transfer arms, enlightened involvement at all these levels is still critical to success.

ACKNOWLEDGEMENTS

I would like to thank Steve Lazarus, CEO, and Neil Wyant, Project Manager, ARCH Development Corporation, and Brian Frost former Director, Technology Transfer Division, Argonne National Laboratory, for their comments on this article.

CRYOGENIC HEAT TRANSFER

A SURVEY OF RECENT DEVELOPMENTS

John C. Chato

Professor of Mechanical, Electrical, and
Bioengineering
Department of Mechanical and Industrial
Engineering
University of Illinois at Urbana-Champaign

INTRODUCTION

Although heat transfer in general and cryogenic
heat transfer in particular are mature fields with
decades of history behind them, new developments and new
applications require a fresh look and new understanding
of the various phenomena influencing the transfer of
thermal energy. This review is intended to highlight
these new areas as reflected in the most recent
technical literature. The emphasis will be on new basic
concepts rather than applications of existing techniques
to new equipment.

GENERAL HEAT TRANSFER STUDIES

Igra et al. (1986) reported results and some
analyses on thermosyphon flows of liquid nitrogen in
rotating, interconnected, radial tubes. The tubes were
heated at an intermediate region along their radial
lengths. As in their previous work with liquid helium,
they suggest the possibility of several double flow
patterns, one radially outwards and the other radially
inwards. The second of these may be caused by an
effective negative buoyancy which the authors attribute
to polytropic heating of the fluid. These results are
not really new but were extended from the previous work
on helium. For effective cooling of a rotating system, a

configuration is proposed which "appears to be the basis
for uniform cooling of the rotor periphery." Narahara et
al. (1988) explored the thermal effects of oscillations
of gas columns for two cases: forced oscillations with
initially uniform temperature and thermally driven
acoustic oscillations with steep temperature gradients.
Two types of heat flows were found: from pressure node
to velocity node and from the warm region to the cold.
Schmidt (1988) investigated the heat transfer from a
copper surface into a closed volume of liquid or
supercritical helium. The transient, pulse-type heat
transfer is reduced in a closed volume as compared to an
open volume. In the closed volume the pressure increases
with time, consequently the temperature is also
increasing.

Yamamoto et al. (1986) studied the effects of a
pulsative heat input on the flow of supercritical helium
in a helically wound tube. They found that (1) the wall
temperatures were strongly dependent on flow velocity,
heat flux, and heating time, (2) inverse flow was
generated with high enough heat fluxes for the helium to
reach the pseudo-critical line, and (3) the energy per
oscillation period depended linearly on the pressure
when thermal oscillations were observed. Kasao and Ito
(1989) reviewed the existing experimental correlations
on forced convection heat transfer to supercritical
helium 4.

HEAT EXCHANGERS

Oonk and Hustvedt (1986) investigated the effects
of variable specific heat, thermal conductivity, and
viscosity on the performance of helium coaxial tube,
counterflow heat exchangers in the 4 K to 20 K range.
They compared the results obtained for effectiveness vs
number of transfer units (Ntu) using a numerical
technique to those obtained by using a simple analytical
model with constant properties evaluated at the mean
fluid temperatures. Their conclusions were that although
the "exact" numerical model predicted higher
effectivenesses, with a maximum increase of 12% at Ntu $\approx$
0.4, in the practical range of interest, Ntu > 3, the
increase was less than 2%; thus, the detailed variations
of the properties need not be considered for the design
of such heat exchangers. Xiao and Guo (1988) analysed
the flow and thermal characteristics of rapid cyclic
flows in cryogenic regenerators using the linear network
theory and perturbation method. Borchi et al. (1989)

found that adding an axial conduction term in the analysis improved the agreement between theory and experiment for matrix type heat exchangers. Kush and Thirumaleshwar (1989) described the design of regenerators. Matsumoto and Shiino (1989) studied the regenerative heat exchanger from the viewpoint of irreversible entropy production. Chato (1989) showed that the second law not only limits the operating range of heat exchangers but can also determine the optimum operating point for the entire thermal system.

INSULATIONS AND SYSTEMS

Because insulations are very design specific, most publications cannot be used for general purposes. Examples are sections of the <u>Advances in Cryogenic Engineering</u> volumes, e.g. 1988, 33:291-348 and Barth et al. (1988). Shu et al. (1986) investigated the effects of polishing or taping with aluminum tape the cold surface of a multilayer insulation system. The results were compared to those obtained with a black painted surface. It was found that the taping improved the performance by a factor of two, i.e. the optimal number of layers was reduced from 60 to 30 at a vacuum level of $1.5*10^{-5}$ torr. It was also found that the vacuum level should be kept below 10^{-4} torr and that cracks in the multilayer insulation seriously degrade the performance, in their case by a factor of about 2.7.

Richardson (1986,1988) has been developing pulse tube refrigeration systems which use acoustic surface energy pumping. Thermal analyses and experiments on the Tokamak Reactor's forced cooled superconducting coils were performed by Kerns et al. (1988) and Volkov et al. (1988). Bunkov (1989) developed a superconducting aluminium heat switch.

TWO-PHASE HEAT TRANSFER IN NORMAL FLUIDS

Lin et al. (1986) examined the enhancement of condensation generated by the surface tension in vertical, V-shaped grooves with saturated nitrogen vapor used as the working fluid. Their analyses and experimental results showed a 5-7 fold increase in the average heat transfer coefficient over those in a smooth tube. However, they did not compare their results with some earlier work and they did not present the results

in a form that would allow for extension to other,
similar configurations. It is to be noted that cryogenic
condensation is not a popular subject because of its
limited use, such as in thermosyphons and heat pipes.

Chen and Van Sciver (1986) conducted an
experimental study of the steady state boiling heat
transfer in rectangular channels submerged in a helium I
bath at a pressure of 100 kPa and at the two
temperatures of 4.2 K and 2.56 K. Although they took
data along the entire boiling curve and even somewhat
beyond it, their discussions and analyses concentrated
on the variations of the peak heat flux (in other
applications called "burnout") with channel orientation
(vertical, inclined, horizontal) and gap width.
Increasing the width increased the peak heat flux, but
increasing the angle from the vertical decreased the
peak heat flux for the narrowest channels, and first
increased then decreased the peak heat flux for the
wider channels with the maximum occuring at increasing
angles of inclination with increasing widths. The
results and correlations proposed apply only to the
specific cases studied and no extrapolations are
possible or suggested. Lezak et al. (1986) measured the
time delay to the onset of film boiling in liquid helium
I in the temperature range from 2.2 K to 4.2 K. The
results were correlated in terms of the the applied heat
flux, $(q/A)_{app}$, the delay time, t_d, and the properties
of liquid helium, specifically the thermal diffusivity,
α, the density, ρ_l, and the latent heat, h_{fg}, as
follows:

$$(q/A)_{app} = (\alpha/t_d)^{1/2} \, \rho_l h_{fg}$$

Vishnev (1988) proposed a molecular and
thermodynamic method to generalize correlations of
experimental data on bubble boiling in tubes and in pool
boiling. The following three correlations were
developed:

For cryogenic liquids

$$Nu = 2.2 M^{-0.25} Pr^{0.35} (Re_{xx})^{0.7} H^{-m} (T_s/T_c)^7$$

For high temperature liquids

$$Nu = 2.2M^{-0.1}Pr^{0.35}(Re_{xx})^{0.7}H^{-m_1}(T_s/T_c)^{10}$$

where for incompletely halogenated refrigerants $M^{-0.25}$ should be used.

For liquid metals

$$Nu = 0.25M^{0.25}Pr^{0.35}(Re_{xx})^{0.7}(T_s/T_c)^8$$

The definitions of the symbols are:

$$Nu = (h/k_l)[\sigma/(g(\rho_l - \rho_v))]^{0.5}$$

M = Relative molecular mass
Pr = Prandtl number

$$Re_{xx} = q(L/d)(h_{fg}\rho_v v_l)^{-1}[\sigma/(g(\rho_l - \rho_v))]^{0.5}$$

g = Gravitational acceleration
h = Heat transfer coefficient
h_{fg} = Latent heat of boiling
H = (Absolute level of hydrostatic pressure)/L; for
  boiling in a free volume H=1 and L/d=80.
k_l = Thermal conductivity of the liquid
L = Length of the tube

$$m = [2900(L/d)^{-1.65}](1 - q/0.1q_c)$$

$$m_1 = [1300(L/d)^{-1.65}](1 - q/0.1q_c)$$

q = Heat flux
q_c = Critical heat flux
T_s = Boiling temperature
T_c = Critical temperature
v = Kinematic viscosity
ρ = Density

and the subscripts l and v refer to the liquid and vapor respectively.

Barron and Dergham (1988) developed a correlation for film boiling of liquid nitrogen on a circular plate facing downward:

$$Nu_D = \left(\frac{0.325}{1+1.486^*10^6\ Ra^{-0.5}}\right)^{0.4} Ra^{0.2}$$

where Nu_D is based on the diameter, D, and

$$Ra = \frac{g(\rho_l - \rho_v)\rho_v(h_{fg} + 0.375c_p\Delta T)D^3}{\mu_v k_v \Delta T}$$

where μ is the dynamic viscosity. The ranges for the variables were:

$$10^{10} < Ra < 5^*10^{12},$$
$$50\ mm < D < 150\ mm,$$
$$50\ K < \Delta T < 210\ K.$$

Beduz et al. (1988) found with liquid nitrogen that, when a smooth surface facing upward is rotated about a horizontal axis, the boiling heat flux, if it is below a critical value, increases with the angle and reaches a maximum when the angle is about 176 degrees. In contrast, rough surfaces did not change very much. The critical heat flux decreased with the angle, but for the rough surfaces the critical heat flux could be greater than predicted by the hydrodynamic theory. Muller-Steinhagen (1988) discussed fouling data for hydrogen, nitrogen, oxygen, and argon observed during pool-boiling. Bodegom et al. (1988) found that both light and sound can substantially lower the superheat required for a given heat flux.

Lutset and Zhukov (1989) correlated experimental data on heat transfer at high centrifugal acceleration fields. At accelerations greater than 200 times gravity, the Nusselt number was found to be constant for developed boiling. The convective Nusselt number followed the turbulent correlation,

$$Nu = c\ Ra^{1/3}$$

where the value of c depended on the range of Ra as
follows:

Ra	10^9-10^{10}	10^{10}-10^{11}	$>10^{11}$
c	0.10	0.13	0.16

Sakurai et al. (1989) found short duration, quasi-steady nucleate boiling in helium I at levels exceeding the steady state value of the critical heat flux (7500 W/m^2). The duration, t_L in seconds, of these high heat fluxes, q_s in W/m^2, were found to be

$$t_L = 9*10^7/q_s^{2.5} \qquad \text{for } q_s \geq 9880 \; W/m^2$$
$$t_L = 8*10^{47}/q_s^{12.5} \qquad \text{for } q_s < 9880 \; W/m^2$$

Shiotsu et al. (1989) developed a correlation for the critical (maximum) heat flux in nucleate boiling of helium I on horizontal cylinders. Unfortunately, two of the symbols (L and D) are not defined properly and the exponent of 10 is different in the text than in the figure.

Tanaka and Kodama (1989) found that, for pool boiling in ^{3}He, a modified form of Kutateladze's correlation, q/A proportional to $\Delta T^{2.5}$, applied to nucleate boiling, while the Brèen and Westwater (1962) correlation applied to film boiling.

Klimenko et al. (1989) investigated the effects of channel orientation and geometry on the heat transfer with two-phase forced flow of nitrogen. For nucleate boiling in channels they recommended the use of the Klimenko formula:

$$(q/\Delta T \; k_l)B = 7.4*10^{-3} \; (qB/h_{fg}\rho_v\alpha_l)^{0.6} \; (pB/\sigma)^{0.5} \; Pr_l^{-0.33}$$
$$*(k_w/k_l)^{0.15}$$

where q is the heat flux, $\Delta T = T_{wall} - T_{sat}$, k_l and k_w are the thermal conductivities of the liquid and the wall, respectively, the length parameter is $B=[\sigma/g(\rho_l-\rho_v)]^{1/2}$, σ is the surface tension, g is the gravitational acceleration, ρ_l and ρ_v are the liquid and vapor densities, h_{fg} is the latent heat of vaporization, α_l is the thermal diffusivity of the liquid, p is the pressure, and Pr_l is the Prandtl number of the liquid. Vaporization became the dominant mechanism of heat transfer for high qualities and mass flow rates, i.e. when

$$\frac{h_{fg}G}{q}\,[1+x(\frac{\rho_l}{\rho_v}-1)]\,[\frac{\rho_v}{\rho_l}]^{1/3} > (1.3 \text{ to } 1.6)*10^4$$

where G is the mass flux and x is the quality. Then the recommended relation was:

$$Nu_B = 0.087\ Re^{0.6}\ (\rho_v/\rho_l)^{0.2}\ Pr_l^{0.17}\ (k_w/k_l)^{0.09}$$

The Nusselt number is based on B and the Reynolds number is defined as

$$Re = \frac{GB}{\rho_l \nu_l}\,[1+x(\frac{\rho_l}{\rho_v}-1)]$$

Xiulin et al. (1989) collected experimental data on pool boiling of nitrogen on smooth and porous coated vertical tubes. Mirza (1990) correlated data for boiling nitrogen using a Rayleigh number, Ra, based on the heat flux, q.

$$Ra = g\beta qx^4/(k\nu\alpha)_l$$

where g is the gravitational acceleration, β is the volumetric expansion coefficient, x is the vertical position, k is the thermal conductivity, ν is the kinematic viscosity, and α is the thermal diffusivity. The most significant correlations were as follows:

In a smooth, vertical, open channel:

$$Nu_x \approx 0.7\ Ra^{0.2} \qquad \text{for the laminar region, } Ra < 10^{12}$$

$$Nu_x = 0.47\ Ra^{0.285} \qquad \text{for the turbulent region, } Ra > 10^{12}$$

In a smooth, 300 mm tall, closed channel:

$$Nu_x = 0.0113\ Ra^{0.4} \quad \text{for the turbulent region, } Ra > 10^{12}$$

In a smooth, 600 mm tall, closed channel:

$$Nu_x = 6\ Ra^{0.155} \quad \text{for the laminar region, } 10^{10} < Ra < 10^{14}$$

$$Nu_x = 0.0011\ Ra^{0.43} \quad \text{for the turbulent region, } Ra > 10^{14}$$

Rohsenow's (1952) boiling surface coefficient, C_{sf}, was also estimated at 0.004-0.006 for the smooth open channel and at 0.002-0.004 for the rough open channel. The boiling heat transfer coefficients were in the ranges of 1400-3900 W/m^2-K in the 300 mm channel and 2100-3100 W/m^2-K in the 600 mm one.

Dress and Kilgore (1988) reviewed the status of cryogenic wind tunnel research, including heat transfer.

HELIUM II HEAT TRANSFER

Breon and Van Sciver (1986) reported on the boiling regimes in saturated helium II confined to a vertical, rectangular (1x10x114 mm) channel. When heated from below, four different boiling regimes were observed at different temperature and heating levels. The first regime occured at the lower temperatures (1.7 and 1.9 K) and lower heat input rates (250-400 mW) and was characterized by relatively high superheat (≈0.1 K) and periodic behavior connected with the formation and rise of single, large vapor bubbles. The second regime occurred at all three temperature levels studied (1.7, 1.9, and 2.1 K) at about the same heat input rates and was characterized by low superheat (0.01-0.05 K) and rather regular temperature oscillations with amplitudes of about 0.02 K and frequencies of 0.01-1 Hz which were related to the formation of several bubbles simultaneously. The third regime occurred at all temperature levels at high heat input rates (400-1500 mW) and was characterized by occasional temperature spikes (≈0.01-0.1 K) followed by small oscillations, which were related to the relatively rapid formation of bubbles. The fourth regime occured only at 2.1 K at the higher heat input rates (300-1000 mW) and was characterized by relatively stable behavior with only very small oscillations in temperature and pressure generated by a steady stream of rising bubbles. Increasing the heat input above those in the last two regimes created dryout. When the channel was heated from above, only two regimes were observed. At the lower heat input rates ($\approx$70-200 mW, depending on the temperature level) the temperature and pressure oscillate with a frequency on the order of 1 Hz and with amplitudes of 0.025 K and 1.5 torr. However, unexplained pressure phenomena were also observed, consisting of an initial spike followed by a reduction to levels 2-3 torrs below expected values. At the higher heat input rates the behavior was periodic at 1 Hz with both temperature and pressure spikes. There was strong hysteresis in the transition between all regimes in this configuration, whereas only the dryout regime exhibited hysteresis when the channel was heated from below.

Fuchino et al. (1986) investigated the effects of heat pulses on a helium II forced circulation loop. The chief motivation of this work was that superfluidity breaks down above some threshold heat flux. This work is very equipment specific and its most significant conclusion seemed to be that its results could not be correlated by previous analyses. Later work on the same equipment by Fuchino and Tamada (1989) developed some analyses based on the Gorter-Mellink energy equation

$$\rho c_p \frac{\partial T}{\partial t} = K \frac{\partial}{\partial x} \left(\frac{\partial T}{\partial x}\right)^{1/3} - \rho c_p V \frac{\partial T}{\partial x} + Q_s$$

where most symbols are standard, $K \approx 9.7$ to 11.5 W/(cm$^{5/3}$K$^{1/3}$) is an apparent thermal conductivity, V is the flow velocity, and Q_s is the heat leak into the system.

Kashani and Van Sciver (1986) measured and analyzed the heat transfer and pressure drop in a 3 mm diameter and 2 m long coiled tube with helium II flowing through it while heat was added at the middle along the length. Their predictions agreed reasonably well with the experimental results. As is to be expected, as the flow velocity increased the importance of axial heat conduction diminished. Gradt et al. (1986) measured the peak and recovery heat fluxes on thin wires in helium II at varying hydrostatic pressures in the laboratory and during a 25 second microgravity flight. The laboratory data showed the heat fluxes increasing linearly with

hydrostatic pressure. The microgravity results were 10%
lower for the peak heat flux and about 50% lower for the
recovery heat flux than the laboratory data at zero
hydrostatic pressure. The 10% reduction was explained by
the authors by taking the Van-der-Waals forces into
account between the heated surface and the helium atoms,
but no explanation was given for the greater deviation
in the recovery heat flux.

Lee et al. (1988) used a finite difference
technique to predict temperature and pressure profiles
for helium II flow. They reported on experimental
results as well and found that, for Reynolds numbers
greater than 10^6, forced flow of helium II can be
treated as classical turbulent flow. For the Fanning
friction coefficient they used Petukhov's correlation

$$f/2 = 2.236 \ln(Re_n) - 4.639$$

Walstrom (1988) investigated the Joule-Thompson
effect and internal convective heat transfer on the flow
of helium II in a tube. Good agreement was found between
theory and experiment using the two-fluid model with
mutual friction and retaining the axial pressure
gradient in the analysis.

Maza et al. (1989) investigated experimentally
"noiseless" film boiling, i.e. with zero mass flux, and
found linear relationships between the heat flux (4-17
W/cm^2) and the bath depth (2-9.5 cm) and also between
the wall temperature (25-55 K) and the bath depth. No
quantitative theoretical predictions were attempted.

Danilchenko et al. (1989) investigated transient
heat transfer from a 9 mm^2 copper film surface deposited
on polished sapphire to both helium I and II. Dresner
(1989) analyzed bubble growth in superheated helium II.
Frederking et al., (1989) conducted experimental and
theoretical studies to determine the critical parameters
for the onset of non-linear counterflow in helium II.
Okamura et al. (1989) examined the behavior of
superfluid helium in a 6 mm i.d., 8 mm o.d., 2.5 m long
helical copper tube, one end of which was attached to a
12 mm diameter, 75 mm long copper rod. The outer end of
the copper tube was heated sinusoidally. The authors
also developed a one dimensional model based on the
"mutual friction counterflow equation" (Gorter-Mellink
correlation),

$$\rho c \frac{\partial T}{\partial t} = \frac{\partial}{\partial x}\left[k(T)\left(\frac{\partial T}{\partial x}\right)^{1/3}\right]$$

where ρ is the density, c is the specific heat, T is the
temperature, t is the time, k(T) is a temperature
dependent coefficient, and x is the axial direction. In
the model the Kapitza conductance was assumed to be
independent of the frequency. Above 40 Hz the system
behaved as if in steady state.

A session was held at the 5th AIAA/ASME
Thermophysics and Heat Transfer Conference in June,
1990, on the heat transfer properties of superfluid
helium. Among the fundamental topics, Kashani (1990)
analyzed forced convective heat transfer to helium II
flowing in a tube in terms of the second law of
thermodynamics, i.e., the generation of entropy by
temperature gradients and by viscous dissipation.
Shiotsu et al. (1990) reported on the transient heat
transfer caused by exponential, step, and exponential-
step heat inputs to a horizontal, 0.08 mm diameter wire
submerged in a pool of saturated helium II. Below the
maximum heat flux the steady state heat transfer was of
the Kapitza type:

$$q/A = 390(T_w{}^4 - T_b{}^4) \quad W/m^2$$

where T_w and T_b are the wall and bulk liquid
temperatures, respectively. The steady state maximum
heat flux was correlated in terms of the liquid head, H,
above the wire in meters:

$$(q/A)_{max} = 5.8{}^*10^4(1 + 5.69H) \quad W/m^2$$

In transient heating the heat flux followed the Kapitza
curve beyond the steady state maximum to a point of
departure from the curve (DKC) and continued to increase
to a higher maximum value. The shape of the curve
depended on the characteristics of the heat input.
Weisend and Van Sciver (1990) investigated heat transfer
in forced flow of helium II and found that the Kapitza
conductance was the controlling mechanism; consequently
the heat transfer was independent of the mass flow rate.
Here the exponent of the temperature used in the Kapitza
equation was 3 as opposed to 4 in the previously
described work.

Yuan et al. (1990) evaluated the overall phonon transmission coefficient, ζ, associated with solid to liquid helium-4 Kapitza resistance with emphasis on copper surfaces:

$$\zeta = 4h/(c_v\rho\ c_D)$$

where h is the heat transfer coefficient, $c_v\rho$ is the specific heat per unit volume, and c_D is the Debye speed of the solid. The value of ζ was found to be approximately 0.1 with variations depending on the temperature level and the difference in temperature between the wall and the fluid.

At the same session, on the topic of thermometry, Brizzi et al. (1990) described two models for niobium surface temperature measurements in superconducting RF cavities. Superfluid systems were presented by DiPirro et al. (1990) on the Superfluid Helium On-Orbit Transfer (SHOOT) flight demonstration facility and by Volz and Ryschkewitsch (1990) on the superfluid helium dewar of the Cosmic Background Explorer (COBE).

THERMAL STABILITY

The stability of superconducting magnets received renewed interest because of the new high temperature units and also because of the use of superfluid helium. Some of the "old," low temperature problems, however, still persist. Examining various aspects of stability were Eckels (1989), Frederking (1989), Hilal et al. (1988), Ito and Kubota (1989), Kerns et al. (1988), Kuroda et al. (1989), Nishijima et al. (1988a & b), Romanovskii (1988), Ogasawara (1989), Tien et al. (1989),and Volkov et al. (1988).

Canavan and Van Sciver (1988) and Waynert et al. (1988) examined stability in helium II; whereas Ohuchi et al. (1988), Phelan et al. (1989), Sekiya and Ichikawa (1989), Tada et al. (1989), van der Linden and Hoogendoorn (1989) used supercritical helium.

THERMAL PROPERTIES AND MEASUREMENT TECHNIQUES

There is always a need for finding the thermal properties of new materials or improving existing data;

consequently, there are always reports on such topics.
Typical examples were published by Asami (1988), Greig
(1988), Hartwig (1988), Iye (1988), Leyarovski et al.
(1988), Schwarz (1988), Tsatis (1988), Asami and Ebisu
(1989), Benda (1989), Bhowmick and Pattanayak (1989),
Bosch et al. (1989), Elsner (1989), Fonteyn and Pitsi
(1989), Gillespie and Ehrlich (1989), Kiselev et al.
(1989), Schwerdtner et al. (1989), Steur and Pavese
(1989), Svoboda (1989), Zimm et al. (1989), and Zych
(1989).

REFERENCES

Asami, T., 1988, New method to determine the BWR
    coefficients in saturated regions, _Cryogenics_,
    28:521-526.
Asami, T. and Ebisu, H., 1989, Thermodynamic properties
    of nitrogen calculated from the BWR equation of
    state, _Cryogenics_, 29:995-997.
Barron, R. F. and Dergham, A. R., 1988, Film boiling to
    a plate facing downward; _in:_ "Adv. in Cryogenic
    Eng.," Vol. 33, R. W. Fast, Ed., Plenum Press,
    N.Y., pp. 355-362.
Barth, W., W. Lehmann, Henry, C., and Walther, R., 1988,
    Test results for a high quality industrial
    superinsulation, _Cryogenics_, 28:607-609.
Beduz, C., Scurlock, R. G., and Sousa, A. J., 1988,
    Angular dependence of boiling heat transfer
    mechanisms in liquid nitrogen; _in:_ "Adv. in
    Cryogenic Eng.," Vol. 33, R. W. Fast, Ed., Plenum
    Press, N.Y., pp. 363-370.
Benda, V., 1989, Thermal resistance between ^{3}He-^{4}He
    mixture and sintered silver, _Cryogenics_, 29:457-
    459.
Bhowmick, T. and Pattanayak, S., 1989, Experimental
    setup for the study of thermal conductivity of
    elastomeric material at cryogenic temperature,
    Cryogenics, 29:463-466.
Bodegom, E., Nissen, J. A., Brodie, L. C., and Semura,
    J. S., 1988, Sound induced enhancement of heat
    transfer from a solid into liquid helium; _in:_ "Adv.
    in Cryogenic Eng.," Vol. 33, R. W. Fast, Ed.,
    Plenum Press, N.Y., pp. 349-354.
Borchi, E., Zoli, M., Frigo, A. and Lombardini, L.,
    1989, Axial conductivity and heat exchanges in a
    hybridized Gifford-McMahon cryocooler, _Cryogenics_,
    29:196-199.

Bosch, W. A., Willekers, R. W., Meijer, H. C., and
    Postma, H., 1989, Heat conductivity below 0.4 K of
    the glass-ceramic Macor and of Staybrite stainless
    steel, _Cryogenics_, 29:982-984.
Breen, B. P. and Westwater, J. W., 1962, Effect of
    diameter of horizontal tubes on film boiling heat
    transfer, _Chem. Eng. Prog._, 58, No. 7:67-72.
Breon, S. R. and Van Sciver, S. W., 1986, Boiling in
    saturated He II; _in:_ "Adv. in Cryogenic Eng.," Vol.
    31, R. W. Fast, Ed., Plenum Press, N.Y., pp. 465-
    472.
Brizzi, R., Fouaidy, M., and Junquera, T., 1990,
    Thermometry of niobium surfaces in superfluid
    helium: a powerful diagnostic technique for
    superconducting RF cavities, 5th AIAA/ASME
    Thermophysics and Heat Transfer Conference,
    Seattle, WA, June 18-20; _in:_ "Superfluid helium
    heat transfer," J. P. Kelley and W. J. Schneider,
    eds., ASME HTD-Vol. 134, New York, 15-22.
Bunkov, Y. M., 1989, Superconducting aluminium heat
    switch prepared by diffusion welding, _Cryogenics_,
    29:938-939.
Canavan, E. R. and Van Sciver, S. W., 1988, Evolution of
    normal zones and temperature in a superconductor
    cooled internally with helium-II; _in:_ "Adv. in
    Cryogenic Eng.," Vol. 33, R. W. Fast, Ed., Plenum
    Press, N.Y., pp. 195-202.
Chandratilleke, G. R., Nishio, S., and Ohkubo, H., 1989,
    Pool boiling heat transfer to saturated liquid
    helium from coated surface, _Cryogenics_, 29:588-
    592.[†]
Chato, J. C., 1989, Second law limitations imposed by
    heat exchangers on the performance of thermodynamic
    cycles, in "Collected papers in heat transfer," M.
    B. Pate, ed., ASME, New York, HTD-Vol. 123:201-210.
Chen, Z. and Van Sciver, S. W., 1986, Channel heat
    transfer in He I - steady state orientation
    dependence; _in:_ "Adv. in Cryogenic Eng.," Vol. 31,
    R. W. Fast, Ed., Plenum Press, N.Y., pp. 431-438.
Danilchenko, B. A., Lutset, M. O., and Poroshin, V. N.,
    1989, Limit of transient heat absorption by
    superfluid helium for very large heat pulses,
    Cryogenics, 29:444-447.
DiPirro, M. J., Boyle, R. F., Figueroa, O., Lindauer,
    D., McHugh, D. C., and Shirron, P. J., The SHOOT
    cryogenic system, 5th AIAA/ASME Thermophysics and
    Heat Transfer Conference, Seattle, WA, June 18-20;
    _in:_ "Superfluid helium heat transfer," J. P. Kelley

and W. J. Schneider, eds., ASME HTD-Vol. 134, New York, 29-36.

Dresner, L., 1989a, Bubble growth in superheated He II, _Cryogenics_, 29:509-512.

Dresner, L., 1989b, Stability margins of superconductors cooled with subcooled He II, _Cryogenics_, 29:668-673.[†]

Dress, D. A. and Kilgore, R. A., 1988, Cryogenic wind tunnel research: a global perspective, _Cryogenics_, 28:10-21.

Eckels, P. W., 1989, Superconductor stability and helium heat transfer: the minimum propagating zone relationship in design, _Cryogenics_, 29:625-629.[†]

Eckels, P. W. and Smith Jr., J. L., 1989, Superconductor stability in the power system environment, _Cryogenics_, 29:651-654.[†]

Elsner, A., 1989, Temperature dependence of energy and entropy of fluid systems, _Cryogenics_, 29:1075-1083.

Fonteyn, D. and Pitsi, G., 1989, Sensitive method for determining thermal conductivity of pure metals at low temperatures, _Cryogenics_, 29:51-54.

Frederking, T. H. K., 1989, Characteristic parameters of superconductor-coolant interaction including high T_C current density limits, _Cryogenics_, 29:610-615.[†]

Frederking, T. H. K., Afifi, F. A., and Ono, D. Y., 1989, Critical transport parameters for porous media subjected to counterflow, _Cryogenics_, 29:498-502.

Fuchino, S., Tamada, N., Sekine, T., Shinada, K., and Tomiyama, S., 1986, Dynamical behaviour of He II circulation loop under heat load; _in:_ "Adv. in Cryogenic Eng.," Vol. 31, R. W. Fast, Ed., Plenum Press, N.Y., pp. 481-487.

Fuchino, S. and Tamada, N., 1989, Characteristics of forced flow superfluid helium circulation system, _Cryogenics_, 29:432-437.

Gillespie, D. J. and Ehrlich, A. C., 1989, Fitting and interpolating calibration data for carbon-glass thermometers, _Cryogenics_, 29:1092-1095.

Gradt, T., Szucs, Z., Denner, H.-D., and Klipping, G., 1986, Heat transfer from thin wires to superfluid helium under reduced gavity; _in:_ "Adv. in Cryogenic Eng.," Vol. 31, R. W. Fast, Ed., Plenum Press, N.Y., pp. 499-504.

Greig, D., 1988, Low temperature thermal conductivity of polymers, _Cryogenics_, 28:243-247.

Hartwig, G., 1988, Thermal expansion of fibre composites, _Cryogenics_, 28:255-266.

Hassenzahl, W. V., 1989, Study of the effects of copper to superconductor ratio on stability, <u>Cryogenics</u>, 29:637-641.[†]

Hilal, M. A., Peck, S. D., and Ibrahim, E. A., 1988, Helium pressure rise of superconducting super collider dipole magnets following a quench; <u>in:</u> "Adv. in Cryogenic Eng.," Vol. 33, R. W. Fast, Ed., Plenum Press, N.Y., pp. 143-148.

Igra, R., Scurlock, R. G., and Wu, Y. Y., 1986, Reverse convection in helium and other fluids in the high speed rotating frame: negative and positive buoyancy effects; <u>in:</u> "Adv. in Cryogenic Eng.," Vol. 31, R. W. Fast, Ed., Plenum Press, N.Y., pp. 447-454.

Iida, F., Tada, N., and Ogata, H., 1989, Stability of 10T-(Nb-Ti)$_3$Sn forced flow cooled superconducting coil, <u>Cryogenics</u>, 29:642-647.[†]

Ito, T. and Kubota, H., 1989, Maddock criterion and minimum propagating zone, <u>Cryogenics</u>, 29:621-624.[†]

Iye, Y., 1988, Small low-cost platinum resistance thermometers for thermometry in magnetic fields, <u>Cryogenics</u>, 28:164-168.

Kasao, D. and Ito, T., 1989, Review of existing experimental findings on forced convection heat transfer to supercritical helium 4, <u>Cryogenics</u>, 29:630-636.[†]

Kashani, A., 1990, Entropy generation in He II forced convection, 5th AIAA/ASME Thermophysics and Heat Transfer Conference, Seattle, WA, June 18-20; <u>in:</u> "Superfluid helium heat transfer," J. P. Kelley and W. J. Schneider, eds., ASME HTD-Vol. 134, New York, 37-43.

Kashani, A. and Van Sciver, S. W., 1986, Steady state forced convection heat transfer in He II; <u>in:</u> "Adv. in Cryogenic Eng.," Vol. 31, R. W. Fast, Ed., Plenum Press, N.Y., pp. 489-498.

Kerns, J. A., Slack, D. S., and Miller, J. R., 1988, Thermal analysis of the forced cooled conductor for the TF (Toroidal Field) superconducting coils in the TIBER (Tokamak Ignition/Burn Experimental Reactor) II ETR design; <u>in:</u> "Adv. in Cryogenic Eng.," Vol. 33, R. W. Fast, Ed., Plenum Press, N.Y., pp. 175-182.

Kiselev, Yu. F., Chernikov, A. N., Gorodishenin, N. L., Evdokimov, V. A., and Katushonok, S. S., 1989, Autocompensating device for ultra-low temperature measurements, <u>Cryogenics</u>, 29:55-58.

Klimenko, V. V., Tyodorov, M. V., and Fomichyov, Yu. A.,
    1989, Channel orientation and geometry influence on
    heat transfer with two-phase forced flow of
    nitrogen, _Cryogenics_, 29:31-36.
Kobayashi, H., Tada, I., Kodama, T., Tanaka, M., and
    Turuga, H., 1989, Stabilization enhancement for
    superconducting magnets by means of hydrostatically
    pressurized He II, _Cryogenics_, 29:674-678.[†]
Kuroda, K., Uchikawa, S., Hara, N., Saito, R., Takeda,
    R., Murai, K., Kobayashi, T., Suzuki, S., and
    Nakayama, T., 1989, Quench simulation analysis of a
    superconducting coil, _Cryogenics_, 29:814-824.
Kush, P. K. and Thirumaleshwar, M., 1989, Design of
    regenerators for a Gifford-McMahon cycle
    cryorefrigerator, _Cryogenics_, 29:1084-1091.
Lee, J. H., Ng, Y. S., and Brooks, W. F., Analytical
    study of He II flow charateristics in the SHOOT
    transfer line, _Cryogenics_, 28:81-85.
Leyarovski, E. I., Leyarovska, L. N. Popov, Chr., and
    Popov, O., 1988, High field magnetocalorimetry
    below 1 K: specific heat of copper at 14 T,
    Cryogenics, 28:321-335.
Lezak, D., Brodie, L. C., Semura, J. S., and Roberts, S.
    M., Temperature dependence of the time delay to the
    onset of film boiling in liquid helium; _in:_ "Adv.
    in Cryogenic Eng.," Vol. 31, R. W. Fast, Ed.,
    Plenum Press, N.Y., pp. 439-446.
Lin, L-h, Hu, L-f, Chen, S-x, Liu, C-y, Sun, Z-f, and
    Chen, H., 1986, Enhancement of condensation heat
    transfer inside V-type corrugated vertical tubes;
    _in:_ "Adv. in Cryogenic Eng.," Vol. 31, R. W. Fast,
    Ed., Plenum Press, N.Y., pp. 423-430.
Lutset, M. O. and Zhukov, V. E., 1989, Heat transfer in
    a rotating cryostat at high centrifugal
    acceleration fields, _Cryogenics_, 29:37-41.
Matsumoto, K. and Shiino, M., 1989, Thermal regenerator
    analysis: analytical solution for effectiveness and
    entropy production in regenerative process,
    Cryogenics, 29:888-894.
Maza, J., Jebali, F., Francois, M. X., and Vidal, F.,
    1989, Temperature and heat flux measurement in
    noiseless film boiling in superfluid helium,
    Cryogenics, 29:200-202.
Mirza, S., 1990, The behavior of boiling of liquid
    nitrogen from aluminum surfaces, _Int. Comm. Heat
    Mass Transfer_, 17:9-18.
Mori, H. and Ogata, H., 1989, Effect of counterflow on
    heat transfer to He II in a channel, _Cryogenics_,
    29:664-667.[†]

Muller-Steinhage, H. M., 1988, Fouling phenomena during
    boiling of cryogenic liquids, Cryogenics, 28:406-
    408.
Narahara, Y., Tominaga, A., Mizutani, F., and Yazaki,
    T., 1988, Thermal effects due to oscillations of
    gas columns, Cryogenics, 28:177-180.
Nemirovskii, S. K. and Tsoi, A. N., 1989, Transient
    thermal and hydrodynamic processes in superfluid
    helium, Cryogenics, 29:985-994.
Nishijima, S., Yamashita, T., Takahata, K., Okada, T.,
    Fukutsuka, T., Matsumoto, K. and Hamada, M., 1988a,
    Effect of epoxy cracking on stability of
    impregnated windings related to thermal and
    mechanical properties; in: "Adv. in Cryogenic
    Eng.," Vol. 33, R. W. Fast, Ed., Plenum Press,
    N.Y., pp. 125-133.
Nishijima, S., Takahata, K., and Okada, T., 1988b, Local
    temperature rise after quench due to epoxy cracking
    in impregnated superconducting windings; in: "Adv.
    in Cryogenic Eng.," Vol. 33, R. W. Fast, Ed.,
    Plenum Press, N.Y., pp. 135-142.
Noto, K., Watanabe, K., Morita, H., Mori, K., Sasakawa,
    M., Isikawa, Y., Sato, K., Ishihara, T., Inukai,
    E., Fujimori, H., and Muto, Y., 1989, Thermal
    conductivity and critical current density in high
    T_c Y_{1-x} Ln_x Ba_2 Cu_3 O_{7-y}, Cryogenics, 29:648-650.[†]
Ogasawara, T., 1989, Stability of composite tape
    superconductors against transient thermal
    disturbances, Cryogenics, 29:825-829.
Ohuchi, N., Makida, Y., Yamamoto, J., and Murakami, Y.,
    1988, AC loss and stability of a force-cooled
    superconducting coil with a bias pulsed
    superconducting magnet; in: "Adv. in Cryogenic
    Eng.," Vol. 33, R. W. Fast, Ed., Plenum Press,
    N.Y., pp. 159-166.
Okamura, T., Yoshizawa, Y., Sato, A., Ishito, K.,
    Kabashima, S., and Shioda, S., 1989, Time dependent
    heat transfer in pressurized superfluid helium,
    Cryogenics, 29:1070-1074.
Onishi, T., Tateishi, H., and Nomura, H., 1989, Dynamic
    behaviour of pulsed magnets, Cryogenics, 29:659-
    663.[†]
Oonk, R. L. and Hustvedt, D. C., 1986, The effect of
    fluid property variations on the performance of
    cryogenic helium heat exchangers; in: "Adv. in
    Cryogenic Eng.," Vol. 31, R. W. Fast, Ed., Plenum
    Press, N.Y., pp. 415-422.

Phelan, P. E., Iwasa, Y., Takahashi, Y., Tsuji, H., Nishi, M., Tada, E., Yoshida, K., and Shimamoto, S., 1989, Transient stability of a Nb-Ti cable-in-conduit superconductor: experimental results, Cryogenics, 29:109-118.

Richardson, R. N., 1986, Pulse tube refrigerator - an alternative cryocooler?, Cryogenics, 26:331-340.

Richardson, R. N., 1988, Development of a practical pulse tube refrigerator: co-axial designs and the influence of viscosity, Cryogenics, 28:516-520.

Rohsenow, W. M., 1952, A method of correlating heat transfer data for surface boiling of liquids, Trans. ASME, 74:969.

Romanovskii, V. R., 1988, Stability of superconducting composites under thermal disturbances with change in the external magnetic field and the critical temperature of the superconductor, Cryogenics, 28:756-761.

Sakurai, A., Shiotsu, M., Hata, K., and Takeuchi, Y., 1989, Quasi-steady nucleate boiling and its life caused by large stepwise heat input in saturated pool liquid helium I, Cryogenics, 29:597-601.[†]

Sato, A., Ishito, K., Kabashima, S., Nakamura, S., and Suzuki, E., 1989, Stability of superconducting magnet indirectly cooled by He II, Cryogenics, 29:655-658.[†]

Schmidt, C., 1988, Transient heat transfer into a closed small volume of liquid or supercritical helium, Cryogenics, 28:585-598.

Schoepe, W., Uhlig, K., and Neumaier, K., 1989, Carbon and germanium resistors in the variable-range hopping regime for thermometry below 1 K, Cryogenics, 29:467-468.

Schwarz, G., 1988, Thermal expansion of polymers from 4.2 K to room temperature, Cryogenics, 28:248-254.

Schwerdtner, M. v., Poppe, W., and Schmidt, D. W., 1989, Distortion of temperature signals in He II due to probe geometry, and a new improved probe, Cryogenics, 29:132-134.

Sekiya, S. and Ichikawa, A., 1989, Numerical analysis of quench transients in forced-flow cooled superconductors, Cryogenics, 29:25-30.

Shiotsu, M., Hata, K., and Sakurai, A., 1989, Effects of diameter and system pressure on critical heat flux for horizontal cylinders in saturated liquid He I, Cryogenics, 29:593-596.[†]

Shiotsu, M., Hata, K., and Sakurai, A., 1990, Transient
    heat transfer from a horizontal wire in superfluid
    helium caused by exponential and step heat inputs,
    5th AIAA/ASME Thermophysics and Heat Transfer
    Conference, Seattle, WA, June 18-20; *in:*
    "Superfluid helium heat transfer," J. P. Kelley and
    W. J. Schneider, eds., ASME HTD-Vol. 134, New York,
    9-14.
Shu, Q. S., Fast, R. W., and Hart, H. L., 1986, An
    experimental study of heat transfer in multilayer
    insulation; *in:* "Adv. in Cryogenic Eng.," Vol. 31,
    R. W. Fast, Ed., Plenum Press, N.Y., pp. 455-463.
Steur, P. P. M. and Pavese, F., 1989, He-3 constant
    volume gas thermometer as interpolating instrument:
    calculations of the accuracy limit *versus*
    temperature range and design parameters,
    Cryogenics, 29:135-138.
Svoboda, P., 1989, Simple device for measurement of
    transport properties of new high temperature
    superconductors, Cryogenics, 29:210-211.
Tada, E., Takahashi, Y., Tsuji, H., Okuno, K., Ando, T.,
    Hiyama, T., Koizumi, K., Nishi, M., Nakajima, H.,
    Yoshida, K., Kato, T., Kawano, K., Oshikiri, M.,
    Yamaguchi, M., and Shimamoto, S., 1989, Downstream
    effect on stability in cable-in-conduit
    superconductor, Cryogenics, 29:830-840.
Tanaka, M. and Kodama, T., 1989, Pool boiling heat
    transfer in ^{3}He, Cryogenics, 29:203-205.
Tarnawski, Z., van der Meulen, H. P., Franse, J. J. M.,
    Kadowaki, K., Veenhuizen, P. A., and Klaasse, J. C.
    P., 1988, New possibility of magnetic ripple
    shielding for specific heat measurements in hybrid
    magnets, Cryogenics, 28:614-616.
Tien, C. L., Flik, M. I., and Phelan, P. E., 1989,
    Mechanisms of local thermal stability in high
    temperature superconductors, Cryogenics, 29:602-
    609.[†]
Tsatis, D. E., 1988, Thermal diffusivity of Araldite,
    Cryogenics, 28:609-610.
Tsukamoto, O., Takao, T., and Honjo, S., 1989, Stability
    analysis of superconducting magnet: an approach to
    quantification of energy disturbance caused by
    conductor motion, Cryogenics, 29:616-620.[†]
van der Linden, R. J. and Hoogendoorn, C. J., 1989,
    Transient cooling of an internally cooled
    superconducting magnet, Cryogenics, 29:179-187.
Van Sciver, S. W., 1989, Forced flow He II in relation
    to stability of internally cooled superconductors,
    Cryogenics, 29:679-682.[†]

Vishnev, I. P., 1988, Molecular and thermodynamic method of heat transfer generalization and classification of boiling substances, <u>Cryogenics</u>, 28:770-778.

Volkov, A. F., Dinaburg, L. B., Kalinin, V. V., Konstantinov, A. B., and Khrushev, V. N., 1988, Simulation of nonstationary thermal regimes in cryogenic loops for tokamak-15 electromagnets; <u>in:</u> "Adv. in Cryogenic Eng.," Vol. 33, R. W. Fast, Ed., Plenum Press, N.Y., pp. 183-186.

Volz, S. M. and Ryschkewitsch, M. G., Ground and early on-orbit performance of the superfluid helium dewar of the Cosmic Background Explorer (COBE), 5th AIAA/ASME Thermophysics and Heat Transfer Conference, Seattle, WA, June 18-20; <u>in:</u> "Superfluid helium heat transfer," J. P. Kelley and W. J. Schneider, eds., ASME HTD-Vol. 134, New York, 23-27.

Walstrom, P. L., 1988, Joule-Thompson effect and internal convection heat transfer in turbulent He II flow, <u>Cryogenics</u>, 28:151-156.

Waynert, J., Eyssa, Y., and Huang, X., 1988, The transient stability of large scale superconductors cooled in superfluid helium; <u>in:</u> "Adv. in Cryogenic Eng.," Vol. 33, R. W. Fast, Ed., Plenum Press, N.Y., pp. 187-194.

Weisend II, J. G. and Van Sciver, S. W., 1990, Surface heat transfer measurements in forced flow He II, 5th AIAA/ASME Thermophysics and Heat Transfer Conference, Seattle, WA, June 18-20; <u>in:</u> "Superfluid helium heat transfer," J. P. Kelley and W. J. Schneider, eds., ASME HTD-Vol. 134, New York,1-7.

Xiao, J. H. and Guo, F. Z., 1988, Analytical network model on the flow and thermal characteristics of cyclic flow cryogenic regenerators, <u>Cryogenics</u>, 28:762-769.

Xiulin, Y., Hongji, X., Yuweng, Z., and Hongzhang, Q., 1989, Pool boiling heat transfer to liquid nitrogen from porous metallic coatings of tube bundles and experimental research of hysteresis phenomenon, <u>Cryogenics</u>, 29:460-462.

Yamamoto, J., Yamamuro, K., Ohuchi, N., and Murakami, Y., 1986, Analysis of flow instability of supercritical helium in curved tubing; <u>in:</u> "Adv. in Cryogenic Eng.," Vol. 31, R. W. Fast, Ed., Plenum Press, N.Y., pp. 473-480.

Yuan, S. W. K., Frederking, T. H. K., and Chuang, T. C., 1990, Thermal contact domain resistance: overall solid helium 4 Kapitza resistance, 5th AIAA/ASME Thermophysics and Heat Transfer Conference,

Seattle, WA, June 18-20; <u>in:</u> "Superfluid helium heat transfer," J. P. Kelley and W. J. Schneider, eds., ASME HTD-Vol. 134, New York,45.

Zimm, C. B., Barclay, J. A., Harkness, H. H., Green, G. F., and Patton, W. G., 1989, Magnetocaloric effect in thulium, <u>Cryogenics</u>, 29:937-938.

Zych, D. A., 1989, Thermal conductivity of a machinable glass-ceramic below 1 K, <u>Cryogenics</u>, 29:758-759.

[†] Papers presented at the Seminar on Basic Mechanics of Helium Heat Transfer and Related Stability of Superconducting Magnets, Fukuoka, Japan, Aug. 29-Sept. 2, 1988.

THE UNIVERSE OF CRYOGENIC STORAGE

FROM HELIUM II TO CO2

Ivan V. (Larry) LaFave

Engineering Consultant
Naperville, Illinois

INTRODUCTION

The purpose of this paper is threefold: (1) to provoke
debate amongst the "cryogentry" by proposing an appropriate
definition for the cryogenic temperature range; (2) to call
attention to the vastness of the universe of cryogenic
storage and to suggest that useful information can be
gleaned from the intertwining of different disciplines; (3)
to discuss a few of the misadventures that have occurred in
the storage and handling of cryogenic fluids.

A DEFINITION FOR THE CRYOGENIC TEMPERATURE RANGE

The definition for cryogenic\cryogenics which appears
in one of Webster's editions is as follows: cryogenic\ of
or relating to the production of very low temperatures;
cryogenics\ a branch of physics that deals with the
production and effects of very low temperatures.

Webster does not quantify "very low temperatures."
In Chicago, -29C (-20F) (244.18K) is very low. Where the
author grew up, -40C (-40F) (233.18K) is very low. In
places humans routinely inhabit on planet Earth, it appears
that somewhere around -57C (-70F) (216.18) is the lowest
naturally occurring temperature. Logically, the upper
boundary for the cryogenic temperature range should have a
value in the vicinity of the lowest ambient temperatures
which occur naturally. However, these naturally occurring
low temperatures are variable and, for a boundary
temperature, a single value is desirable.

Applications of Cryogenic Technology, Vol. 10
Edited by J.P. Kelley, Plenum Press, New York, 1991

Therefore, it is proposed that a <u>unique</u> value be
selected for the upper boundary of the cryogenic
temperature range and further that this unique value be the
<u>triple point of CO2</u> which is 216.6K (-56.6C) (-69.9F). In
the opinion of the author, the storage of CO2 at its triple
point should be considered cryogenic storage.

The lower boundary for the cryogenic temperature range
is <u>absolute zero.</u> It is presumed that there will be little
debate regarding the value for the lower boundary. It will
be left to others to install any intermediate boundaries
which become necessary to separate the low, the very low,
the ultra low and the super ultra low temperature ranges.

A DEFINITION OF A CRYOGEN

With the cryogenic temperature range defined, a method
for deciding what substances are cryogens and what ones are
not is needed. It is proposed that all fluids which have a
<u>critical</u> temperature which lies <u>below</u> or <u>within</u> the normal
ambient temperature range be defined as cryogens. The upper
limit of the normal ambient temperature range is selected
to be 323.18K (50C) (122F). This is arbitrary but rational.
Based on this concept, cryogens then, are those gases which
cannot exist as liquids, without refrigeration, regardless
of the amount of pressure which is applied.

It follows then, that the family of cryogens will
begin with Helium II near the lower boundary and include
carbon dioxide at the upper boundary. The critical
temperature of CO2 is 304.18K (31C) (87.8F). It is fitting
that these two cryogens with such unique properties exist
at the extremes of the cryogenic range.

Helium II is probably the most unusual of the family
of cryogenic fluids. Its unique internal convection
mechanism leads to an apparent thermal conductivity and
superfluid behavior that is truly phenomenol.

CO2 is also a very unique material. It can be stored
as a liquid at an infinite number of combinations of
temperature and pressure, a common one being about -18C
(0F) (255.18K) at approximately 2.07 Mp (300 psi) of
pressure. Under these storage conditions, CO2 does not fit
the usual perception of a cryogen, but when it is released
from storage pressure to atmospheric pressure, desirably in
a controlled manner, a substantial portion of the liquid
becomes the familiar solid (dry ice) at a temperature of
194.68K (-78.5C) (-109.3F).

THE UNIVERSE OF CRYOGENIC STORAGE

One of Webster's editions defines "universe" as <u>the whole body of things and phenomena observed or postulated.</u>

That definition anticipates much more than could be included in a brief paper. However the chosen title is not meant to convey that the entire universe will be discussed, but rather to suggest the magnitude of that universe and to point out the possible benefits for some if they will expand their cognizance of the overall cryogenic field, so as to take advantage of the potential cross fertilization which can occur when disciplines are intertwined and the art of one field is applied in others.

Now that a cryogenic temperature range has been defined, and a method has been proposed for deciding what a cryogen is, a table listing the cryogenic fluids which fall within the proposed definition can be constructed. Table 1 lists the common cryogens along with their boiling point temperatures.

Short Term Cryogenic Storage

What is short term storage? It is not necessary to be specific with regard to time. It is only necessary to grasp the concept. Some examples will illustrate.

The transportation of cryogenic fluids by ship, rail, barge, airplane, or truck trailer falls in the category of short term storage. During transit, there is heat gain into the storage container. This must be provided for in the overall system design. Either the bulk temperature of the cryogenic fluid increases and the pressure goes up or the pressure is controlled by removing vapor and the bulk temperature does not increase. Combinations of some pressure rise, some temperature rise and some vapor removal are also possible.

An LNG ship experiences heat gain during a sea voyage. This heat gain is translated into boiloff, which is used as part of the ship's fuel requirements. In the case of a barge, the heat gain is generally translated into an average bulk temperature rise in the stored liquid. In some cases, venting can be used to maintain the pressure and the temperature.

The concepts of aircraft transporting LNG have generally considered that the boiloff would fuel the aircraft engines. Note that this is conceptual only.

Table 1 THE FAMILY OF CRYOGENIC FLUIDS

CRYOGENS	CRITICAL TEMPERATURE			NORMAL BOILING POINT		
	DEG K	DEG C	DEG F	DEG K	DEG C	DEG F
Carbon dioxide	304.2	31.0	87.8	273.2	* -78.5	* -109.3
Acetylene	309.5	36.3	97.4	189.8	-83.3	-118.0
Ethane	305.5	32.3	90.1	184.6	-88.6	-127.5
Nitrous Oxide	309.7	36.5	97.7	183.7	-89.5	-129.1
Ethylene	283.1	9.9	49.8	169.5	-103.7	-154.7
Xenon	289.8	16.6	61.9	165.1	-108.1	-162.6
Krypton	209.5	-63.7	-82.7	120.3	-152.9	-243.2
Methane	190.7	-82.5	-116.5	111.7	-161.5	-258.7
Oxygen	154.8	-118.4	-181.1	90.2	-183.0	-297.4
Argon	150.7	-122.5	-188.5	87.4	-185.8	-302.4
Fluorine	144.2	-129.0	-200.2	85.1	-188.1	-306.5
Carbon monoxide	134.2	-138.9	-218.1	81.2	-192.0	-313.6
Air	132.5	-140.7	-221.3	78.9	-194.3	-317.7
Nitrogen	126.1	-147.1	-232.8	77.4	-195.8	-320.4
Neon	44.5	-228.7	-379.7	27.3	-245.9	-410.6
Hydrogen	33.3	-239.9	-399.8	20.4	-252.8	-423.0
Helium	5.3	-267.9	-450.2	4.3	-268.9	-452.0
Helium II **	5.3	-267.9	-450.2	4.3	-268.9	-452.0

* The sublimation point is given for Carbon dioxide.
** Helium II properties begin at 2.17 degrees Kelvin.

Truck and truck/trailer transportation of cryogenic fluids is well developed. There are in place federal regulations, which govern the transport of cryogens by truck and truck/trailer.

The use of a laboratory container for bringing small quantities of a cryogen to the point of use is another example of short term storage. The container can be a styrofoam cup, a urethane foam insulated aluminum or stainless steel cooking utensil, a vacuum bottle, or a fine quality commercially available laboratory container designed specifically for such use.

Long Term Cryogenic Storage

For those applications where cryogenic fluids are employed for the maintenance of cryogenic temperatures on essentially a continuing basis, refrigeration is generally required for reliquefying the cryogen vapor.

Bulk storage of cryogens for later use in partial quantities also fits the concept of long term storage. Again, time is not as relevant as the concept itself. One example of long term storage is peakshaving with LNG. In some instances, the LNG is stored for months and is not used until the peak demand for gas requires that the LNG be vaporized and fed into the gas distribution system.

Cryogenic fluids have been and are now employed in space applications. It is envisioned that long term storage of cryogenic fluids will be required for future space missions. Remember, cryogenics got us to the moon. And back!

The challenge for long term storage is to design an insulation system which will limit the rate of heat input into the boiling cryogenic fluid to a value which can be tolerated. Generally it all boils down to economics.

There is, however, always room for improvement in the design of cryogenic insulation systems. Reliquefiers can play a role for those cases where continuous cryogenic cooling is required or where preservation of the cryogen is paramount. This is a likely requirement for many space applications and as the space voyages get longer, cryogenic storage system design challenges will become greater. Some new breakthroughs would be welcome in the area of insulation system design for cryogenic storage throughout the cryogenic temperature range.

Sizes and Cryogenic Storage Vessels

The size range for cryogenic storage vessels covers
about eight orders of magnitude. The smallest laboratory
vessel is somewhere around a liter in capacity. At
present, the largest storage vessel for LNG is about
144,000,000 liters in capacity. The diameter of a vessel
this size may be in excess of 75 meters (246 feet).

Insulation space thicknesses vary from 12 millimeters
to as much as 1.8 meters, or in English units, from about
1/2 inch to about six feet.

The large LNG ships have a capacity of about 125,000
cubic meters (800,000 bbls).

The Many Disciplines of Cryogenic Storage

The disciplines which come into play in the practice
of cryogenic engineering as related to cryogenic storage
are many and diverse. Those that quickly come to mind in
no special order are:

Heat transfer	Fluid mechanics
Insulation systems	Structural design
Vapor barriers	Seismic design
Thermodynamics	Concrete design
Materials	Physical properties
Metallurgy	Refrigeration
Welding	Reflectivity
Pressure vessels	Emissivity
Foundations	Vacuum
Soil mechanics	Leak detection
Frost susceptibility	Mass spectrometry

There are many more which could be added to the list.
The intent is not to make an exhaustive list but to point
out that the field of cryogenics throughout the cryogenic
range requires a multi-discipline approach and that the
design and construction of cryogenic storage systems
utilizes a variety of disciplines.

It is most desirable to keep in touch with the
directions new developments are taking in related and non-
related fields and observing carefully how technological
advances in these other fields may be useful to you in your
chosen field. The author's personal experience is that
there are many opportunities to apply knowledge gained from
one field to another field which is completely unrelated to
the first.

A few words are in order about just two of the
disciplines on the list. Reference is made to <u>vacuum</u> and
to <u>leak detection,</u> which are employed in the construction
of cryogenic storage vessels which have vacuum insulation
systems.

There should be a special memorial erected in some
appropriate place dedicated to all those scientists,
engineers, and technicians who have had to struggle through
the process of learning the intricacies of vacuum and leak
detection without benefit of a teacher. Those who are self
taught in these two disciplines will likely understand. A
leak in a vacuum system which cannot be located immediately
comes very close to producing the highest level of
frustration one can experience in the cryogenic field or
perhaps in any field.

The Many Shapes of Cryogenic Storage

There are spheres inside of spheres; cylinders inside
of cylinders; spheres inside of cylinders; flat bottom
tanks inside of flat bottom tanks; flat bottom tanks with
dome roofs; flat bottom tanks with suspended insulation
decks; vertical cylinders and horizontal cylinders. There
is almost always a good reason for the specific selection
of tank configuration for the cryogenic storage vessel.
Almost always it boils down to economics.

CRYOGENIC STORAGE MISADVENTURES

Few technologies escape an occasional misadventure and
the cryogenic field is no exception. The author has been
involved, at times during the last three decades, in
dealing with the results of some of these misadventures.
Dealing, as used herein, means either analyzing what
happened or developing procedures for making repairs, or
both. Most of the incidents characterized as misadventures
have not been serious but rather have been nuisances. They
are excellent lessons for all to learn from.

Improper Gas In Insulation Space

This misadventure occurs, on the average, about once a
decade, and is a classic. It first occurred because of the
inadvertent admission of a mixture of gaseous nitrogen and
gaseous oxygen into the annular space of a double wall
liquid nitrogen tank. The temperature of the outer surface
of the inner tank shell was cold enough to condense some of
the "air" in the annular space.

The liquid formed was rich in oxygen. As the liquid formed on the outside of the shell of the inner vessel, it ran down the outer surface where it then entered the perlite insulation underneath the bottom of the inner tank. When liquid first entered the perlite mass, it vaporized and cooled the perlite. Succeeding drops of liquid could penetrate further into the perlite. The process was a continuing one inasmuch as the vapor would recondense and add to the liquid supply. This is a classic case of very efficient reflux heat transfer and should not be allowed to happen in the insulation space of a storage vessel.

Permitting it to occur not only increases the heat leak into the tank but, also, when the liquid finally reaches the outer carbon steel container, it can cause the carbon steel to get cold enough to lose its ductility and become a brittle material subject to brittle fracture.

The annular insulation space in a double wall non-vacuum cryogenic storage tank must have a gas in the insulation space which, under the specific storage conditions, is not subject to condensation at the temperature of the outer surface of the inner tank shell with which it is normally in contact.

Brittle Fracture

Another aspect of the cryogenic universe is the phenomenon of brittle fracture. All it takes is mild carbon steel and a cryogenic fluid spilled on the carbon steel. A very rapid series of events take place.

The cryogen boils vigorously. The carbon steel experiences a very rapid change in temperature downward. The change in temperature produces two distinct effects. The mild carbon steel changes from a tough ductile material to a brittle material with reduced toughness. In addition, thermal shrinkage takes place and, if the area which gets cooled is restrained, rapidly changing patterns of stress are established. Almost always there will exist a stress raiser within the brittle area. Almost always cracking will occur. The cracks will generally initiate at one of the stress raisers. Some observers will erroneously conclude that the flaw, if that happens to be the stress raiser, is to blame for the event. Not so! The cause is the change in properties because of the lowering of the temperature of the carbon steel. The flaw is of no significance until the temperature has been reduced to levels way outside the design temperature range for the carbon steel.

170

There have been a number of incidents of brittle
fracture resulting from accidental spills of cryogenic
fluids. The more common ones have been oxygen, nitrogen
and liquid natural gas spills on the carbon steel outer
container of the storage vessel, which is not designed to
be spilled upon by a cryogenic fluid.

The Case of the Frozen Foundation

Flat bottomed cryogenic tanks frequently rest directly
on the grade and the foundations for these tanks are fitted
with heating systems which maintain a plane of temperature
a few degrees above freezing so that the 0C (32F) isotherm
does not penetrate the grade.

There have been a few cases where the heating system
was not kept working properly and the freezing isotherm did
penetrate the grade several feet. If the soil within the
grade is frost susceptible, i.e. if ice lenses form, it is
quite possible to experience an uplifting of the tank.
This is usually not an uplifting experience for the tank
owner. If not detected early enough, expensive repairs may
be required.

DO'S AND DON'TS

The misadventures experienced can and should be used
as object lessons and learned from so that the number of
such misadventures can be reduced.

The Don'ts

Do not allow the storage container to become
overfilled.

Do not allow a grade foundation to freeze.

Do not allow the pressure in the tank to exceed the
range of pressure the vessel was designed for.

Do not spill cryogenic fluids on carbon steel
structures.

Do not permit a gas in the annular insulation space of
a double wall tank which can condense on the outer
surface of the inner tank and set up a reflux heat
exchange. This can lead to brittle fracture of the
outer tank shell and this must be avoided. It is a
nuisance to decommission a tank for a crack repair.

<u>The Do's</u>

<u>Do</u> monitor the condition of the tank on a planned basis and note the changes which take place over time. Generally, a tank which is experiencing a problem will show signs of the problem developing over a long period.

<u>Do</u> maintain the appurtenances associated with the tank, particularly the safety relief valves, the liquid level gaging equipment, and the pressure controlling and indicating equipment.

CONCLUSIONS

Cryogenics covers a lot of territory these days and will continue to grow in stature as a very useful science, art, and technology. In the vernacular of the teenager, "Cryogenics" is cool. But, those who work in the field, whether it be with Helium II or CO2, or anywhere in between, are the "cryogentry." They are the ones who really know what cool is.

CRYOGENIC LIQUID CYLINDER DEVELOPMENT

James P. Eaton

Minnesota Valley Engineering, Inc.
407 Seventh Street N.W.
New Prague, Minnesota 56071

ABSTRACT

The evolution of liquid cylinders has resulted in the
present, durable, long lasting tough design. All cylinders are
built to the DOT-4L code, with exacting specifications for
materials, pressure, and capacities. They use superinsulation
(a combination of aluminum foil and fiberglass paper) which will
not support combustion.

FIGURE 1. Modern Industrial Liquid Cylinder

The modern vessel configuration, (Figure 1) used with gas and liquids, has a complex construction consisting of: pressure building coil, pressure building regulator, vaporizer, pressure building valve, economizer regulator, liquid valve, vent valve, gas use valve, pressure gauge, relief valve, and rupture disk.

Non-Code design considerations include minimizing heat loss by minimizing radiation, conduction, and convection heat losses and constructing a neck tube strong enough to support the tank. Overriding all other concerns is that the tank must be designed to take abuse.

The cryogenic container design has development through three generations. As they evolved, these tanks have incorporated more and more elements that provided for safety and ruggedness. Examples of tanks from various major manufacturers and how they developed over the years are presented.

CYLINDER DESIGN CODE AND FUNCTIONAL CHARACTERISTICS

Before reviewing the history of cryogenic liquid cylinder development, an explanation of the cylinder's design code and functional characteristics is in order.

The industrial type liquid cylinders were first introduced in the 1950's. At that time, the design was governed by the Bureau of Explosives of the Federal Government under the DOT-4L code. The DOT-4L code, with some minor revisions over the years, is still the governing specification. DOT-4L covers the design of "Welded Cylinders - Insulated" for the industrial gases Argon, Nitrogen, and Oxygen, plus Helium, Hydrogen, and Neon.

In the DOT-4L code, design of the inner vessel is specific (Figure 2). The heads must be 2:1 elliptical. The shell and heads must be Type 304SS with certain limitations on allowable stress. Testing of materials and welds are also listed. The insulation must be fire resistant. Additionally, the maximum heat transfer rate is limited to 0.0005 BTU/HR (0.00015 W) per degree Fahrenheit differential in temperature per pound of water capacity. For a typical 160 liter cylinder in liquid nitrogen service, this would result in a total maximum heat transfer rate of 75.5 BTU/HR (22.1 W) or a boiloff rate of approximately 7.2% per day. The minimum shell thickness is 0.60 inches (0.0152 m) for stainless steel and 0.70 inches (0.0178 m) for aluminum. The inner vessel size is limited to 1,000 pounds (454.5 kg) of water capacity - approximately 120 gallons (454 liters) - and the service pressure must be over 40 psig (377 KPa ABS) and not exceed 500 psig (3.55 MPa ABS). The relief devices are specified per CGA S-1.1. The filling limits are tabulated in Paragraph 173.316 of the Code of Federal Regulations.

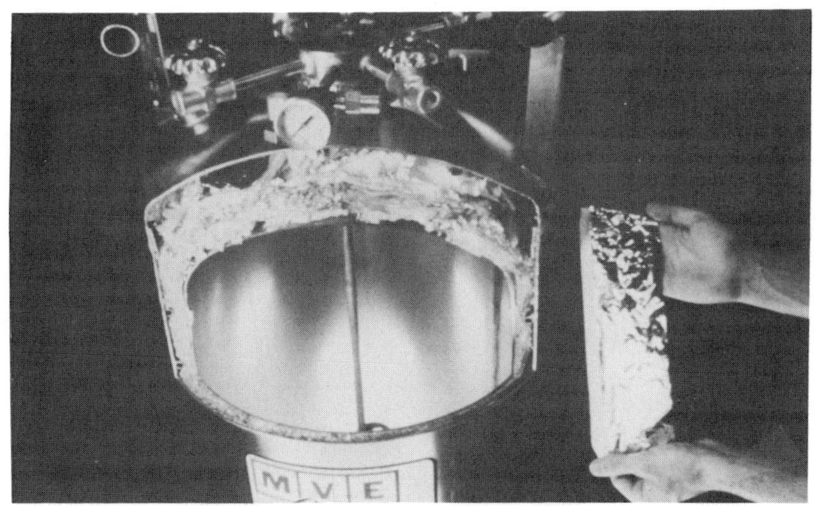

FIGURE 2. Cross Section of DOT-4L Liquid Cylinder

The functional components of a typical cryogenic liquid
cylinder include:

. A liquid fill and withdrawal valve.
. A dip-tube.
. A vent valve to relieve pressure, primarily during the fill
  process.
. A gas-use system, including a valve and an internal
  vaporizer coil to warm the liquid during gas withdrawal.
. A pressure building system, including valve, regulator, and
  an internal coil to maintain a preset cylinder operating
  pressure.
. An economizer regulator to release excess ullage space
  pressure into the gas-use system.
. A pressure gauge, a spring loaded relief device, and a
  rupture disc for the inner vessel.
. A float type liquid contents indicator.
. A vacuum casing rupture disc.

Figure 3 presents a cut-away drawing of a modern cryogenic
liquid container and labels some of the pertinent components.
Figure 4 displays the external plumbing of a container.

The design changes that the cryogenic liquid cylinder went
through were motivated by the fact the cylinder depends on the
bottom support and neck tube for durability.

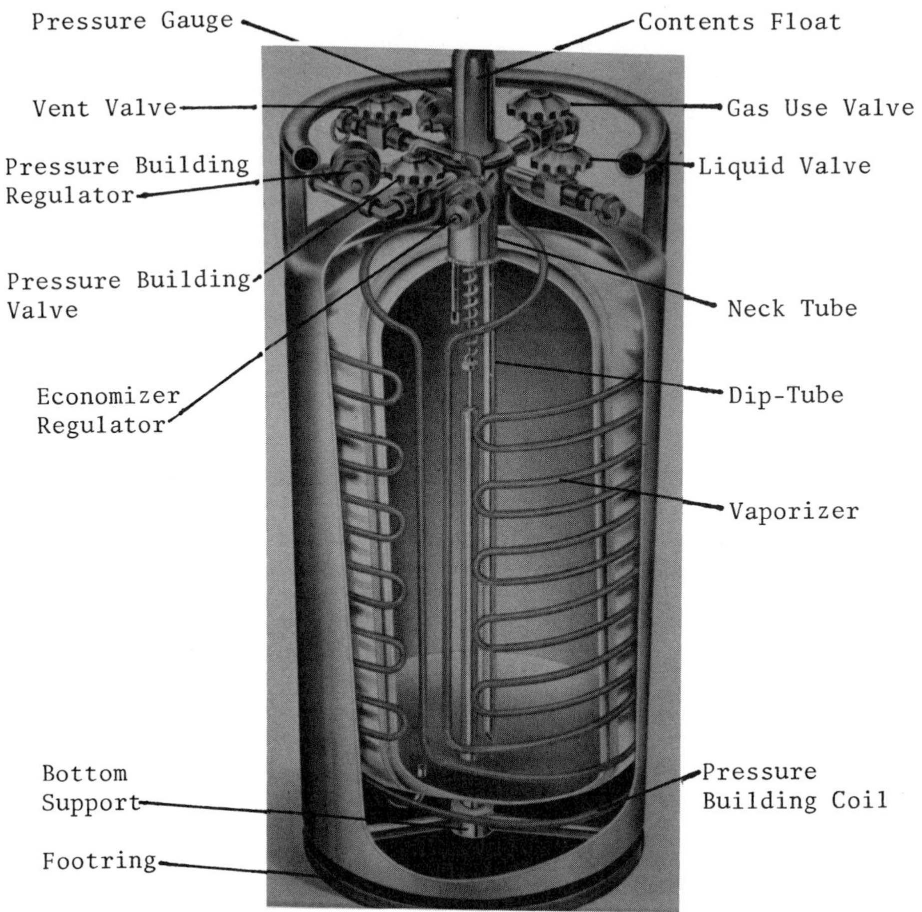

FIGURE 3.   Components of the Modern Liquid Cylinder

HISTORY

The liquid cylinder history can be classified into three generations. Generation I, from 1950-1965, could not survive normal use. Generation II, from 1965-1982, could not survive normal abuse. Generation III, from 1982 to the present, can successfully survive normal abuse.

The Generation I design was characterized by a 110 liter capacity, small diameter thin walled, brittle neck tubes, weak bottom supports, specialized function (depended on saturated liquid), and poor vacuum technology. The primary models available were the Linde LC-3 (Figure 5) and the Ryan 110 (Figure 6).

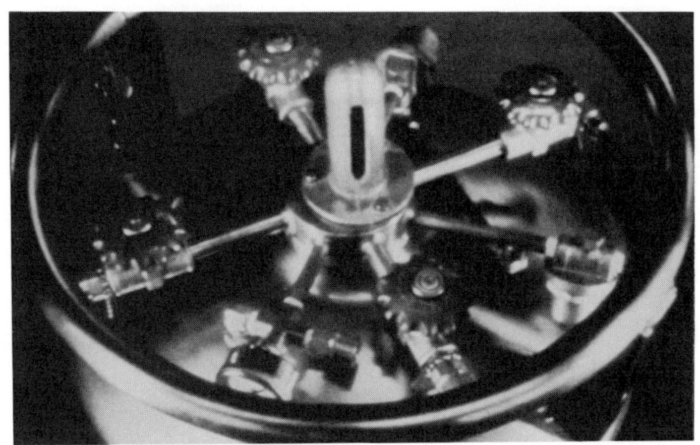

FIGURE 4. All Function External Plumbing of a Liquid Cylinder

FIGURE 5. Linde Model
LC-3

FIGURE 6. Ryan Model
110

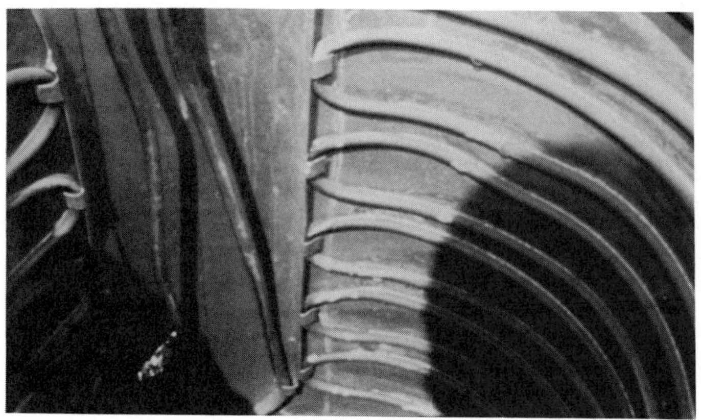

FIGURE 7. Clips Wore Through Copper Vaporizer Tubes on LC-3

The Linde LC-3 incorporated a powder type insulation which
required a large annulus. It also utilized a tube of fiberglass
epoxy for a bottom support. It had weak copper external plumbing
lines and a small diameter dip-tube fill line. The internal
vaporizer used clips and solder to attach the lines to the outer
shell. The clips would wear through the copper lines, resulting
in a vacuum loss (Figure 7).

The Ryan 110 had a type of crude multi-layer insulation (Figure
3). The bottom support was a metal cup attached to the outer shell
with small diameter rods. This model also had weak copper lines
but did incorporate a pressure building system.

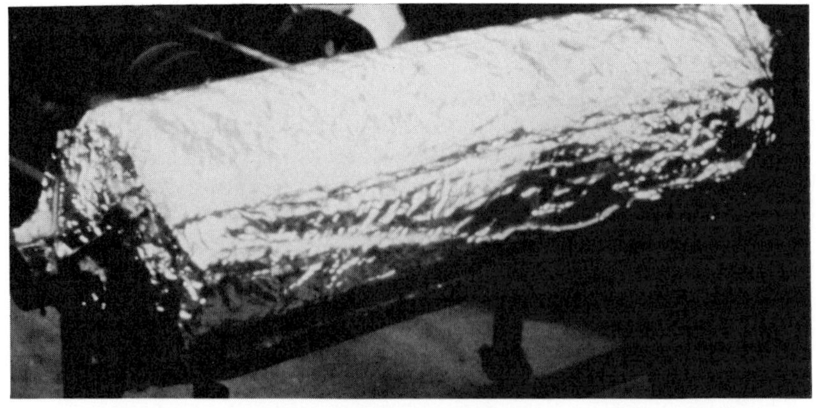

FIGURE 8. Multi-Layer Insulation on Ryan 110

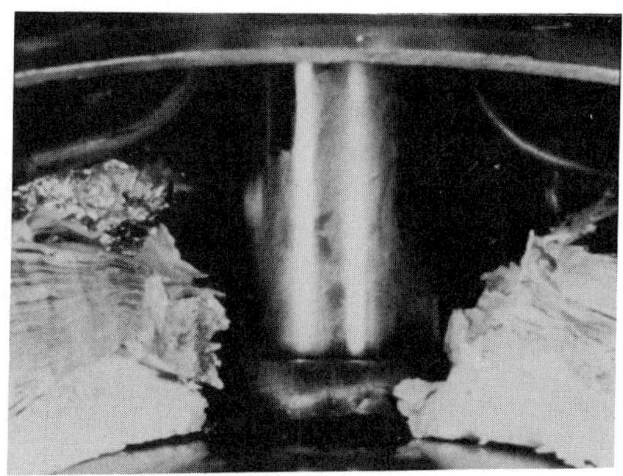

FIGURE 9.   Larger Neck Tube

    The Generation II design included a 160 liter capacity,
all-function sturdy plumbing, larger diameter neck tubes (Figure 9),
stronger bottom supports (Figure 10), and acceptable vacuum and
insulation technology.   The models in this group were Linde's PGS-45
and LS-156, MVE's VGL-160, and the Ryan LG-45.

    The inner shell diameter of the Generation II cylinder was
larger (Figure 11) for the same diameter outer shell as in the Generation I cylinder because of the improved insulation type and
method of application.   The all-function plumbing provided easy
access to the operating controls and the main valves were attached
to pipe nipples welded to the knuckle.   Most of these plumbing
components are in use on current cylinders.

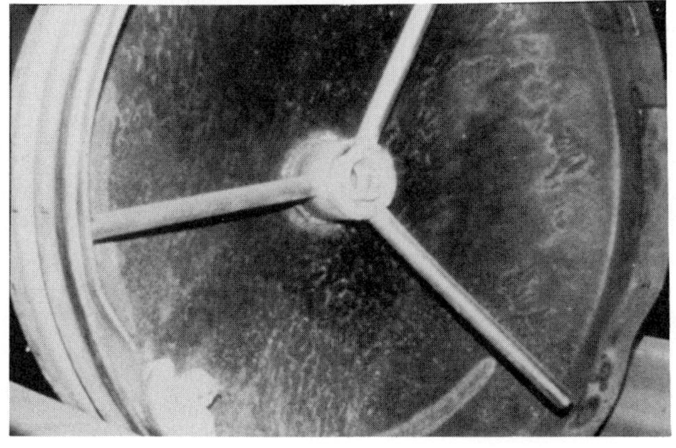

FIGURE 10.   Stronger Bottom Support

FIGURE 11.   110 Liter and 160 Liter Inner Vessel

However, external evidence of abuse usually resulted in a
wrinkled neck. After one tip-over a neck would wrinkle, and continued use after multiple tip-overs would result in a total neck
failure (Figure 12). In addition to neck failure, the bottom
support would fail by shearing, in the case of the fiberglass
tube, or deforming, in the case of the metal support. The metal
support did not stop the outer shell's top head from caving in
(Figure 13). To eliminate the structural damage of severe vertical loads, some Generation II cylinders added rubber mounts
between the shell and footring (Figure 14).

This helped, but bolts would rust and the mounts would
delaminate over time. In locations with high humidity and a salt
atmosphere, the carbon steel footring rusted through and the
cylinder could fall over or lose vacuum (Figure 15). The metal
clips were still being used to hold the vaporizer line and, after
many thermal cycles, would wear through the copper tubing.

FIGURE 12.   Total Neck
             Failure

FIGURE 13.   Deformed Outer
             Top Head

180

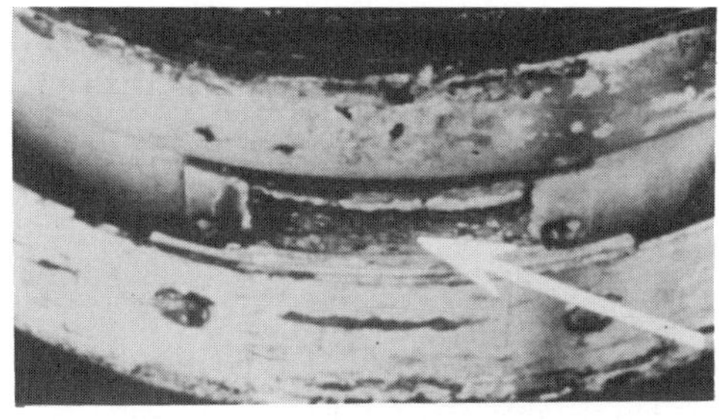

FIGURE 14. Rubber Shock Mount

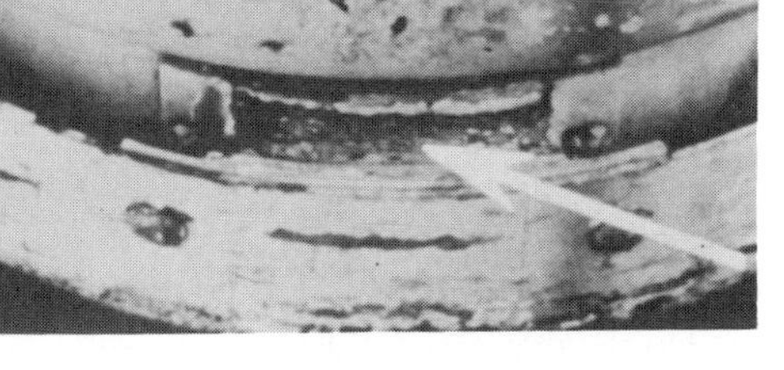

FIGURE 15. Rusted Out Carbon Steel Footring

Generation III is the latest and current design of liquid cylinders and is characterized by an almost indestructible support system, a stainless steel outer shell, choice of operating pressures, and state of the art vacuum technology coupled with the most efficient multi-layer insulation and a vacuum maintenance system. After a detailed 2-1/2 year study of cylinder repair records (Table 1) to determine the major failure causes, a number of design changes were made to make the Generation III liquid cylinder able to "Survive Normal Abuse" (Table 2). Generation III models include the MVE Dura-Cyl and CSI super tank.

TABLE 1

2-1/2 Year Study of Causes for Vacuum Failure

45% Rusted Outer Bottom Head
28% Burst Disk Leak
20% Tired Vacuum
 7% Inner Leak (Wrinkled Neck)

TABLE 2

Changes Made to Enhance Service Life of Liquid Cylinders

1. Redesign to protect the neck tube.
2. Promote stainless steel outer jacket.
3. Change material and add thickness to vacuum jacket rupture disk.
4. Change rupture disk cap to polyethelene.
5. Add protective cover to burst disk to protect from atmosphere and tampering.
6. Maintain a constant vacuum system technology program to produce longer lasting and safer vacuum system.

To protect the neck tube, the wall thickness of the neck tube wall was increased, the diameter of the plumbing protective ring was made slightly larger than the cylinder outside diameter to act as a shock absorber in case of a tip-over, the footring thickness was increased to act as a spring, and the bottom support was changed to eliminate any deflection which could cause the neck tube to wrinkle. In addition, a vinyl cap and protective cover were added to prevent tampering or corrosion from damaging the vacuum casing's rupture disk. An angle ring was welded to the inside top of the jacket with protective sleeves to prevent the annular lines from being crushed during tip-over. The bottom head was reinforced and the support stud changed to a solid rod. The vaporizer and pressure building coil are soft soldered to the inside of the outer shell without using clips. Finally, major improvements were incorporated in the vacuum techniques, insulation type and application method. The use of getters in the annular space for long term vacuum maintenance and thermal efficiency was improved.

TESTING

To affirm that the design changes made would increase the durability of the Generation III liquid cylinder, the following tests were performed with a cylinder filled with liquid nitrogen.

1. Vertical drop tests from 2(0.6), 3(0.9), and 4(1.2) feet (meters).
2. The cylinder was tipped over eleven times.
3. The cylinder was loaded onto a hand-cart and pushed from a 4 foot (1.2 meter) high loading dock.
4. The cylinder was pushed from the rear of a delivery truck traveling at 30 mph (40 km/h).

After each test the cylinder was checked for performance of the pressure building system, flow through the internal vaporizer and boiloff rate. The cylinder was then inspected for any damage to the neck tube or the bottom support.

The only negative performance change between the Generation I and Generation III cylinder is an increase in the boiloff rate, a consequence of the change in design of the bottom support and neck tube. For a cylinder in nitrogen service, this increase in boiloff rate is from 1.5% to 2.5% per day, which is still well within the 7.2% per day allowed by the DOT-4L code.

CONCLUSION

The Generation III liquid cylinder incorporates the design features necessary to make it the durable cryogenic storage container needed to absorb, without failure, the abuses to which

it is subjected during normal industrial gas handling and
transport. Additional design changes using advanced materials and
techniques will make it even more durable and efficient in service
for the industrial gas industry.

ACKNOWLEDGMENT

The author would like to extend his thanks to Rosann Blaschko
for her patience and help in typing this paper.

COMPOSITES FOR CRYOGENICS

Michael S. Kramer

ACPT, Inc.
Huntington Beach, CA

ABSTRACT

Composite materials have been used in cryogenic applications for many years. Typically, cryogenic composites consist of G-10 type fiberglass material which has limited mechanical and thermal properties. More advanced applications with stringent design requirements require material properties beyond the capability of generally used composites. Composites utilizing advanced materials and fabrication methods provide designs with properties unobtainable with G-10. This paper gives mechanical and thermal design data and comparisons of various composite materials for cryogenic use.

INTRODUCTION

Composites are becoming important materials for use in cryogenic applications. The unique mechanical, thermal and electrical properties have made possible cryogenic designs such as the Superconducting Super Collider and vehicles utilizing cryogenic fuels.

Composites do not benefit from a detailed specification system such as is available for metals. What makes detailed specification difficult is the large number of different combinations of composites available, each with unique material properties. There are few commonly used specifications for composites. Of note to the cryogenic designer is G-10, G-11 and the cryogenic derivatives.[5,12] G-10 has been very well characterized at room temperature and cryogenic temperatures. The National Bureau of Standards (now NIST) has the basis of a more complete specification system, however the system has not found wide acceptance with industry.

COMPOSITE SPECIFICATION

Advanced composites consist of a fibrous material in a matrix binder. There are a number of parameters that determine the resulting mechanical and thermal properties of a composite, including, fiber, matrix, the ratio of fiber to matrix, fiber orientation and the form of the fiber. Determining the material properties can be difficult for composites using published data for the following reasons: Fiber manufacturers may present data for the fiber alone without any matrix. The "dry" fiber properties are significantly higher than can be expected from a composite. Composites are anisotropic; the material properties are dependent on the orientation of the fiber reinforcement. A unidirectional oriented composite (all fibers aligned) will have significantly different properties if tested with the fiber oriented axially or transversely. Composites may contain a number of different orientations or may consist of cloth with its' particular orientation. Published data must be analyzed to determine the direction of the reinforcement in relation to the direction of testing.

Data presented herein is given for unidirectional composites and is referenced from the fiber axial and the fiber transverse direction. Cloth based laminates are referenced to the warp and fill direction of the cloth. In an actual composite, a number of different orientations are combined to achieve particular mechanical properties. The unidirectional information should be used as a bench mark comparison of different composite materials.

The type of resin selected for the matrix and its' percentage in the composite determines many of the properties of the composite. Composites for cryogenics typically use an epoxy resin matrix. Epoxy is used for its' processing characteristics and mechanical properties at cryogenic temperatures.[13] Toughened epoxy resins show tensile strain capability of 10% or more.[14] Resin content in the range of 30 to 40 percent by volume is optimum. A composite with higher resin content will have lower stiffness and strength. Too little resin and the composite will be weak in shear. The resin content is dependent on the fabrication technique and the viscosity of the resin selected. Resin should be selected for its' processing characteristics in addition to mechanical properties. There are a number of good cryogenic resins available. Markley et al.[14], have characterized a number of resins.

There are two primary methods of placing the resin onto the fibers: preimpregnation, or " wet resin" processing. Epoxy resins begin to cure when the epoxide is mixed with a suitable hardener. During the cure, the viscosity of the resin increases until it is an amorphous solid. The amount of time available for processing can be adjusted by adjusting the chemistry and keeping the resin cool (typically -20 C) until ready for use. The resin may be placed on a fiber and then stored frozen until ready to use. This process is referred to as preimpregnation. Fibers of cloth that have been preimpregnated are called prepreg. Prepreg is convenient for fabrication, however, prepreg parts have complex curing procedures and the preimpregnation process degrades the mechanical properties of the composite. An alternative is to mix the components of the resin at the time of processing. This technique is referred to as " Wet Resin Processing." Wet resin provides better mechanical properties[4] but extra care must be taken to insure the proper resin content.

There are a number of mechanical or thermal properties that may be needed to characterize an application. Presented are Tensile Strength, Tensile Modulus, Thermal Conductivity, and Thermal Expansion. These properties are highly dependent on the fiber selected. The main classifications are as follows:

High Strength Graphite	Y-300 (Toray), AS-4 (Hercules), XA-S (Courtaulds)
High Modulus Graphite	GY-70 (BASF), P-75 (Amoco), P-100 (Amoco)
Medium Modulus Graphite	HM-S (Courtaulds), G-50 (BASF), T-50 (Amoco)
Aramid	Kevlar 29,49,149
Fiberglass Cloth Composite	G10, G11, CR Grades

Data is given for the underlined materials as typical for each class. The other materials in the classification have similar properties, however, since mechanical properties may vary, data for the particular material should be used for designing.

Figures 1 and 2 show Young's modulus and tensile strength for a number of materials. The data is for the fiber direction in the unidirectional composites, and the warp direction in cloth based composites. Figures 3 and 4 show the thermal properties, comparing graphite with fiberglass. Figure 5 presents a summary of composite property data.

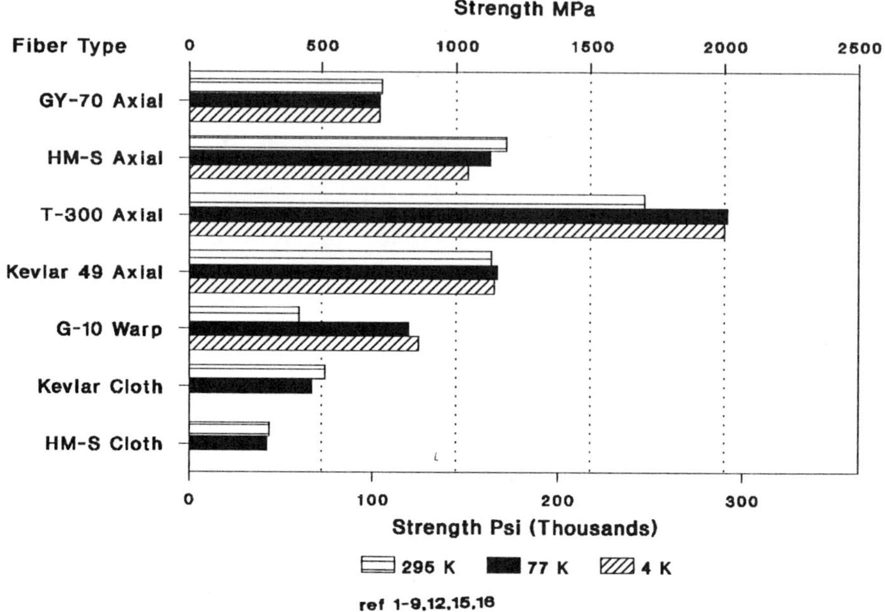

Figure 1 Strength of Composites

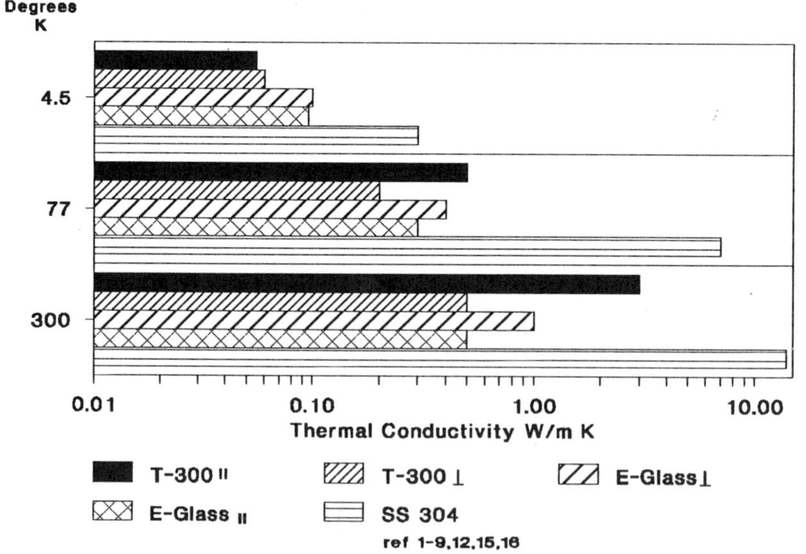

Figure 2 Tensile Modulus of Composites

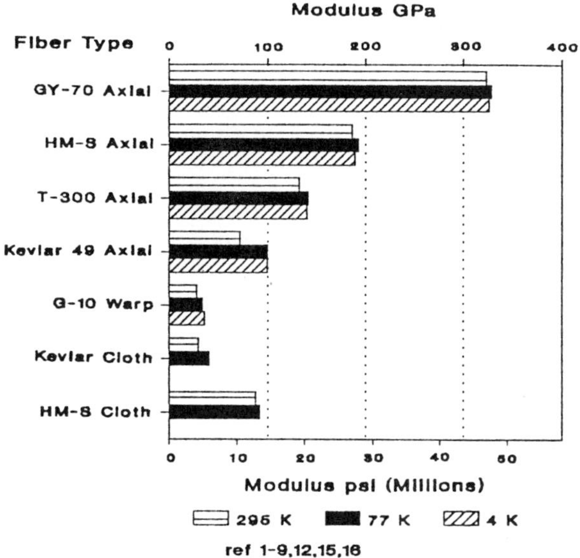

Figure 3 Thermal Conductivity of Composites

Change in Temperature, degrees K

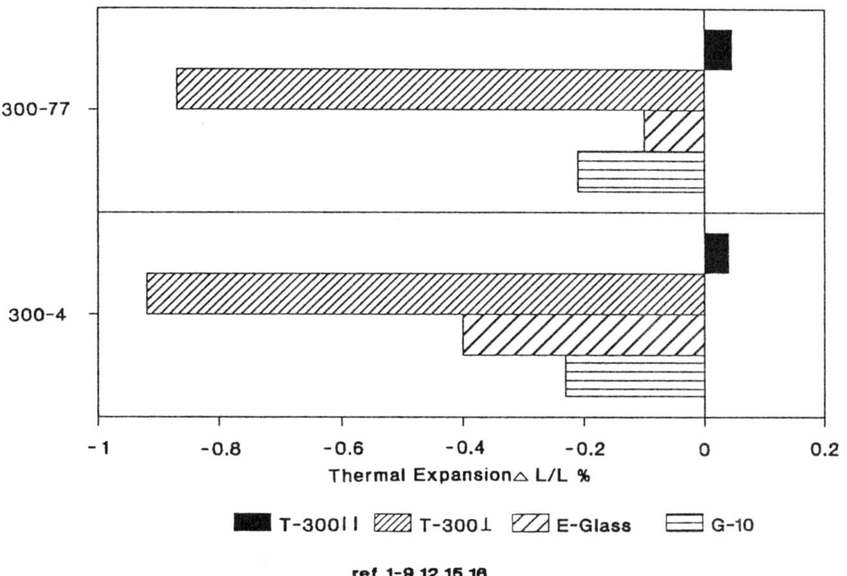

Figure 4 Thermal Expansion of Composites

Material Property	T-300 ref(2,4,5,6,7,15)	HM-S ref(1,8,16)	GY-70 (1)	Kevlar 49 ref(1,9)	G10 CR ref(3,12)	Kevlar Cloth ref(6,15)	HM-S Cloth (6)
Axial (warp) Young's Modulus (GPa)							
4 (K)	140	189	326	99.4	35.9		
77(K)	140	192	328	99.4	33.7	40.09	91.40
295 (k)	132	186	323	71.4	28.0	29.77	87.08
Transverse (fill) Young's Modulus							
4 (K)	13	12.1	8.94	4.56	29.1		
77 (K)		12.0	8.70	3.59	27.0		
295 (K)	9.6	8.62	6.70	2.51	22.4	26.1	
Tensile Strength (MPa)							
4 (K)	2000	1045	716	1142	862		
77 (K)	2010	1127	717	1154	825	463	293
295 (K)	1700	1187	726	1132	415	513	304
Thermal Conductivity W/mK							
4 (K)	.043	.025			.06		
77 (K)	.09	.11			.24		
295 (K)	5.8				.6		
Thermal Expansion L/L%							
295 - 4 (K)	.045			.09	-.23		
295 - 77 (K)	.039			.08	-.21		

Fig. 5 Cryogenic Composite Properties

COMPOSITE DESIGN EXAMPLE: SSC SUPPORT POST

The Superconducting Super Collider (SSC) requires materials with very specific properties. The support post was particularly challenging in its' requirements. The support post is a reentrant design consisting of two concentric composite cylinders connected by a metallic tube in the annular space. Figure 6 shows a support post. The support post is a load bearing thermal insulator. It must provide flexural and torsional stiffness and strength, yet limit the heat leak to the magnet. The design criteria for the post is as follows:

Structural Requirements (@ 300 K)[10]	Inner Tube	Outer Tube
E (GPa)	68.9 min	27.5 min
Axial Stress allowable ± MPa	413 min	275 min
Hoop Stress allowable ± MPa	275 min	172 min

Thermal Conductivity in the Axial direction, W/m-K[10]		
@ 300 K	3.5 max	.81 max
@ 77 K	.67 max	.36 max
@ 4.5 K	.06 max	.09 max

The heat leak rate of the post is a function of both the thickness of the tubes and the thermal conductivity of the material. As the tubes become thinner, buckling failure becomes critical. This is more evident with the inner tube due to its smaller diameter. Raising the modulus of the tube raises the critical buckling stress.

Graphite Composite was chosen of the inner tube. The inner tube operates between 4 and 80 degrees K. In this temperature range, graphite is a good insulator, and is stiff enough to meet the modulus requirements.

Fiberglass composite was chosen for the outer tube. Fiberglass has good insulating capability over the wide span of temperatures the outer tube will see in service. Graphite could not be used because of its higher thermal conductivity. Fiberglass has sufficient mechanically properties to resist buckling without an excess wall thickness.

The tubes were manufactured by filament winding. The winding process wraps fibers in a tow form around a mandrel. The angle of the fibers in relation to the mandrel may be adjusted to match the mechanical requirements. Filament winding has an advantage over

Figure 6 SSC Reentrant Post

cloth wrapped tubes in that there is no seam. A seam would have induced thermal stresses that could have caused the tube to fail prematurely.

The tubes have slightly higher strength requirements in the axial direction than in the hoop. The strength of a composite is higher in the direction of the fibers. For the tubes a combination of axial fibers and hoop fibers was used, with a higher percentage of axial fibers to match the higher strength requirement. Wet resin processing was chosen over preimpregnation for its superior strength. The filament wound tubes have been tested and have shown to meet all the mechanical and thermal requirements.

The SSC support post is a good example of utilization of particular composites for a specialized purpose. The support post efficiency would have been severely limited if it would have been fabricated entirely of G-10.

CONCLUDING REMARKS

Composites encompass a number of different materials and combinations. By carefully selecting the proper material and form, very efficient components may be designed. The data presented showed some of the material properties available with composites. There are many more materials available for composite designers to use. Some promising matrix materials are thermoplastics such as poly ether ether keton (peek) which could provide very high elongation composites. Alumina fibers show good insulating properties and are beginning to be used in cryogenics.[11]

The data herein is intended as a reference to be used for preliminary material selection. The data is obtained from a number of sources and each may have used slightly different sample geometries, or matrix materials.

REFERENCES

1. M.B. Kasen, Mechanical Performance of Graphite Reinforced Composites at Cryogenic Temperatures, in: " Advances in Cryogenic Engineering Materials, Vol. 28," R.P. Reed and A.F. Clark, eds., Plenum Press, New York (1982), pgs. 165-177.

2. G. Hartwig, Reinforced Polymers At Low Temperatures, in: " Advances in Cryogenic Engineering Materials, Vol. 28," R.P. Reed and A.F. Clark, eds., Plenum Press, New York (1982), pgs. 183-186.

3. S.S. Wang and E.S.-M Chim, Degradation of Fiber-Reinforced Composite Materials, in: " Advances In Cryogenic Engineering Materials, Vol. 28," R.P. Reed and A.F. Clark, eds., Plenum Press, New York (1982), pg. 192.

4. G. Hartwig and S. Knaak, " Fiber-Epoxy Composites at Low Temperatures," Cryogenics, 24:639-647 (1984).

5. M. Takeno et al., Thermal and Mechanical Properties of Advanced Composite Materials at Low Temperatures, in: " Advances in Cryogenic Engineering Materials, Vol 32," R.P. Reed and A.F. Clark, eds., Plenum Press, New York (1986), pgs. 217-224.

6. K. Dahlerup-Peterson, Test of Composite Materials at Cryogenic Temperatures: Facilities and Results, in: " Advances in

Cryogenic Engineering Materials, Vol. 26," A.F. Clark and R.P. Reed, eds., Plenum Press, New York (1980), pg 303.

7. W. Weiss, Low Temperature Properties of Carbon Reinforced Epoxide Resins in: "Non-metallic Materials and Composite at Low Temperatures 2," Plenum Press, New York (1979), pg. 303.

8. K.A. Philpot and R.E. Randolph, The Use of Graphite/Epoxy Composites in Aerospace Structures Subject to Low Temperatures, in: "Non-metallic Materials and Composites at Low Temperatures 2," Plenum Press, New York (1982), pgs. 314-319.

9. G. Hartwig, "Thermal Expansion of Fiber Composites," Cryogenics, 28:257 (1988).

10. M. Kramer and T. Nicol, Composite Materials for SSC Dipole Magnet Cryostats, in: "Composites in Manufacturing 9," SME (1990), pg. EM-90-101-4.

11. R.D. Kriz and L.L. Sparks, Performance of Alumina/Epoxy Thermal Isolation Straps, in: "Advanced in Cryogenic Engineering Materials, Vol. 34," R.P. Reed and A.F. Clark, eds., Plenum Press, New York (1988), pgs. 107-114.

12. M.B. Kasen et al., Mechanical, Electrical and Thermal Characterization of G11CR Glass-Cloth/Epoxy Laminates Between Temperature and 4 K, in: "Advances in Cryogenic Engineering Materials, Vol. 26," A.F. Clark and R.P. Reed, eds., Plenum Press, New York (1980), pgs. 237-241.

13. H. Hacker et al., Epoxies for Low Temperature Application Technology, in: "Advances in Cryogenic Engineering Materials, Vol. 30," A.F. Clark and R.P. Reed, eds., Plenum Press, New York (1984), pg. 54.

14. F.W. Markley, J.A. Hoffman, D.P, Hoffman and D.P. Muniz, Cryogenic Properties of Basic Epoxy Resin Systems, in: "Advances in Cryogenic Engineering Materials, Vol. 32." A.F. Clark and R.P. Reed, eds., Plenum Press, New York (1986), pg. 119.

15. H.D. Neubert, "SQ5N Program Input Data", Hans D. Neubert and Assoc., Anaheim Hills, CA.

16. D.J. Radcliffe and H.M. Rosenberg, " The Thermal Conductivity of Glass-Fiber and Carbon Fiber/Epoxy Composites From 2 to 80 K," <u>Cryogenics</u> 22:85 (1982).

DEVELOPMENT OF 3-D COMPOSITES

J. R. Benzinger

Spaulding Composites Co.
310 Wheeler Street
Tonawanda, New York 14151-5101

INTRODUCTION

Interlaminar shear strength is one of the most important
properties of laminated composites, not only for fusion magnets, but for
many other applications. Several processes that have been developed
over the years to improve interlaminar shear strength include 3-D
woven orthogonal fabrics, multiple warp weaving, stitch bonding of
multi-layers of 2-D fabrics, and needle punching of non-woven and
woven fabrics.

Fabrics woven in three orthogonal directions have been in
existence since 1966 or earlier. However, they have not been
commercialized to any great extent except in the aircraft and the
aerospace industries because of economics or the lack or a market.
Recently, however, there has been considerable interest in the use of
3-D Fabric Reinforced Plastics (FRP) composites, loaded biaxially in
shear and compression, as insulators for superconducting cryogenic
fusion magnets.

There are several manufacturing companies in the U.S.A. capable
of weaving 3-D orthogonal fabrics. They include Advanced Textiles,
Fabric Development, Techniweave, Textile Products, Textile Technology
and Hitco's Woven Structures. Recently Shikishima Canvas Co. of
Japan introduced 3-D glass fabrics in a series of reinforced plastics
(3DGFRP). This development program was done in cooperation with
1S1R Osaka University and supported in part by the Ministry of
Education in Japan[1]. In the U.S.A., Avco Co., NASA, Drexel

University, Hercules Aerospace and the University of Delaware among others have either developed or tested 3-D composites but no one has manufactured a 3-D glass reinforced flat sheet for commercial applications.

The first composite materials fully characterized at cryogenic temperatures for fusion magnets were the CR grades developed by Spaulding Composites Co. in cooperation with NIST (Formerly NBS).[2] Designated G-10-CR and G-11-CR, they are composite epoxies reinforced with 2-D E-glass woven fabric and laminated in sheet and tube forms.[3]

Radiation effect studies by Coltman et al. at Oak Ridge National Laboratory (ORNL) showed that at 1×10^{10} rads, the properties of G-10-CR and G-11-CR were reduced to barely useful levels.[4] Since future fusion magnet designs would require better resistance to radiation, Spaulrad®-E and Spaulrad®-S laminated composites were subsequently developed. Both are 2-D glass reinforced bismaleimides (Kerimid 601). Spaulrad®-E uses E-glass and after irradiation is less damaged and remains 5 to 10 times stronger than its epoxy variants at 77 K.[5] Spaulrad®-S uses boron free high strength S-2 glass which yields better radiation resistance and higher mechanical strengths.

MANUFACTURING CAPABILITIES

In the development of 3-D semi-cured resin impregnated woven fabric (prepreg) for making panels using the high pressure "laminating" technique, consideration must be given to the maximum thickness of the fabric that can be saturated in a continuous treater. Since the 3-D fabrics are tightly woven to maintain integrity, saturation with resins presents a problem as thickness increases. As a consequence, only thin composites can be economically manufactured by this method.

Another problem presents itself in that for each thickness gradient, a different weight fabric is needed. This may not be a big problem with fusion magnets because projected thicknesses are mainly in the range of 0.5 to 1.6 mm. However, thicker panels, tubes and shapes would have to be made by some other process such as resin transfer molding (RTM).

DEVELOPMENT OF 3-D FRP IN JAPAN

The 3-D fabrics reported by Y. Iwasaki et al. are woven from T glass (S-glass). The orientation of the glass fibers in the fabric were reported as 34% in the X direction, 57% in the Y direction and 9% in the Z direction. The epoxy 3-D FRP contains 54% glass fibers by volume and the BT Bismaleimide contains 50%. Properties were compared with Spauldite 2-D FRP G-10-CR and G-11-CR made with E-glass. The data reported by them shows improved tensile and flexural strength at 300 K and 77 K for the 3-D composites. Compressive values at 300 K are also better than the 2-D G-10-CR and G-11-CR, but at 77 K the values are very close. The improved strengths of 3-D FRP are attributed by Iwasaki to the use of T-glass which has higher strength than E-glass and to the straight arrangement of the fibers in 3-D FRP; those in 2-D are waved.

APPLICATIONS IN FUSION MAGNETS

<u>Alcator C-MOD</u>

The Alcator C-MOD is a compact Tokamak designed to study plasma physics related to fusion. The basic design requirement is survival for 50,000 operational cycles under 9 tesla. The superstructure is cooled to 77 K before each operational pulse which lasts for about 3 seconds. The insulating material selected for the copper coil central core of the Alcator C-MOD was Spaulrad®-S. The horizontal and vertical legs of the device were insulated with G-10-CR.

The copper coil central core, and the horizontal and vertical legs were manufactured in Japan by Mitsubishi. The small coils were made in the U.S.A. by MIT in cooperation with Everson Electric. The 316LN steel forgings, 3.65 to 4.57 meters tall, were made by VSG and are said to be the largest ever made in the world. The assembly of the tokamak is being done at MIT.

As a prelude to the use of Spaulrad®-S for insulating the Alcator C-MOD, H. Becker et al. discovered that shear strengths of Spaulrad®-S increased rapidly with pressure and leveled off at pressures on the order of 700 MPa. Laminated panels of Spaulrad®-S were machined into disks 0.5 mm thick and 11 mm in diameter and thickness, were subjected to face loading at room temperature without prior irradiation. The 0.5 mm thick specimens, loaded only statically, revealed strengths greater than 800 MPa.

Thin discs had been irradiated in two facilities, the Advanced Test Reactor (ATR) at the Idaho National Engineering Laboratory and the Intense Pulsed Neutron Source (IPNS) at Argonne National Laboratory.[7] Neutron fluences attained in these irradiations were 3.41 x 10^{23} n/m^2 (E>1 MeV) and 3.17 x 10^{24} n/m^2 (total) for the ATR and 1.9×10^{21} n/m^2 (E>0.1 MeV) and 2.8×10^{21} n/m^2 (total) for the IPNS. In addition, a gamma dose of 3.9×10^9 Gy was obtained in the ATR irradiation. Temperatures of 325 K and 4.2 K were used for the ATR and IPNS irradiations, respectively. Post-irradiation tests included static compression, compression fatigue, and electrical resistance. No failures had been observed in static compression tests when unirradiated specimens were loaded to stress in excess of 7000 MPa. Only one of five specimens from each ATR irradiation group failed in fatigue after cycling to 640 MPa for 277,500 cycles. No failures were observed for the IPNS specimens when cycled to the same maximum stress for 646,540 cycles. No change in electrical resistance was observed for the IPNS irradiation group when compared to the control specimens (6×10^{11} Ohms) whereas a decrease by a factor of ten was found for ATR group.

The irradiation dose obtained by specimens placed in the ATR was higher by a factor of one hundred than that anticipated for most fusion reactor magnets, and the testing stresses were conservatively high, as well. No failures of any kind were observed for the specimens irradiated in IPNS, and it is thought that a damage threshold had not been reached for those specimens, whereas the damage threshold probably had been reached by those specimens irradiated in the ATR.

Based upon the irradiation tests, compression fatigue, compression strength, electrical resistance, and other factors, it was concluded that Spaulrad®-S should be used in the Alcator C-MOD.

<u>Compact Ignition Tokamak</u>

The Compact Ignition Tokamak (CIT) is a fusion reactor with copper alloy coils cycled from 77 K to over 300 K.[8] The turn-to-turn coil insulation can be exposed simultaneously to a neutron radiation dose of 10^{10} rads, with high shear and compressive loads. The insulation loading, in effect, is somewhat analogous to the loading of the insulators in the central coil core of the Alcator C-MOD. The insulators are normally 1 mm thick in the CIT and 0.5 mm thick in the Alcator C-MOD.

Preliminary screening tests were made on fifteen materials from six vendors at the Material Testing Laboratory of Princeton Plasma Physics Laboratory.[9] The preliminary screening tests were combined shear and compression loading. In general, all of the materials demonstrated very high interlaminar shear strengths with high compressive loads in the biaxial test device. At 345 MPa compression, the average shear strength of Spaulrad®-S was 118 MPa. Nearly pure interlaminar shear failures were found with high compression of Spaulrad®-S as compared to a tearing failure with the 3-D materials.

MATERIALS SELECTED FOR IRRADIATION BY ORNL

Three of the fifteen screened materials were selected for irradiation testing. Spaulrad®-S, a 2-D laminated composite which has been shown to be resistant to radiation damage[10], and two 3-D composites from Japan[11] were the selected materials. All three materials contain boron-free glass. Their compositions are listed in Table I.

<u>Radiation Dose</u>

The three materials were irradiated at room temperature in the Advanced Technology Reactor and tested by EG&G at the Idaho National Engineering Laboratory. Test specimens were subjected to radiation does at two levels. The lower dose was calculated at approximately 5×10^9 rads and the higher dose at 3×10^{10} rads. The ratio of gamma to neutron dose was on the order of 3 to 2. The two levels of fast fluence (10^{18} neutrons/cm^2) were 0.5 and 2.5. Details of the dose and fluence levels are reported in references 12 and 13.

Table I

Composition of Candidate Materials for the CIT
Selected for Irradiation Tests

Description	Spaulrad®-S	PG5-1	PG3-1
Manufacturer	Spaulding[a]	Shikishima[b]	Shikishima
Resin	Bismaleimide (Kerimid 601)	Bismaleimide triazine	Bisphenol-A
Hardener	None	None	Aromatic amine
Fiber			
Material	S-2 glass	T-glass	T-glass
Preparation	Silane	Epoxy silane	Epoxy silane
Form	2-D Fabric	3-D weave	3-D weave
Resin, wt %	26	33	29

a Spaulding Composites Company, Tonawanda, N.Y.
b Shikishima Canvas Co., Ltd., Shiga, Japan

TEST RESULTS

Listed in Table II are the properties of the three materials after irradiation at the two levels. The test data was reported by T. J. McManamy et al. in reference 14.

Large reductions in the flexural strength samples after irradiation were observed. Better flexural strengths were expected for the 3-D over the 2-D material. However, only the PG5-1 was significantly higher. The PG3-1 was reported to be warped after irradiation. The most significant finding was a general trend for the samples to shrink after irradiation. There was no great difference except in thickness where the PG5-1 was somewhat better than the other two. Weight changes were insignificant.

Table II

Irradiation Effects on Properties

Properties of a 1 mm Thick Specimen	Spaulrad®-S	PG5-1	PG3-1
Flexural Strength Flatwise, MPa (Change, %)			
Control	675	946	697
5×10^9 Rad	531(-21)	831(-12)	548(-21)
3×10^{10} Rad	362(-46)	723(-24)	699(0)
Shear Stress with 345 MPa Compression, MPa (Change, %)			
Control	125	135	123
5×10^9 Rad	129(+2)	134(-1)	119(-4)
Post-Fatigue			
Control	125(0)	130(-3)	111(-10)
5×10^9 Rad	133(+6)	130(-3)	113(-8)
3×10^{10} Rad	136(+9)	129(-4)	133(+8)
Dimensional Changes, %			
Thickness, 5×10^9 Rad	-1.5	-0.7	Warped
3×10^{10} Rad	-2.6	-1.7	-2.4
Length, 5×10^9 Rad	-0.2	-0.1	-0.0
3×10^{10} Rad	-0.4	-0.3	-0.4
Width, 5×10^9 Rad	-0.4	-0.4	-0.8
3×10^{10} Rad	-0.7	-0.7	-0.9
Weight Changes, %			
5×10^9 Rad	-0.8	0.0	-2.2
3×10^{10} Rad	-0.2	+1.4	-1.2

After irradiation, the shear/compression strength of all three materials remained high, even after the dose of 3×10^{10} rad. There is little correlation between the reduction in flexural strength due to irradiation and the interlaminar shear strength, when samples are tested with applied compression.

The epoxy 3-D material warped slightly and was more sensitive to fatigue than the 2-D and 3-D bismaleimides. Spaulrad®-S and PG5-1 were tested with 30,000 cyclic shear loads at 90% of the static strength at constant compression without any failures. The percentage changes given are compared to the unirradiated static control averages.

CONCLUSIONS

Only in the aerospace industry have 3-D composites been used to any great extent and then only in shapes and structures. There are no standard "off the shelf" 3-D fabrics available in either E or S-2 glass in the U.S.A. For each significant panel thickness a new 3-D fabric must be designed, woven and inventoried. Thickness of panels is limited owing to saturating problems. RTM should be considered to process thick panels. Cost and availability factors will escalate with 3-D composites.

Spaulrad®-S was the insulating material selected for the copper coil central core of the Alcator C-MOD and is now under consideration for the CIT fusion device. Other than shrinkage, which may or may not be a factor, the use of 3-D composites did not show any advantage over 2-D Spaulrad®-S as used in fusion magnets loaded under shear/compression.

Development of 3-D radiation resistant composites is continuing in the U.S.A. through a cooperative program consisting of the National Institute of Standards and Technology, Spaulding Composites Co. and Composite Technology Development Co. There are plans to irradiate and test lap shear strength of the 3-D composites at 4 K.

ACKNOWLEDGEMENT

The author wishes to gratefully thank T. McManamy, ORNL and H. Becker, MIT for their permission to reproduce their data and test results.

REFERENCES

1. Y. Iwasaki, S. Nishiima, J. Yasuda, T. Hirokawa, and T. Okada, Development of 3DGFRP for cryogenic use, in: "Advances in Cryogenic Engineering, Vol. 36," R.P. Reed ann F.R. Fickett, Eds., Plenum Press, NY (1990).

2. M. B. Kasen, G. R. MacDonald, D. H. Beekman and R. E. Schramm, in: "Advances in Cryogenic Engineering, Vol. 26," A.F. Clark and R.P. Reed, Eds., Plenum Press, New York (1980), pp. 235-244.

3. J. R. Benzinger, Manufacturing capabilities of CR grade laminates, in: "Advances in Cryogenic Engineering, Vol. 26," A.F. Clark and R.P. Reed, Eds. Plenum Press, New York (1980), pp. 252-258.

4. R. Coltman, Jr. and C. Klabunde, "Mechanical strength of low temperature-irradiated polymides: a five-to-tenfold improvement in dose-resistance over epoxies," J. Nuc. Mat., 103 & 104 (1981), pp. 717-722.

5. J. R. Benzinger, The manufacture and properties of radiation resistant laminates, in: "Advances in Cryogenic Engineering, Vol. 28," R.P. Reed and A.F. Clark, Eds., Plenum Press, New York (1982), p. 231.

6. H. Becker and T.T. Cookson, "Effect of Normal Pressure on Interlaminar Shear Strength" NBS-DOE Workshop, 1985.

7. R. E. Schmunk, H. Becker, "Irradiation and Testing of Spaulrad®-S for Fusion Magnet Applications" Fusion Energy, p. 282.

8. T. J. McManamy, J. E. Brasier and P. Snook, Insulation interlaminar shear strength testing with compression and irradiation, in: "Proc. IEEE 13th Symposium on Fusion Engineering Vol. 1," IEEE, New York (1990), pp. 342-347.

9. H. Blake, Stress Analysis of the Princeton Plasma Physics Laboratory 1/2 x 1/2 in. Shear/Compression Test Fixture, Oak Ridge National Laboratory Report ORNL/ATD-23, 1989.

10. R. Schmunk, L. Miller, and H. Becker, "Tests on irradiated magnet insulator materials," J. Nuc. Mat., 122 & 123 (1984), pp 1381-1385.

11. S. Nishiima et al., "Development of radiation resistant composite materials for fusion magnets," presented at ICMC Conference, Los Angeles, July 1989.

12. M. Jacox, Neutron and Gamma Dose Calculation for CIT Materials in the ATR |1 Position, Idaho National Engineering Laboratory Report NRRT-N-90-003, 1990.

13. J. Rodgers and R. Anderi, Neutron Monitoring Measurements for the CIT Materials Irradiation in the ATR |1 Position, EG&G Report EGG-cS-8821, 1989.

14. T. J. McManamy, G. Kanemoto, P. Snook, "The Insulation Irradiation Test Program for the Compact Ignition Tokamak" 1CMC Low Temp., V, Heidelburg, May 1990.

CRYOBIOLOGY - TWO SIDES OF THE SAME COIN?

Sidney A. Wilchins

Clinical Associate Professor of Obstetrics and
Gynecology, New Jersey Medical School, Newark, N.J.
Adjunct Research Professor, New Jersey Institute of
Technology

Stan D. Augustynowicz

SSC Laboratory, Dallas, Texas

INTRODUCTION

The purpose of a meeting of this sort is to fuel the development of
communication so that physicians know what engineers have available and
hence, can determine different ways to use what is available. The conference also shows a need for physicians to define what their needs are
and see if there is a way of either improvising or developing new
equipment that they can apply biologically. Thus, will cryogenics be
both more useful and meaningful in the medical field?

TECHNICAL ASPECTS

The International Institute of Refrigeration Dictionary defines
cryobiology as "a domain of biology where refrigerating techniques
are applied". Usually, cryobiology applies refrigeration techniques
with the temperature close to the boiling point of liquid nitrogen,
77K (-320 $^\circ$F).

Using cryogenics for biology and medicines allows preservation, for
unlimited periods of time, of some type of tissues (like frozen blood
banks) (Fig.1). Cryogenics can also be used to destroy tissues (like
cancer tissue) during a cryosurgery procedure. These two fascinating
but distinctly opposite advantages can be attained by varying the cooling
rate (Fig.2). It is how fast we cool or freeze a biological substance
that dictates whether preservation or destruction occurs. Like the
fingerprint, which is a unique characteristic of each person, so too is
the cooling rate required for the preservation of each tissue a unique
feature of that tissue.

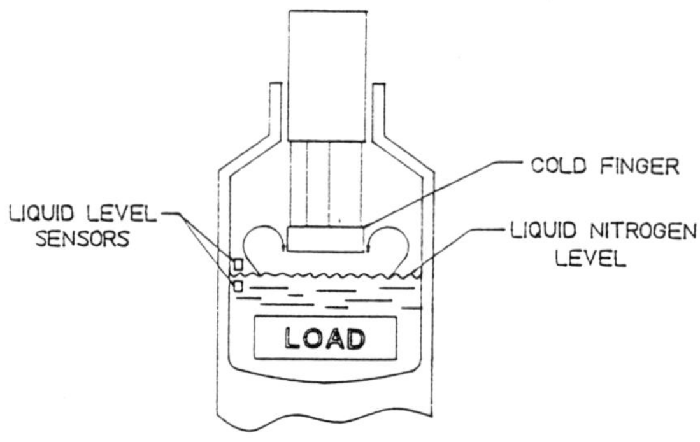

Possible Loads

Bank Of Blood
Bank Of Organs
Bank Of Sperm
Bank Of Tissues

Bank Of Animal Tissue
Insemination

Fig. 1 Cryopreservation

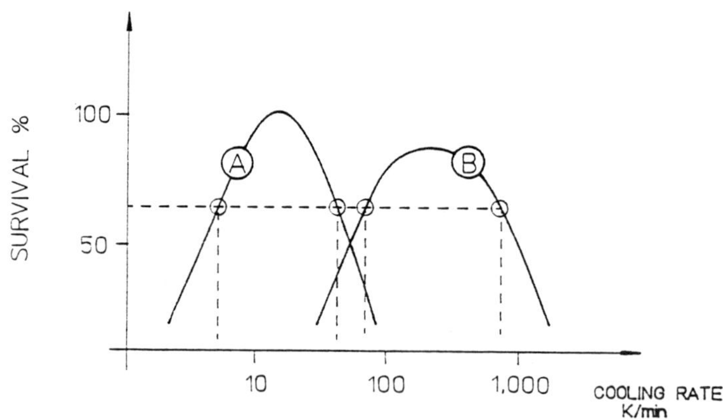

A --- Preservation: A Few Of Deg/Min.

B --- Cryosurgery: Hundreds Of Deg/Min.

Fig. 2 Cooling Rates

The best cooling rate for preservation of a particular tissue takes place
when the maximum percentage of cell survival, after thawing, can be
obtained.

Temperature reduction of a biological substance is accompanied
by physical-chemical processes in the cells which, in most cases,
leads to necrosis of the substance after freezing. At a temperature
of a few degrees below 273K (32 $^{\circ}$F), water inside the cell forms ice
crystals. This results in an increasing concentration of the inner
components, which decreases the freezing temperature, and increases
the pressure. During the freezing process, the formation of ice
crystals can result in the damage of the cell membrane. Ice crystals
can be formed inside or outside the cell. Their growth can destroy the
membrane. The greatest danger of damaging the cell occurs at the freezing point, during the transition from liquid to solid state, when the ice
increases its volume vs. the liquid state. The formation of ice crystals
is strongly influenced by the cooling rate. The formation of big crystals
destroys the cell and is due to the slow removal of heat with a low
cooling rate. Removing heat too quickly from a cell results in rapid
temperature reduction of a substance already solid and damages it
(cryosurgery) by the generation of thermal stresses. (Fig. 3). In practical procedures for preservation, freezing of tissue is always done with
protective fluids, which is necessary for survival of highly organized
cells. In this case, freezing is done in a program way and cooling rates
are low, close to 1 deg/min (Fig. 4).

Reasons that tissues are destroyed during freezing are as follows:

        *formation of ice crystals
        *denaturation of protein
        *dehydration
        *mechanical deterioration

Cryosurgery is the destruction of tissue by freezing "in site". The
technique requires the use of cryogenic agents to freeze the tissue, which
can be done either by direct application of the cryogen or by the use of
closed probe systems (cryoprobes). In deep tissues, the use of metal
cryoprobes is required.

Cryosurgical procedures for the destruction of tissue depends on
efficient heat transfer from the tissues, which in turn is dependent upon
well-known thermodynamic laws. For this reason, the performance characteristics of cryosurgical probes are a matter of substantial importance.
The amount of heat extracted from the tissue via cryoprobes depends upon a
number of factors, including the temperature of the probe, the surface
contact area with the tissue, and the heat supplied to the area. Cryosurgical techniques require fast freezing rates and the attainment of a
lethal temperature in the target tissue. The functional performance of
cryosurgical probes, with regard to these requirements, should be known
prior to clinical use.

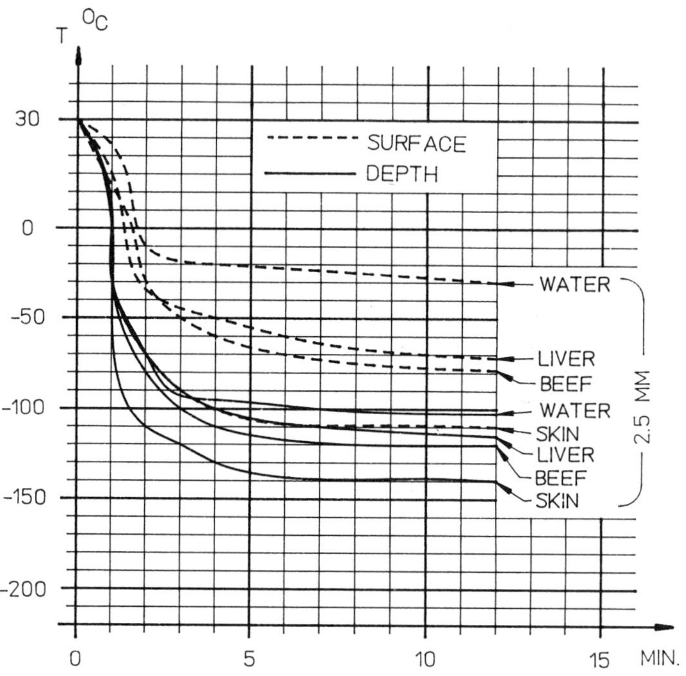

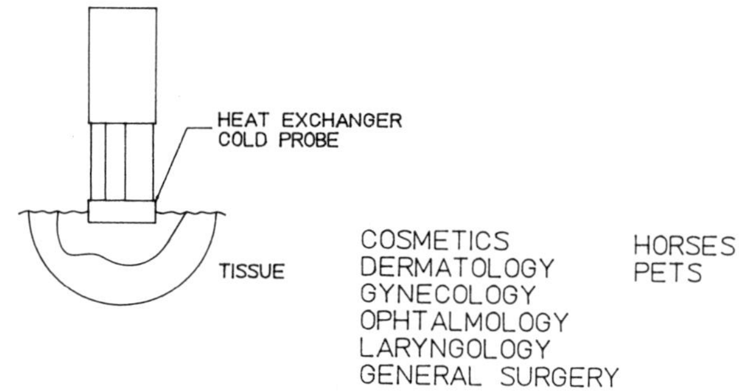

Fig.3   Cryosurgery

BIOLOGICAL ASPECTS

In the biological realm, there are really only four
circumstances that have cryogenic application. These involve
"Preservation by Intent", "Preservation by Default," "Destruction
by Intention" and "Destruction by Default". Preservation by
virtue of intention involves the entire area of taking a cell and
keeping that cell alive so that at some future time it will be
functional. This applies for ova, sperm, embryos, and zygotes.
The questions are, how long can they be preserved, and how long
should they be preserved? Is it appropriate to have a frozen ova
utilized by another recipient? There are tremendous issues in
gynecology involved in these very types of questions. There are
other issues: the use of frozen plasma, the use of frozen cells,
and the use of replacement cells and replacement organs. How
long can each survive? Is it appropriate to bank or not? Should
we take normal pancreas from someone and freeze it for 30 years
so that a person who has a carcinoma of the pancreas and has had
that pancreas removed can have the frozen organ transplanted?
This represents the generalized category of "Preservation by
Intention".

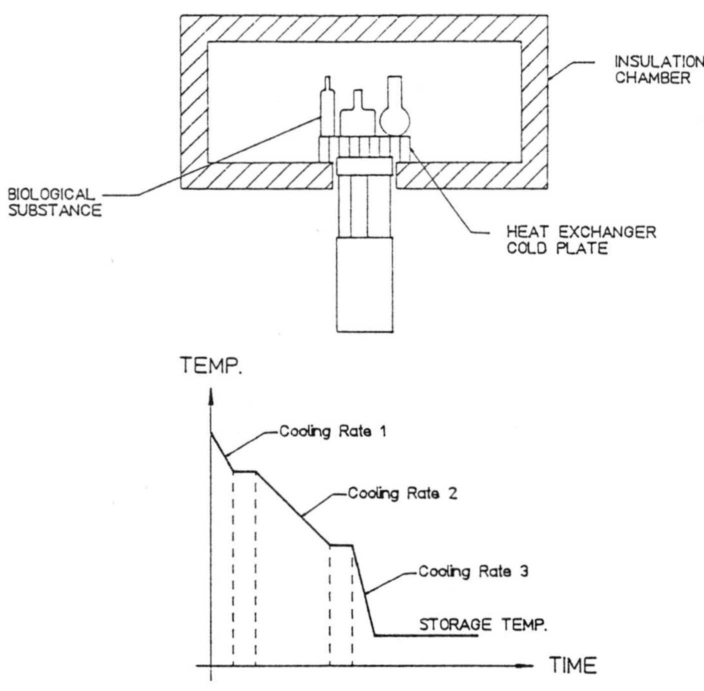

Fig. 4 Programmed Freezing

The next area to question is "Preservation by Default". A
Gynecologist will utilize cryosurgical techniques to destroy
malignant cells and, to all apparent appearances, that particular
organ appears to be free of disease when the patient is
re-examined with a PAP Smear. Then, two to three years later,
the patient has an abnormal PAP Smear which, following biopsy,
again demonstrates malignancy. Does this represent a new
malignancy? Does this represent the presence of those environmental circumstances which help provoke that malignancy in the
first place, or by default, did the Gynecologist manage to
preserve one of the malignant cells which had been there from the
first cancer? Instead of assisting this individual, the surgeon
has created the patient's own organ storage bank where this cell
had lingered in a state of dormancy and then kicked over to an
active state. This would represent the category of "Preservation
by Default".

"Destruction by Intention" - Cryosurgery represents a viable
method of cell destruction which is both inexpensive and pain-
less. It is frequently used by dermatologists, proctologists,
and gynecologists. By the same token, it also has several
drawbacks. The drawbacks are basically the following: Given the
circumstances of a very superficial lesion, something that has
the surface depth of 1 mm, from a technological point of view in
terms of being able to utilize cryosurgery, how can one only
freeze to 1.1 mm. (i.e.: 1 mm and .1 mm beyond the margin of the
depth of the lesion). Suppose this lesion exists on the surface
of the bowel, if the surgeon freezes any lower and creates a
tissue necrosis, a hole in the bowel wall will develop, which in
turn will lead to spillage of contaminant from bowel contents,
which will produce a condition of peritonitis. So crucial
margins become very, very critical in terms of efficacy of the
therapy and in terms of the avoidance of the side affects of the
therapy. One of the major things that has occurred in medicine
is the replacement of cryosurgery by laser surgery. Laser
surgery is capable of being delivered in two ways. It can be
delivered by virtue of actual contact with a probe, or it can be
delivered, if you will, through the air. In other words, you do
not need a contact point to have the ultimate role of cellular
destruction achieved. Of course, the down side of cryosurgery is
you must have contact. But, even that would be overcome if there
were a way of controlling that contact, if there were a way of
controlling the margins in a horizontal and vertical fashion. Is
there a way to control depth of penetration?

Is there a way of being able to measure the peripheral heat input
because Cryosurgery is performed in a field where the surrounding
environment is 37 $^{\circ}$C (98.6 $^{\circ}$F), i.e., body temperature? Is there
a way of being able to tell when a normal cell is destroyed

versus when a malignant cell is destroyed, even if you are able
to define lateral margins of the freeze zone, and even if you are
able to define the depth of a freeze zone?   Is there a way of
differentiating that freeze so that it is limited just to those
cells which represent the malignant portion of the lesion while
leaving behind the perfectly normal cell? The idea of
destruction is to have control of the structure to only remove
those basic areas that involve the lesion itself. This, then,
represents the definition of the boundary of what, hopefully, can
be done in the future to swing the performance of surgery back in
the direction towards cryosurgery.

Now let us move to the area of "Destruction by Default."
What occurs is the following: an ova is obtained for a donor and
fertilized. There is no take on implantation. This failure
sometimes indicates damaged genetic material and, if it were to
take, in essence what you would be growing and maturing would be
an aberration. Fortunately, in the course of gestation, the body
has a recognition or an "in grown filtration system" whereby
damaged genetic material is eliminated. It is estimated that
between 15% and 25% of all pregnancies end as spontaneous
miscarriages with a shedding of material, which had it been
permitted to continue to grow, would have provoked an
aberration. It is an essential obligation that this material not
be damaged by default, for the obvious reasons that the implications are earth-shaking. In essence, in each one of these four
categories, what we are ultimately talking about is "what do we
do with the double helix," this combination of deoxyribonucleic
acids? Do we preserve it? Do we destroy it? And, when we
follow the former category is it by intention or is it by
default? With this definition of objectives, thus can the
equipment and the technology be designed for these goals. So,
the intent defines the equipment and the equipment defines the
intent.

EQUIPMENT COST

The other important difference between laser surgery and
cryosurgery is the issue of cost. Commercially available today
for office utilization, you can buy a cryosurgery machine from
$3,000 to $4,000. The average outpatient office laser unit will
cost approximately $70,000. In the realm of health care cost,
these are very realistic implications if the pendulum can be
swung from laser back to the cryosurgery.

ACKNOWLEDGMENT

A special thanks to James Takacs of CEBAF, for his figures and
Marva Watkins, Union Hospital for clerical support.

VITRIFICATION CAPABILITY OF METAL MIRROR ULTRA-RAPID

COOLING APPARATUS: A THEORETICAL EVALUATION

ZhaoHua Chang and John G. Baust

Center for Cryobiology Research
State University of New York
Binghamton, New York, 13901

ABSTRACT

A theoretical heat sink model was developed to assess the
limitations and potential of the "bounce-free" metal mirror cooling
methodology now used to vitrify biological materials. Calculations
indicated that the current copper and ˉliquid nitrogen (LN_2)-based
slamming apparatus can vitrify a water sample to approximately 11
microns in depth. Liquid helium (LH_e) can enhance the vitrification
depth another 4 microns. Approximate 18 microns can be vitrified by
virtue of liquid helium and silver. The vitrification efficiency of the
apparatus is directly related to the initial temperature and the
penetration coefficient of the heat sink. The best combinations able to
effect maximum heat transfer for the purpose of vitrification appears to
be in the following ordering: Tin (LHe) > Lead (LHe) > Silver (LHe) >
Aluminium (LHe) > Copper (LN_2). A maximum vitrification depth of
approximately 23 microns can be achieved when pure tin is used as the
heat sink material, and the sink's initial temperature is maintained at 4
K or below. Gold and sapphire are less efficient than pure copper as a
heat sink at any bath temperature. Diamond is less efficient than copper
if its initial temperature is lower than 120 K.

Key Words: Metal Mirror Slamming, Ultra-rapid Cooling, Vitrification.

INTRODUCTION

A glass is referred to as a solid in which the molecules retain
restricted rotational and vibrational freedom but effectively lose

translational (diffusion) motion (Franks, 1985). Such an acrystalline state is often essential to avoid various physical and chemical injuries typically caused by the formation of ice crystals and, therefore, to preserve the ultrastructure of the biological specimens at low temperatures. Ultra-rapid cooling due to its presumed capability of attaining a glassy state is, at present, considered ideal for morphological studies at the electron microscopic level and for cryopreservation. It can be conveniently achieved by different techniques, which include plunging, spray freezing, jet freezing, and metal mirror slamming. The first three techniques typically use those cryogens with very low freezing points and considerably high boiling points, such as ethane, propane, Freon 22, Freon 12, etc., as secondary cryogens. Liquid nitrogen is often used as a primary cryogen to pre-cool the secondary cryogen, since direct contact of the specimen with boiling liquid nitrogen would, inevitably, create a vapor layer, blanketing the surface of the specimen and reducing the heat transfer. Except for "spray freezing", which is most suitable for low viscosity specimens composed of droplets no larger than 50 microns, all other techniques can be used for larger specimen.

When a tissue specimen at room temperature is suddenly exposed to a polished surface of a metal block immersed in a cryogen (liquid nitrogen or liquid helium), it experiences an unsteady temperature variation in the form of a rapidly changing temperature gradient. An ultra-rapid cooling rate capable of arresting water's molecular motion could be expected in the tissue boundary close to the polished surface of the metal block. This method of "bounce-free" metal mirror cooling has received particular attention in the past primarily because the metal heat sink may have potentially more choice in a materials search than the secondary liquid cryogens in terms of the vitrification efficiency and operation safety, and because liquid helium can be used to pre-cool the heat sink. It has been suggested (Chang and Baust, 1990a) that liquid helium can neither be effectively used as a primary cryogen nor as a secondary cryogen in spray/plunge techniques. Since Van Harreveld and Crowell (1964) first successfully adopt this metal-mirror slamming method for cryofixation in electron microscopy (EM), it has been commonly used in the fields of cryobiology and EM and appears to vitrify a variety of thin tissue samples. Its feasibility in preserving the ultrastructure has been demonstrated in numerous mammalian tissue systems (Philips et al, 1984; Linner et al, 1986; Sitte et al, 1987; and Baust et al, 1987). It was found, however, that only 10-15 microns of amorphous phase tissue water is formed by using a copper mirror-liquid nitrogen slamming apparatus (Dempsey and Bullivant, 1976; Philips et al, 1984., and Linner et al, 1986). The maximum depth of vitrification appears to be 15-20 microns as reported by Phillips et al (1984), using a pure copper block at liquid helium temperature (4 K).

There have been many theoretical analyses regarding the mechanism of heat transfer associated with the "metal mirror"

216

slamming apparatus (Bald and Crowley,1979; Kopstad and Elgsaeter, 1982; Jones, 1984; and Bald, 1985). It was suggested most recently (Diller, 1990) that the rate of heat transfer depends upon the thermal driving potential, the heat flow impedance, and the internal impedance inherent to the thermophysical properties - of the specimen. The present work is intended to further analyze the heat transfer process and evaluate the vitrification efficiency, the potential for and limitations of vitrifying biological materials associated with different metal heat sinks and different primary cryogens (LN_2 or LH_e), which are, in effect, the only potential factors that can be significantly improved.

THEORETICAL MODEL

If there is no thermal disturbance of the liquid cryogen when the specimen is exposed to the polished surface, the cooled metal block can be viewed as a heat sink, whereas, the cryogen serves to pre-cool the metal block. This will occur if the metal block above the cryogen is longer than a critical value, X_C (Figure 1). Then, the portion of the metal block near the cryogen will maintain its temperatures at a minimum of 99.9% of its initial value, as the specimen is cooled from its initial temperature down to its glass transition temperature (T_g).

When the specimen and heat sink come into sudden thermal contact at t = 0 (Figure 1), it is reasonable to assume that their mutual interface temperature would immediately assume the same value T_m, i.e. T_m (x=-0) = T_m (x=+0).

For t > 0, the temperature changes in the specimen and heat sink, based upon the heat equations for transient conduction in a semi-infinite solid (Incropera et al, 1981), can be expressed as:

Specimen (X > 0): $T = T_m + (T_1 - T_m)$ erf $[0.5 (\alpha_w t)^{-1/2} X]$

$$(1)$$

Heat sink (X < 0): $T = T_m + (T_2 - T_m)$ erf $[-0.5 (\alpha t)^{-1/2} X]$

$$(2)$$

in which T is temperature (K), t is time (s), erf is the Gaussian error function, α_w and α are, respectively, the thermal diffusivity (m^2/s) of the specimen and heat sink, T_1 and T_2 are, respectively, the initial temperatures of the specimen and heat sink, and x denotes the distance (m) from the mutual interface. Both equations 1 and 2 are derived on the assumption that the specimen is effectively a semi-infinite body and there is no heat transfer from the specimen holder through the specimen to the metal block. This is a reasonable assumption, since the biological sample generally has a low thermal diffusivity and consequently the thermal influences are only confined in a limited

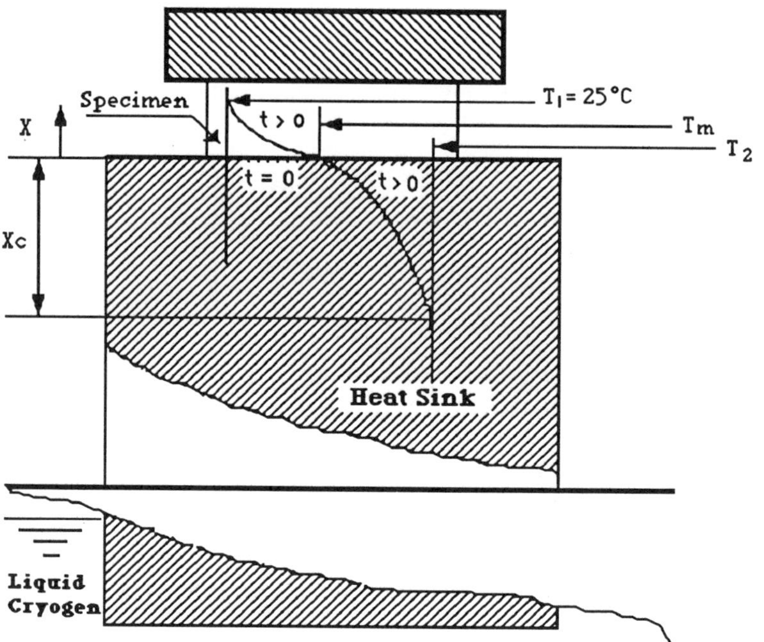

Figure.1. Schematic Representation of the Heat Sink Model

x --   The distance of the specimen or heat sink from the
            mutual contacting surface.
Xc--   critical height of the metal block
T(x,t)--   Temperature of the specimen at position x and time t.
T_1 --   Initial temperature of the specimen
T_2 --   Initial temperature of the heat sink
T_m --   Mutual interface temperature of the specimen and the
            heat sink
t --   Time

zone near the contacting surface within a short time interval after
slamming. Additionally, the following assumptions which have been
discussed in the previous thermal analyses (Bald and Crowley, 1979;
Kopstad and Elgsaeter, 1982; Jones, 1984; and Bald, 1985) also have been
made: 1). constant material thermophysical properties, 2). negligible
thermal radiation from surroundings, and 3). perfect thermal contact
between the specimen and the heat sink.

Since the boundary between the specimen and the heat sink
cannot store heat, for t > 0,

$$K \left. (\partial T/\partial x) \right|_{x=-0} = K_w \left. (\partial T/\partial x) \right|_{x=+0} \tag{3}$$

By solving equations (1) - (3), we obtain

$$(T_m - T_1)/(T_2 - T_1) = b/(b_w + b)$$

(4)

$$b = (K\rho C_p)^{1/2} \qquad b_w = (K\rho C_p)_w^{1/2}$$

Here, K is thermal conductivity (W/m·k), ρ is mass density (K_g/m^3) and C_p is the specific heat at constant pressure (J/K_g·k). The subscript w denotes the specimen. The values b and b_w ($J/m^2ks^{1/2}$) are defined as the penetration coefficients of the heat sink and the specimen, respectively. A penetration coefficient measures the quantity of heat which penetrates into a body during a given time interval when its surface is suddenly cooled by a given amount to some final temperatures. The cooling rate of the specimen actually depends, in the first instance, on the value of b, since the value of b_w is inherent to the specimen. The larger the value of b, the larger the amount of heat absorbed by the heat sink from the specimen within a definite time, and, as such, a greater depth of vitrification can be realized.

RESULTS OF CALCULATIONS AND DISCUSSION

The Efficiency of Various Metals as Heat Sink at Low Temperature in Comparison with Copper

From the discussion above, we know that the temperature distribution and cooling rate in the specimen strongly depend on the penetration coefficient, b, of the heat sink. The choice of material is, therefore, crucial in designing a slamming apparatus. An ideal material should have a very high value of b at low temperatures. Copper has a relatively large b value at various temperatures and is the material most widely used at present. To compare the efficiency of other materials with that of copper as a heat sink, we define a new parameter, ξ, to measure the potential gain from using other materials:

$$\xi = b/b_c$$

(5)

Where b_c is the penetration coefficient for copper.

The results of calculation for several possible candidate materials for a heat sink are illustrated in Figure 2. The thermophysical properties were obtained from Touloukin, et al. (1970), and Field (1979). It can be seen from the figure that pure tin, lead, silver and aluminium all have a penetration coefficient larger than that of the pure copper at liquid helium temperature (approximately 4 K). The penetration coefficients of gold and sapphire are both smaller than that of the pure copper at any temperature. Diamond has a smaller penetration coefficient than does pure copper when the temperature is lower than 120 K. Contrary to a relatively complicated analysis by Bald (1983), we did not see that the thermal diffusivity of the heat sink plays a significant, if any, role in determining the cooling rate of the specimen. If the penetration coefficient is the only important parameter for choosing the heat sink material, as proven to be true by our analysis, then Figure 2 actually presents the efficiency of various materials as heat sinks as compared to pure copper. The best material for a heat sink at liquid helium temperature (about 4 K) is, therefore, pure tin followed by lead, and then silver and aluminium. Silver is superior to copper if the heat sink temperature is below 25 K, which is in agreement with Bald's analysis (Bald, 1983 and 1985). With the assumption that the thermal inertia (penetration coefficient) changes linearly with temperature, Bald (1983) found an optimum operating temperature of 15.6 K for a silver heat sink. According to Figure 2, however, there is only a slight difference (about 12%) in the penetration coefficient between a silver heat sink and a copper heat sink at this so-called optimum temperature. Such a marginal benefit

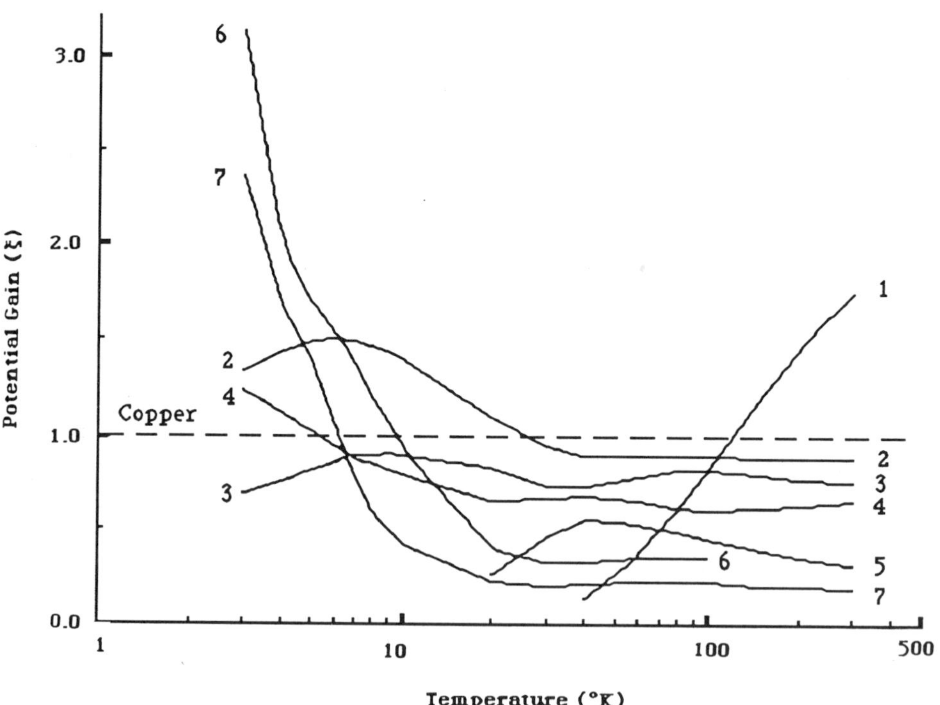

Figure 2. The Potential Gain from Using Type ∏a Diamond (1), Silver (2), Gold (3), Aluminium (4), Sapphire (5), Tin (6), and Lead (7) at Different Temperatures.

could be well compensated for by other factors, such as the cost of construction and the maintenance of the apparatus. If the initial heat sink temperature can be maintained below 6 K, we suggest that pure tin or lead be the first choice for the heat sink material. Pure copper is no doubt the best choice at liquid nitrogen temperature, as indicated in Figure 2.

Many investigators (Field, 1979) have concluded that the heat dissipation from solid state semi-conductor devices can be almost doubled between 200 K and 300 K if the Type ∏a diamond is placed on the top of a copper block to form a diamond heat sink as compared with a copper sink. This is not, however, the situation for the slamming apparatus, which operates at much lower temperatures. In fact, diamond is less efficient than pure copper as a heat sink at any temperature lower than 120 K (Figure 2).

<u>The Calculation of Cooling Rate in a Pure Water Specimen</u>

As indicated in equation 1, the temperature of the specimen changes in a time (t) and position (x) dependent manner. At a specific position, the temperature decreases sharply at first and then slowly as cooling continues. In other words, the lower the temperature, the

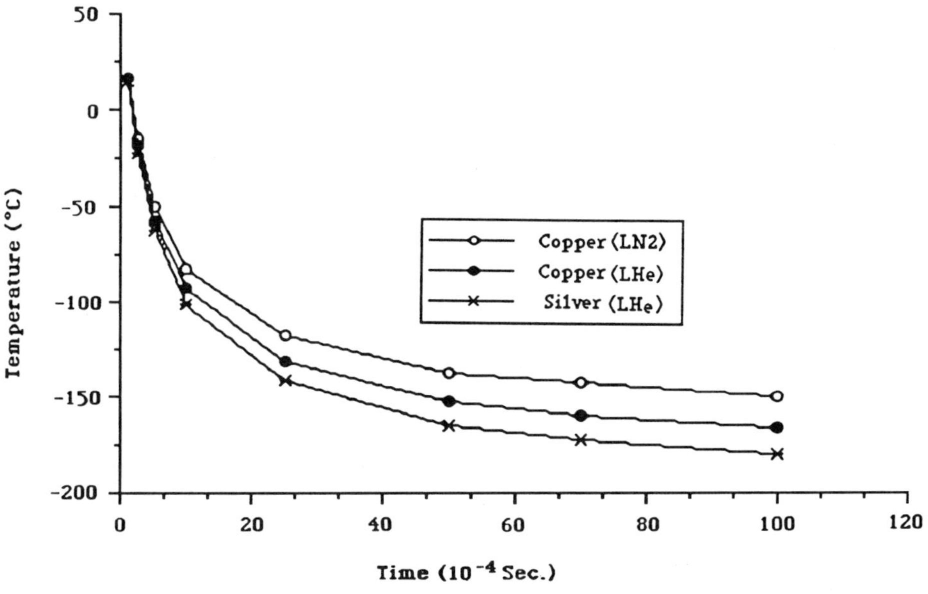

Figure 3. Temperature Changes in a Water Specimen at a Position X=20 mm.

smaller the cooling rate. Such a cooling patten is exemplified in Figure 3 by the cooling history at a position of 20 microns for a pure water specimen. To facilitate the calculation, the heat capacity (C_p) of the water has been assigned the mean value of the amorphous solid water (Sugisaki et al, 1968) between 0°C and -137°C (an accepted glass transition temperature of pure water) which otherwise would vary appreciably with temperature changes. Other thermophysical properties of the water, such as mass density and thermal conductivity, are not readily available in the literature. They are here assumed to be constant and equal to the values at room temperature (Incropera et al, 1981). Such involvement of elements of uncertainty as to the thermophysical properties of water in the calculation seems to be reasonable and does not have significant effect on the final results. The difference in the cooling rate at different temperatures may vary by many orders of magnitude. The definition of cooling rate is therefore meaningless if it is not specified at a particular temperature.

The cooling rates of the specimen using different heat sink materials at different initial temperatures have been calculated at temperatures near the freezing point of water (Bald, 1985). The results thus obtained can be used to compare the differences in cooling rates at different conditions, but are not sufficient to directly tell the vitrification capability of the apparatus. According to Franks (1985), the liquid within the zone from T_c (crystallization temperature) to T_g (glass transition temperature) is vulnerable to freezing. It has been experimentally verified (Chang and Baust, 1990b) that the time lapse or average cooling rate in this temperature zone is critical to vitrification. The cooling rate beyond this critical zone has no apparent effect on vitrification. To form a glass and avoid crystallization, the metastable zone between T_c and T_g must be crossed quickly. To be convenient, we define the critical rate as the average cooling rate in the temperature range of 0°C to -137°C. The calculation of the average cooling rate from 0°C to -137°C can then be performed by using equation:

$$E = 137/t_g \qquad (6)$$

Here we denote E to be the average cooling rate between 0°C and -137°C, and t_g is the time required to reach -137°C from 0°C.

The calculated results are illustrated in Figure 4. As expected, the maximal cooling rate is attained by utilizing a tin block and liquid helium. There has been a wide divergence of opinion as to the specific value of the critical cooling rate for vitrifying pure water, which ranges from 10^4 to 10^{7}°C$\cdot$Sec.$^{-1}$ (Luyet, 1956; and Uhlmann, 1972). If a moderate value of 10^5°C$\cdot$Sec.$^{-1}$ is assumed, then the vitrification depth is approximate 11 microns using pure copper and liquid nitrogen. This result agrees well with the electron microscopy data reported by Linner

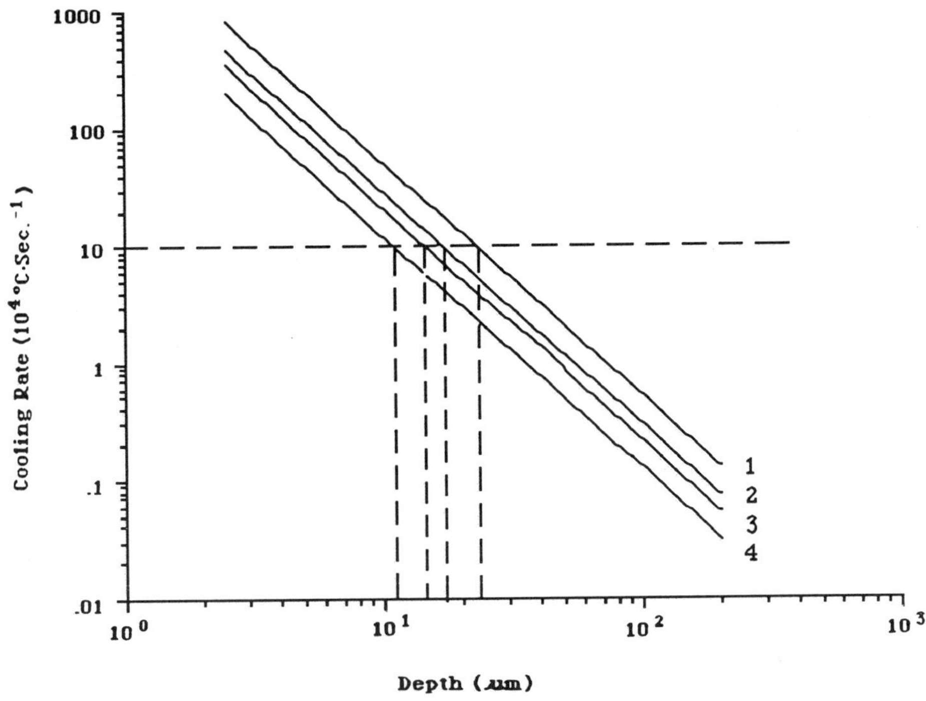

Figure 4. Cooling Rate Distribution in a Pure Water Specimen, with Material and Cryogen as Parameters. Curve 1- Tin (LHe); 2- Silver (LHe); 3- Copper (LHe); and 4- Copper (LN$_2$).

et al. (1986) and Phillips et al. (1984). Surprisingly, the liquid helium can enhance the vitrification depth by only approximate 4 microns. Such an unexpected limited value is mainly due to the great decrease in the penetration coefficient for copper as it approaches liquid helium temperature. This compensates for the contribution of the low initial temperature, which is expected to increase the cooling rate. Silver, which has a higher penetration coefficient value at liquid helium temperature than copper, can improve the vitrification depth by another 3 microns. The combination of silver and liquid helium can thus only achieve a marginal improvement in vitrification, which is in agreement with the conclusion made by Bald (1985). When tin is used in combination with liquid helium, a maximum vitrification depth of approximately 23 microns can be achieved.

In conclusion, the most important parameters that directly determine the thermal driving potential and, therefore, the vitrification efficiency of the metal-mirror slamming apparatus are the penetration coefficient of the heat sink and its initial operating temperature. According to our analysis, the best combination of the heat sink material and initial temperature in which the apparatus will achieve maximum vitrification capability seems to be in the following ordering: Tin (LHe) > Lead (LHe) > Silver (LHe) > Aluminium (LHe) > Copper (LN2). Such an ordering is derived purely from a theoretical analysis and certainly still remains to be evaluated experimentally.

REFERENCES

Bald, W.B., and Crowley, A.B., 1979, On Defining the Thermal History of Cells during Freezing of Biological Materials, J. Microsc., 117: 395.

Bald, W.B., 1983, Optimizing the Cooling Block for the Quick Freeze Method, J. Microsc, 131: 11.

Bald, W.B., 1985, The Relative Merits of Various Cooling Methods, J. Microsc, 140: 17.

Baust, J.G., May, S.R., Wolanczyk, J.P., Livesey, S.A., and Linner, J.G., 1987, Differential Scanning Calorimetric and Mass Spectrometric Analysis of Mammalian Tissue Samples, Cryobiology, 24: 548.

Chang, Z.H., and Baust, J.G., 1990 (a), Ultra-Rapid Cooling by Spraying/ Plunging: Pre-cooling in the Cold Gaseous Layer, J. Microsc., (in Press).

Chang, Z.H., and Baust, J.G., 1990 (b), Further Inquiry into the Cryobehaviour of Aqueous Glycerol Solutions, Cryobiology, (in press).

Dempsey, G.P., and Bullivant, S., 1976, A Copper Block Method for Freezing Non-Cryoprotected Tissue to Produce Ice-Free Regions for Electron Microscopy. I. Evaluation Using Freeze-Substitution, J. Microsc., 106: 251.

Diller, K., 1990, A Note on Optimal Techniques of Rapid Cooling for Low Temperature Preparation of Electron Microscopy Specimens, Cryo-Letters, 11: 75.

Field, J.E., 1979, "The Properties of Diamond," Academic Press, London,

Franks, F., 1985, "Biophysics and Biochemistry at low Temperatures," Cambridge University Press, Cambridge.

Incropera, F., and DeWitt, D.P., 1981, "Fundamentals of Heat Transfer," John Wiley & Sons, New York.

Jones, G.J., 1984, On Estimating Freezing Times during Tissue Rapid Freezing, J. Microsc., 136: 349.

Kopstad, G., and Elgsaeter, A., 1982, Theoretical Analysis of Specimen Cooling Rate during Impact Freezing and Liquid-Jet Freezing of Freeze-etch Specimen,. Biophysics. J., 40: 163.

Linner, J.G., Livesey, S.A., Harrison, D.S., and Steiner, A.L., 1986, A New Technique for Removal of Amorphous Phase Tissue Water Without Ice Crystal Damage: A Preparative Method for Ultrastructural Analysis and Immunoelectron Microscope. J. Histochemistry and Cytochemistry, 34: 1123.

Luyet, B.J., 1956, "Cytology of the Blood and Blood Forming Organs," Bessis, M. ed. Grune and Stratton, New York, 1956.

Phillips, T.E., and Boyne, A.F., 1984, Liquid Nitrogen-Based Quick Freezing: Experiences with Bounce-free Delivery of Cholinergic Nerve Terminals to a Metal Surface, J. Electron Microscopy Technique, 1: 9.

Sitte, H., Edelmann, L., and Neumann, K., 1987, Cryofixation Without Pretreatment at Ambient Pressure, in: "Cryotechniques in Biological Electron Microscopy," R.A. Steinbrecht and K. Zierold, ed. Springer-Verlag, Berlin.

Sugisaki, M., Suga, H., and Seki, S., 1968, Calorimetric Study of the Glass State IV. Heat Capacities of Glass Water and Cubic Ice. Bull. Chem. Soc. Jap., 41: 2591.

Touloukin,Y.S., Powell, R,W., and Klemens, P.G., 1970, "Thermophysical Properties of Matter," Vol. 1, IFI/Plenum, New York.

Touloukin,Y.S., Powell, R,W., and Klemens, P.G., 1970, "Thermophysical Properties of Matter," Vol. 2, IFI/Plenum, New York.

Touloukin, Y.S., and Buyco, E.H., 1970, "Thermophysical Properties of Matter," Vol. 4, IFI/Plenum, New York.

Touloukin,Y.S., and Buyco, E.H., 1970, "Thermophysical Properties of Matter," Vol. 5, IFI/Plenum, New York.

Uhlmann, D.R., 1972, A Kinetic Treatment of Glass Formation, J. Non-Crys. Solids, 7: 337.

Van Harreveld, A., and Crowell, L., 1964, Electron Microscopy after Rapid Freezing on a Metal Surface and Substitution Fixation, Anat Rec., 182: 377.

MICROFRACTURES IN CRYOPRESERVED HEART VALVES:

VALVE SUBMERSION IN LIQUID NITROGEN REVISITED

L. Wolfinbarger, Jr., M. Adam, P. Lange, and J-F Hu

Center for Biotechnology, Old Dominion University
Norfolk, VA 23529 and LifeNet: The Virginia Tissue
Bank, 5809 Ward Court, Virginia Beach, VA 23455
USA

INTRODUCTION

Cryopreservation of human tissues is finding increased usage as a means of increasing the availability and quality of clinically usable biomaterials. Tissues being cryopreserved include heart valves, vein segments, articular cartilage, tendons, ligaments, bone, and cellular tissues such as the pancreas. Cryopreservation of organs has met with more limited success than cryopreservation of "acellular" tissues, i.e. tissues with minimal cell populations. Tissues such as a heart valve consist primarily of a collagen/proteoglycan matrix in which small numbers of fibroblast type cells are sparsely distributed. In it's native state, the heart valve also contains a surface coating of metabolically active endothelial cells whose primary function is to provide a smooth surface to the blood streaming across its' face and to prevent blood clotting.

Heart valves are typically obtained postmortem (or, infrequently, from organ donors), surgically trimmed of excess tissues, immersed into tissue culture media amended with cryoprotectant (dimethylsulfoxide, DMSO), sealed in an appropriate "plastic" container and control rate frozen (Lange, 1989). These cryopreserved tissues are stored below 143K (-130° C) (vapor phase) or directly in liquid nitrogen (liquid phase) until needed, at which time they are rapidly thawed, the cryoprotectant removed by step-wise dilution into a tissue culture medium, and implanted by a cardiovascular surgeon. These implanted allograft heart valves are remarkably durable and are vastly superior to other mechanical or bioprosthetic valves currently in use.

It is generally agreed that properly cryopreserved tissues stored below 143 K (-130° C) experience minimal recrystallization complications and thus, efforts are normally focused on maintaining cryopreserved heart valves below this temperature. Complications arise when tissues need to be transported from one storage container to another, transported between storage facilities, and/or during refilling of the liquid nitrogen storage containers. Valves stored in the vapor phase of liquid nitrogen may become submerged during an automatic refill cycle of a storage chamber or valves may be submerged in liquid nitrogen as a "routine" aspect of transport. Indeed, the authors became aware of potential problems with valve submersion into liquid nitrogen when a valve storage system in a hospital overfilled during one of its' automatic refill cycles. Valves taken after this "accident" were discovered to have numerous full thickness fractures in the valve conduit, following normal thawing procedures in the operating room. Experiments were immediately begun to assess the effects of valve submersion into liquid nitrogen on valve structure. Adam, et al (1990) reported that valves stored in the vapor phase of liquid nitrogen experienced extensive surface fracturing when immersed into liquid nitrogen. Similar fractures were

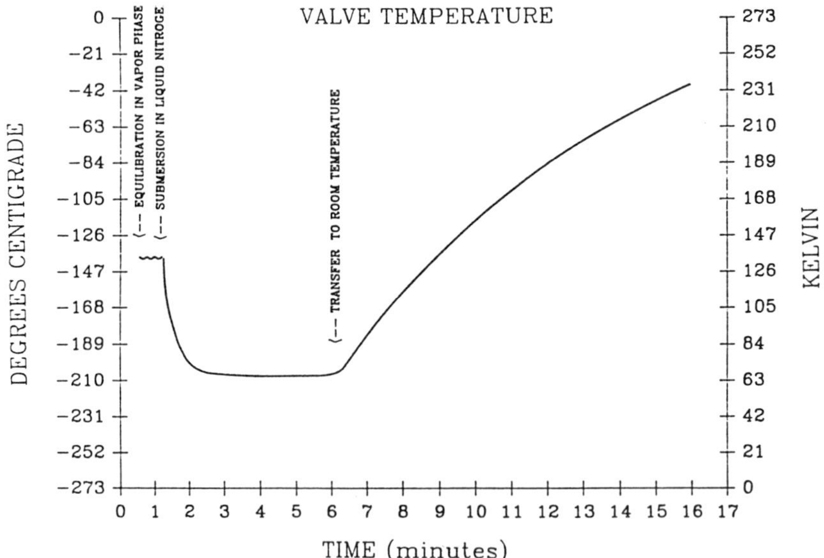

Figure 1   Temperature, in degrees centigrade and Kelvin, in close proximity to the valve leaflets as a function of time.

not present in valves stored in the vapor phase of liquid nitrogen, but not immersed into liquid nitrogen. The fracture lines were shown to penetrate into the collagen matrix, running parallel and perpendicular to the orientation of the collagen fiber bundles in that matrix. Figure 1 represents the temperature profile of a typical valve cryopreserved in tissue culture media amended with 10% DMSO such that the total volume of material frozen constituted 100 mls. The temperature probe from a Cryomed Controlled Rate Freezer is implanted in close proximity to the leaflet tissues of the valve and temperatures, consistent with limitations imposed by the invasive nature of the temperature probe, in this region are measured. As may be seen in figure 1, the temperature within the probe region dropped within 2.5 minutes from 143K to 77K (-130° C to -196° C) for an "average" temperature change of 26.4 K/minute. It is important to note that this temperature change is not uniform for the entire volume of the valve package and that mass near the surface of the bag cools more rapidly than mass in the bag interior. These differences may be important to the introduction of fractures in tissues frozen in these matrices in that these temperature differentials may create stress within the solid matrix of the bagged material. Research by Kroener and Luyet (1966a, 1966b) may suggest mechanisms for tissue damage. They reported that 97% glycerol solutions contracted by 2.8 cubic millimeters/gram for every 10 degrees above Tg (glass transition temperature) and by 0.4 cubic millimeter/gram for every 10 degrees below Tg. As temperatures in these solutions approached 158K (-115° C) cracks developed in the specimens and their readings became "irratic". Their second study (Kroener and Luyet, 1966b) suggested that the formation and disappearance of cracks depended on "the interaction of several factors, in particular, the mechanical properties of the material, the concentration of solute in it, the temperature gradients through it, the overall temperature, and the rate of change of temperature". At cooling rates on the order of 1 to 3 K/minute (with temperature gradients of 6.5 to 8.5 K/cm) cracks appeared. When higher cooling rates and larger temperature gradients were attempted, the cracks appeared "earlier". When the solutions were warmed, the cracks disappeared, but only at temperatures above that of the glass transition. Cryopreservation solutions typically contain quantities of salts and cryoprotective agents. During the freezing process, crystalline ice forms resulting in an unfrozen fraction which becomes increasingly concentrated with solute (salts, cryoprotectant). These highly concentrated solute solutions may eventually form an amorphous solid (at sufficiently low temperatures) and thus the solid matrix in a valve package will consist of different forms of solid matrix each with its own coefficients of expansion/contraction. Kroener and Luyet (1966a, 1966b) suggested that changes in the coefficient of expansion and their correlatives, the density and specific volume, involve generation of stress. Relief of this stress could presumably occur through formation of "cracks", and rapid temperature changes or mechanical impact to the matrix could initiate the process (Adam, et al. 1990).

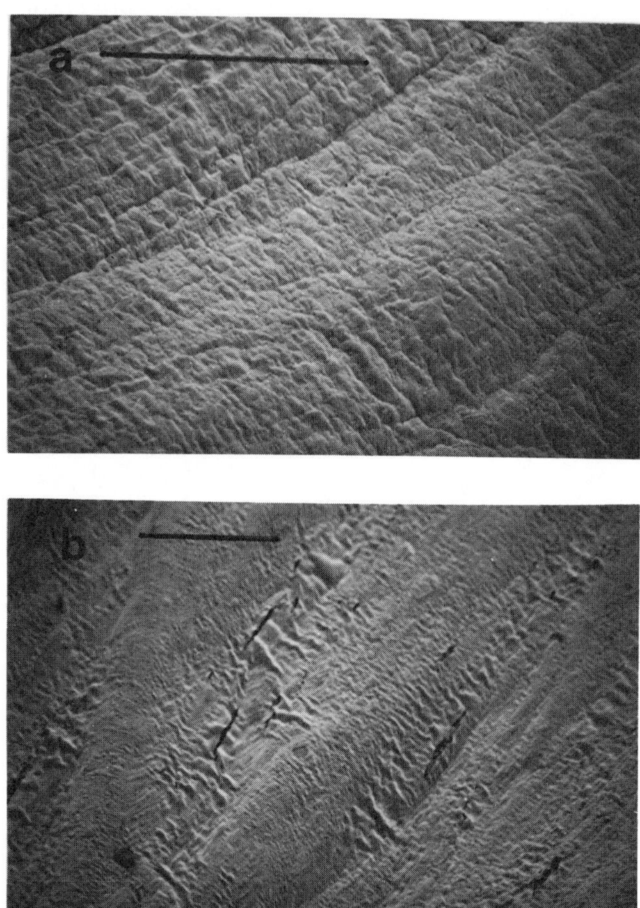

Figure 2 Scanning electron photomicrograph of a control valve leaflet which was not submerged into liquid nitrogen. a) Magnification: 219X with the bar indicating 200 um b) Magnification: 121X with the bar indicating 200 um, note that this photograph was taken of a valve suffering typical fixation damage to demonstrate the different kind of damage than observed in experimental valve leaflet tissues.

Cracks formed in the matrix could cause deep or surface fractures to tissue embedded within that matrix provided the fracture line (crack) extended within or through the valve tissue matrix. Figure 2a shows a representative scanning electron micrograph of the surface of a cryopreserved human valve which was stored in the vapor phase of liquid nitrogen, but not immersed into the liquid phase of liquid nitrogen. It was thawed and fixed for viewing with the scanning electron microscope as described by Adam, et al. (1990). The surface of the valve is essentially unscarred. Figure 2b shows a similar representative photomicrograph of the kind of damage to a nonimmersed valve usually attributed to fixation damage. Here, the "basement membrane" (to which endothelial cells would normally attach) appears "blistered" and contraction of the surface material results in small-short "tears", usually parallel to the orientation of the underlying collagen fibers.

Similar scanning electron micrographic analysis of valves immersed into liquid nitrogen (after storage in the vapor phase for varying periods of time) reveal a more dramatic blistering and tearing of the surface material (figure 3a). In general, this blistering and tearing is more severe with valves immersed into liquid nitrogen, but it is unknown if the rapid immersion is responsible for the damage. The initial study by Adam et al. (1900) revealed the presence of larger fracture lines in the surfaces of the valve leaflet material (on both surfaces). Additionally, these fracture lines were shown to penetrate into the collagen matrix of the tissue. Photomicrographs similar to those shown by Adam, et al. are shown in figure 3b. These fracture lines typically run both parallel and perpendicular to the orientation of the underlying collagen fibers (perpendicular in figure 3b) and appear to cut across the underlying collagen fibers (figure 3c).

Problems associated with mechanical cracking of vitrified solutions are not new and Bank & Brockbank (1987) have discussed the use of vitrification solutions and the damage that can occur by spontaneous devitrification and phase separations. At present, there is minimal data regarding the mechanism(s) which may result in mechanical damage to cryopreserved tissues. The temperature of storage of these tissues may be unimportant. However, it would appear that thermal perturbations within tissues, which may occur during subsequent handling, may be important and researchers should consider the damage which may be occurring to tissues whenever they are subjected to dramatic and sudden temperature changes or rough handling.

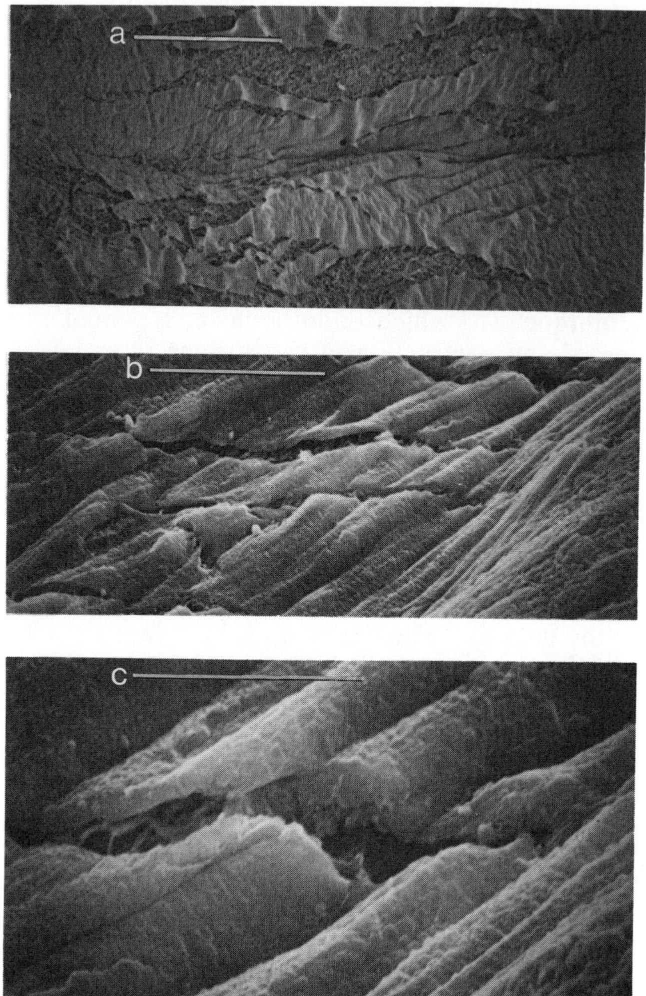

Figure 3a   Scanning electron photomicrograph of a submerged valve to demonstrate that experimental valves experience more dramatic effects with fixation damage than control valves.   Magnification: 121X with the bar indicating 200 um.

Figure 3b   Scanning electron photomicrograph of a submerged valve demonstrating the damage occurring to valves submerged into liquid nitrogen. Magnification: 604X with the bar indicating 50um.

Figure 3c   Scanning electron photomicrograph of a submerged valve demonstrating the damage occurring to valves submerged into liquid nitrogen. Magnification: 1.90KX with the bar indicating 50 um.

REFERENCES

Adam, M., Hu, J-F, Lange, P., and Wolfinbarger, L., Jr., 1990, The Effect of Liquid Nitrogen Submersion on Cryopreserved Human Heart Valves. Cryobiology 27:605-614.

Lange, P., Hopkins, R.A., 1989, Allograft Valve Banking; Techniques and Technology, in: "Cardiac Reconstructions with Allograft Valves," Hopkins, R.A., Springer-Verlag Publishers, New York, pp 37-63.

Kroener, C. and Luyet, B., 1966, Discontinuous Change in Expansion Coefficient at the Glass Transition Temperature in Aqueous Solutions of Glycerol, Biodynamica, 10:41-45.

Kroener, C. and Luyet, B., 1966a, Formation of Cracks During the Vitrification of Glycerol Solutions and Disappearance of the Cracks During Rewarming, Biodynamica, 10:47-51.

Bank, H.L. and Brockbank, K., 1987, Basic Principles of Cryobiology, J. Card. Surg., 2:137-143.

ACKNOWLEDGEMENTS

This work was supported in part by grants from the Virginia Chapter of the American Heart Association and LifeNet.

AN ESTIMATION OF THERMAL STRESS INDUCED BY THE FREEZING

PROCESS FOR BIOLOGICAL CELL PRESERVATION

Sui Lin

Department of Mechanical Engineering
Concordia University
Montreal, Quebec, Canada H3G 1M8

ABSTRACT

Thermal stresses in a frozen solution, induced by the freezing process, are related to the cryoinjury of biological cells which are suspended inside a physiological medium for the purpose of preservation. Experimental and analytical results indicate that the circumferential compressive stresses inside the frozen solution are much higher than the radial and axial stresses. The maximum circumferential compressive stress is located at the interface position between the frozen and the unfrozen solution.

In this paper, an estimation of the maximum circumferential compressive stress is presented and is based on the one-dimensional experimental freezing model for simulation of cell freezing preservation taking place in a long cylindrical test tube. The result of the analysis indicates that a decrease of the temperature at the outside surface of the tube results in an increase of the maximum thermal stress in the frozen solution. For freezing preservation, a slow freezing process having a relatively high surface temperature has the benefit of reducing the cryoinjury of biological cells caused by the thermal stress.

NOMENCLATURE

C, C_1, C_2	constants
E	Young's modulus
h_{LS}	latent heat of water freezing
k	thermal conductivity
kR	dimensionless parameter defined in Eq. (8)
K_1, K_2	constants defined in Eqs. (20) and (21), respectively
L	length of the tube
q	heat transfer rate

r	radial coordinate
R_o	outside radius of the tube
R_i	inside radius of the tube
S	dimensionless interface position between the ice and water, defined in Eq. (6)
t	time
T	temperature
T_o	outside surface temperature of the tube
T_f	freezing temperature of water
ΔT	temperature difference $(T - T_f)$
u	displacement
X	interface position between the ice and water
z	axial coordinate

Greek Symbols

α^*	linear thermal expansion coefficient
β	initial strain of the ice
ε	strain
ν	Poisson's ratio
ρ	density
σ	stress
τ	dimensionless time defined in Eq. (7)

Subscripts

1	tube
2	ice
r	radial direction
t	circumferential direction
z	axial direction

INTRODUCTION

In the preservation of biological cells, freezing techniques have been used to keep the cells dormant, but potentially alive. However, there is an apparent contradiction from the experimental findings that a part of the living cells often experiences injury during the freezing process. From a survey of the literature, the generally acceptable reasons for the cryoinjury of the cells can be conceptualized as "solution effect"[1], "intracellular ice-crystallization"[2] and "packing effect"[3]. Recently, a new hypothesis[4] concerning one mode of the cryoinjury of cells suspended in a physiological medium has been raised. This hypothesis states that the thermal stress in the frozen-cell-suspension could cause mechanical damage to the cell membrane/structures. An experimental and numerical investigation of thermal stresses related to cryoinjury of biological cells in freezing preservation[5] was conducted. The results indicate that the circumferential compressive stresses inside the frozen solution are much higher than the radial and axial stresses. The maximum circumferential compressive stress is located at the interface position between the frozen and unfrozen solution.

The purpose of the present paper is to estimate the maximum circumferential stress induced by the freezing process during preservation of biological cells. The paper confirms that the maximum thermal stress in the frozen solution induced by a slow freezing process is smaller than that induced by a quick freezing process.

MATHEMATICAL FORMULATION

For the present analysis, as a first approximation, pure water is considered as the ·suspension medium of the cells (actually, the major component of the physiological medium is water) and the effect of the suspended cells on the thermal stresses in the frozen medium may be neglected. Because the cell suspension is usually contained in a long cylindrical test tube, the freezing process of the suspension takes place mainly in the radial direction. A one-dimensional freezing model is used in the thermal stress analysis.

For the approximation, the following assumptions are made:

1. The physical properties of the tube material, the ice and water are constant except that the change of the density of the ice is considered in the thermal stress analysis.
2. The temperature distributions in the tube wall and in the ice are allowed to be determined by the quasi-steady state approximation.
3. The ice is assumed to be the isotropic polycrystalline ice.
4. The shear stress between the tube wall and the ice may be neglected.

For the isotropic polycrystalline ice, the brittle behavior is characterized by an elastic deformation, followed by a sudden fracture. Therefore, the elastic theory may be applied to the polycrystalline ice before the fracture/crush occurs. After the crush of the ice, the strains and stresses in the ice and in the tube wall will be redistributed. These redistributions will not be considered by the present analysis.

Heat Transfer Process

Using the above assumptions, the one-dimensional freezing process taking place inside the tube, shown in Fig. 1, and initially at the freezing temperature of water, can be formulated as follows:

The heat transfer rate is described by

$$q = \frac{2\pi L \, (T_f - T_o)}{\dfrac{ln(R_i / X)}{k_2} + \dfrac{ln(R_o / R_i)}{k_1}} \tag{1}$$

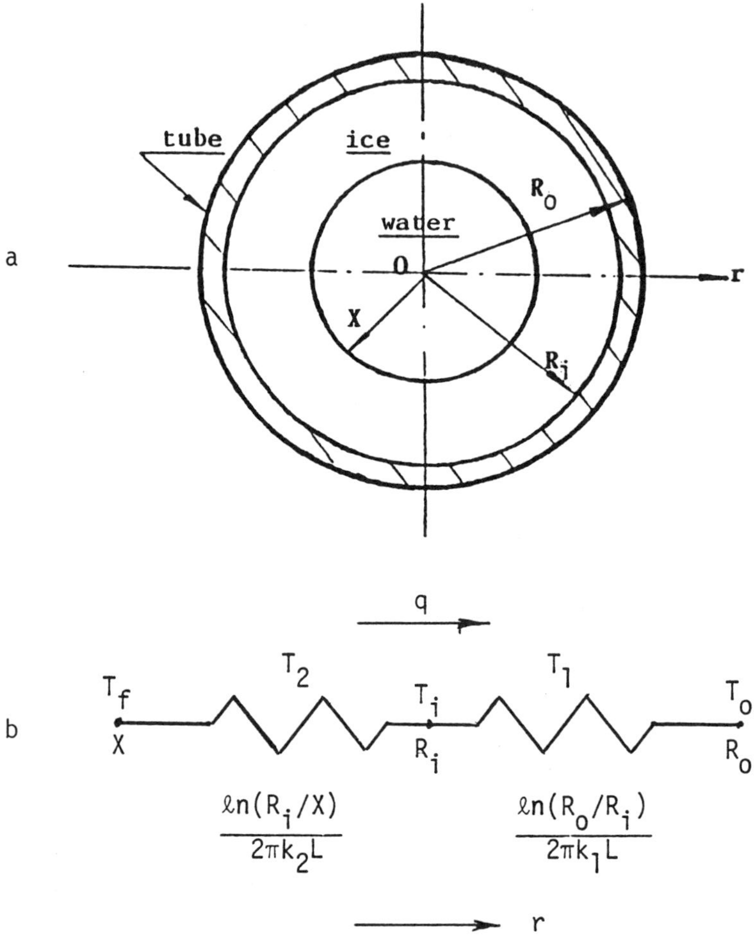

Fig. 1 (a) Schematic diagram of one-dimensional water freezing process in a tube. (b) Heat flow network for the one-dimensional water freezing process.

The temperature distributions in the tube wall, T_1, and in the ice, T_2, are as follows:

$$T_1 = T_o + \frac{q}{2\pi L} \bullet \frac{ln(R_o/r)}{k_1} \tag{2}$$

$$T_2 = T_f - \frac{q}{2\pi L} \bullet \frac{\ln(r/X)}{k_2} \tag{3}$$

The interface position, X, between the ice and water is determined by using the heat balance equation,

$$h_{LS}\, \rho_2 \frac{dX}{dt} = k_2 \left. \frac{\partial T_2}{\partial r} \right|_{r=X} \tag{4}$$

Substituting Eq. (3) into Eq. (4) and utilizing Eq. (1) yields

$$\frac{dX}{dt} = - \frac{T_f - T_o}{h_{LS}\, \rho_2\, X\left[\dfrac{\ln(R_i/X)}{k_2} + \dfrac{\ln(R_o/R_i)}{k_1}\right]} \tag{5}$$

Integrating Eq. (5) from R_i to X and introducing the following dimensionless representations,

$$S = X/R_i \tag{6}$$

$$\tau = \frac{k_2(T_f - T_o)\, t}{h_{LS}\, \rho_2\, R_i^2} \tag{7}$$

and

$$kR = \frac{k_2}{k_1} \ln \frac{R_o}{R_i} \tag{8}$$

we obtain

$$(1 - S^2)(0.5 + kR) + S^2 \ln S = 2\tau \tag{9}$$

Note that X represents the interface position between the ice and water. $(R_i - X)$ represents the thickness of the ice formed inside the tube. Hence, $(R_i - X)/R_i = (1 - S)$ represents the dimensionless thickness of ice formed inside the tube. Fig. 2 shows the dimensionless thickness of the ice, $(1 - S)$, as a function of the dimensionless time, τ, with the dimensionless group kR as a parameter.

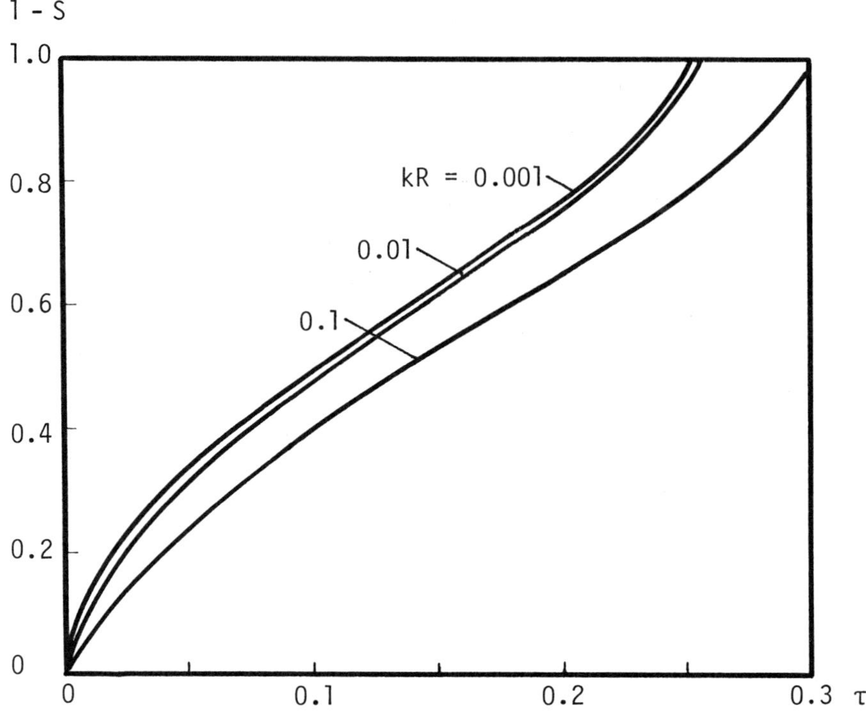

Fig.2. Dimensionless thickness of ice, (1 - S), as a function of dimensionless time, τ, with the dimensionless group kR as a parameter.

<u>Thermal Stress Analysis</u>

For the thermal stress analysis, we consider only the circumferential compressive stress in the ice, because it is much higher than the radial and axial stresses. The circumferential compressive stress can be expressed by[5]

$$\sigma_{tt2} = \frac{E_2}{1 + \nu_2} \left[\frac{1 - \nu_2}{1 - 2\nu_2} \varepsilon_{tt2} + \frac{\nu_2}{1 + 2\nu_2} (\varepsilon_{rr2} + \varepsilon_{zz2}) \right] - \frac{\alpha_2^* E_2}{1 - 2\nu_2} \left(\Delta T_2 + \frac{\beta}{\alpha_2^*} \right)$$

(10)

where

$$\Delta T_2 = T_2 - T_f$$

(11)

240

$$\varepsilon_{tt2} = \frac{u_{r2}}{r} \tag{12}$$

$$\varepsilon_{rr2} = \frac{\partial u_{r2}}{\partial r} \tag{13}$$

$$\varepsilon_{zz2} = \frac{\partial u_{z2}}{\partial z} = C = \text{constant} \tag{14}$$

and

$$u_{r2} = \frac{1 + \nu_2}{1 - \nu_2} \frac{\alpha_2^*}{r} \int_X^r (T_i + \frac{\beta}{\alpha_2^*}) \, r\,dr + C_1 r + \frac{C_2}{r} \tag{15}$$

The maximum circumferential compressive stress is located at the interface position, $r = X$, where $\Delta T_2 = 0$. Under this condition, Eqs. (15), (12) and (13) become,

$$u_{r2} = C_1 X + \frac{C_2}{X} \tag{16}$$

$$\varepsilon_{tt2} = C_1 + \frac{C_2}{X^2} \tag{17}$$

$$\varepsilon_{rr2} = \frac{1 + \nu_2}{1 - \nu_2} \alpha_2^* (T_f + \frac{\beta}{\alpha_2^*}) + C_1 - \frac{C_2}{X^2} \tag{18}$$

Substituting Eqs. (14), (17) and (18) into Eq. (10) gives

$$\sigma_{tt2,\, r=X} = K_1 - \frac{K_2}{X^2} \tag{19}$$

where

$$K_1 = E_2 \left[\frac{C_1 + \nu_2 C + (\frac{1 + \nu_2}{1 - \nu_2}) \nu_2 \alpha_2^* T_f}{(1 - 2\nu_2)(1 + \nu_2)} - \frac{\beta}{1 - \nu_2} \right] \tag{20}$$

$$K_2 = \frac{E_2 C_2}{1 + v_2} \; . \tag{21}$$

The values of the constants K_1 and K_2 are dependent on the material properties of the ice and the boundary conditions of the problem. The constant K_2 is always positive.

For single ice crystals, a stress-strain relation was given by Higashi[6]. The characteristic of the relation is that the yield stress of the ice increases with an increase of the strain rate. The rate of the maximum circumferential compressive stress of the ice inside the tube can be obtained by differentiating Eq. (12) with respect to time, t,

$$\dot{\varepsilon}_{tt2,\, r=X} = \left. \frac{d\varepsilon_{tt2}}{dt} \right|_{r=X} = \frac{d}{dt}\left(\frac{u_{r2}}{r}\right)_{r=X} = -\frac{2C_2}{X^3}\frac{dX}{dt} \; . \tag{22}$$

Substituting Eq. (5) into Eq. (22) yields

$$\dot{\varepsilon}_{tt2,\, r=X} = \frac{2C_2(T_f - T_o)}{X^4\, h_{LS}\, \rho_2 \left[\dfrac{ln(R_i/X)}{k_2} + \dfrac{ln(R_o/R_i)}{k_1}\right]} \tag{23}$$

where C_2 is a positive constant.

DISCUSSION AND CONCLUSION

For the one-dimensional freezing process, the change of the dimensionless thickness of the ice, $(1 - S)$, formed inside the tube, is a function of the dimensionless time, τ, with the dimensionless group kR as a parameter, as shown in Fig. 2. The parameter kR represents the dimensionless thermal resistance in the tube wall, as shown in Fig. 1. Hence, it is expected that the smaller the value of the thermal resistance, kR, the faster the growth of the ice.

From the literature, the crush strength of the ice varies in a range from 4×10^5 to 130×10^5 Pa[7]. When the maximum circumferential compressive stress inside the ice induced by the freezing process reaches a value lying in the above range, the ice may crush.

From Fig. 2 it can be seen that the dimensionless ice thickness $(1 - S = 1 - X/R_i)$ increases (or the interface position, X, decreases) with an increase of the dimensionless time, τ. For water, the freezing temperature, T_f, is equal to zero at atmospheric pressure. The temperature at the outside surface of the tube, T_o, has a negative value. From the definition, Eq. (7), for a fixed time t, the dimensionless time, τ, increases with a decrease of the surface temperature, T_o. Therefore a decrease of the surface temperature, T_0, results in an increase of the ice thickness (1 - X/R_i) or a decrease of the interface position, X, as shown in Fig. 2. The decreases of the values of T_0 and X result in a significant increase of the strain rate, as shown in Eq. (23), because the strain rate is proportional to the surface temperature, T_0, and nearly inversely proportional to the fourth power of the interface position, X. For such a case, the yield stress of the ice increases correspondingly. On the other hand, the reduction of the value of X gives an increase of the maximum circumferential compressive stress, as shown in Eq. (19). When the circumferential stress reaches the crush strength of the ice, the ice crushes.

From the above discussion, it can be concluded that a decrease of the temperature at the outside surface of the tube, T_0, results in an increase of the maximum thermal stress in the ice. Therefore, for freezing perservation, a slow freezing process having a relatively high surface temperature, T_0, will have the benefit of reducing the cryoinjury of biological cells caused by the thermal stress.

REFERENCES

1. H.T. Meryman, R.J. Williams, and M.St. J. Douglas, Freezing injury from solution effects and its prevention by natural and artificial cryoprotection, <u>Cryobiology</u>, 14:287-302 (1977).
2. P. Mazur, The role of intracellular freezing in the death of cells cooled at supraoptimal rates, <u>Cryobiology</u>, 14:251-272 (1977).
3. D.E. Pegg, M.P. Diaper, H.LEB. Skaer and C.J. Hunt, The effect of cooling and warming rate on the packing effect in human erythrocytes frozen and thawed in the presence of 2M gylcerol, <u>Cryobiology</u>, 21:491-502 (1984).
4. B. Rubinsky, E.G. Cravalho, and B. Mikic, Thermal stresses in frozen organs, <u>Cryobiology</u>, 17:66-73 (1980).
5. S. Lin and D.Y. Gao, Thermal stresses related to cryoinjury of biological cells in freezing preservation, <u>in</u>: "Heat Transfer with Phase Change," I.S. Habib and R.J. Dallman, ed., HTD-Vol. 114, ASME, New York (1989).
6. A. Higashi, Mechanical properties of ice single crystals, <u>in</u> "Physics of Ice," N. Riehl, B. Bullemer and H. Engelhardt, ed., Plenum Press, New York (1969).

ROLE OF MEMBRANE PHOSPHOLIPIDS IN NONFREEZING COLD INJURY

Dipak K. Das, Swapna Maity and Dan Lu

Surgical Research Center, Department of Surgery
Cardiovascular Division
University of Connecticut School of Medicine
Farmington, Connecticut 06030 U.S.A.

INTRODUCTION

Nonfreezing cold injury (NFCI) represents a potential threat
to infantry and marine operations carried out in inclement weather
conditions.[1,2] The term "NFCI" is used to identify the syndrome that
results from damage to tissues that have been cooled, usually for
prolonged periods, at temperatures between about 288°K (15°C) and
their freezing point (272.5°K) (-0.5°C). NFCI reduces man's
mobility at the time, but through cold sensitization it may
compromise his ability to fight under similar conditions in the
future.

During recent years, increased incidences of cold injury have
been encountered among civilians.[3,4] In addition to accident victims
who might have been exposed to cold for several hours for various
reasons, people without adequate clothing and shelter or people
involved in winter sports are frequently subjected to cold injury.
Unfortunately, the pathogenesis of cold injury still remains unknown
and, as a result, it remains a crippling problem.

Recent studies from our laboratory have indicated that cold
injury is associated with an ischemic episode, as evidenced by the
reduction of blood flow.[5] Upon rewarming, ischemia is relieved and
revascularization occurs. Our results further indicated that
rewarming of cooled tissue was associated with the generation of
oxygen-derived free radicals, causing additional tissue injury
similar to that occurring during reperfusion of ischemic tissue.[5-9]

It is known that free radicals can attack lysosomal membrane releasing phospholipases, and activation of Ca^{2+}-dependent phospholipases has been suggested as one of the mechanisms by which free radicals potentiate the breakdown of membrane phospholipids during reperfusion of ischemic tissue.[10] The breakdown of phospholipids is likely to cause an accumulation of nonesterified fatty acids (NEFA). Free radicals can also directly attack polyunsaturated fatty acids (PUFA) of membrane phospholipids, causing lipid peroxidation which can further promote additional tissue injury.[11,12] In this study, the fatty acid profiles of membrane phospholipids were examined during NFCI induced in a rabbit leg.

MATERIALS AND METHODS

New Zealand white rabbits of about 2.5 kg body weight were anesthetized with xylazine (5 mg/kg) and ketamine (30 mg/kg) and were maintained under anesthesia during the entire experiment. An electric clipper was used to remove hair from the hind limb as well as from the front of the neck area. A tracheostomy was performed, and the rabbits were ventilated by a Harvard ventilator (15 ml volume and 50 strokes/min). Femoral arteries and veins of both sides were exposed and dissected free of tissue. A flow probe (Transonic, Ithaca, NY) was placed around the femoral artery in the leg to be cooled, which in turn was connected to a six-channel simultrace chart recorder (Honeywell Inc., Pleasantville, NY). The femoral vein of the same side was cannulated with an I.V. placement catheter for withdrawal of blood samples. Another such catheter was placed in the femoral artery of the other limb. A continuous display of electrocardiogram (EKG) was obtained on Lead II by connecting the limb leads to the same recorder. Baseline values were established by measuring EKG and blood flow, as well as by estimating the arteriovenous difference of creatine kinase (CK), lactate dehydrogenase (LDH), and malonaldehyde (MDA) from the blood samples. One of the legs was then perfused with either 10 μM quinacrine (experimental) or saline (control) for 10 min and then cooled down to 275°K (2°C) with a freezing mixture containing ice and salt. Continuous monitoring of the interstitial temperature was achieved by inserting a thermocouple probe (Omega Engineering, Inc., Stamford, CT) into the limb. Tissue temperature was maintained for 20 min at 275°K (2°C). The ice was then removed and the limb allowed to rewarm to room temperature. During the experiment, blood samples were withdrawn at regular intervals of time from both the vein and artery for the subsequent assay of CK, LDH, and MDA formation. Blood flow was also continuously monitored. At the end of the experiment, the rabbits were sacrificed by an overdose of sodium pentobarbital. Tissue biopsies were obtained from the leg.

Lipids were extracted with chloroform-methanol mixture by the method of Folch et al.[13] Phospholipids, except for lysophosphatidylcholine (LPC), were separated on silica K6 plates (Whatman, Clifton, NJ) using a mixture of chloroform-methanol-petroleum ether-acetic acid-boric acid (40:20:30:10:1.8, vol/vol/vol/vol/wt) as a solvent system.[14] LPC was separated on silica gel H plates (Analtech, Newark, DE) using a mixture of chloroform-methanol-acetic acid-water (75:25:3:4, vol/vol/vol/vol) as a solvent system.[15] The phospholipids on the silica gel plates were identified by cochromatography with authentic phospholipid standards after a brief exposure to iodine vapor, scraped off, and quantitated by the method of Bartlett.[15]

Lipids were extracted as described above after including 20 nmol of heptadecanoic acid (17:0) as the internal standard. Lipids were subjected to silica gel G thin layer chromatography (TLC) using hexane:ether:acetic acid (80:20:1, v/v/v) containing 25 mg butylated hydroxytoluene/100 ml as solvent system to separate FFA. The spots corresponding to NEFA were scraped off and converted to methyl esters as described elsewhere.[16] The methyl esters of fatty acids were separated by gas chromatography using a gas chromatograph (model HP5890A, Hewlett Packard) equipped with a 5% diethylene glycol succinate-phosphoric acid-coated column.[16] An initial temperature of 393°K (120°C) was used for 2 min, followed by two successive temperature programs: (i) a linear increase of 393-453°K (120-180°C) at a rate of 283°K/min (10°C/min), and a 25 min period at 453°K (180°C); and (ii) a second linear temperature gradient of 453-493°K (180-220°C) at a rate of 278°K/min (5°C/min), and a 10 min period at 493°K (220°C). The fatty acids were identified by their equivalent chain lengths. The quantitation of 20:4 and other tissue fatty acids was performed on the basis of 17:0 (10 nmol), which was added during the extraction of lipids.

Malonaldehyde was measured as an index for lipid peroxidation. Plasma (0.5 ml) was added to 0.5 ml ice-cold perchloric acid (15%) and then treated with 0.75% thiobarbituric acid (TBA) as described previously.[17] Samples were boiled for 20 min and centrifuged to remove the pellet. The color of the supernatant was read at 535 nm. The concentration of MDA (nmol/ml) was calculated by using a molar extinction coefficient of 156 mM^{-1} cm^{-1}. CK and LDH were assayed in plasma samples obtained from the femoral artery and vein using an assay kit obtained from Sigma Chemical Company (St. Louis, MO) as described elsewhere.[18]

All measurements are expressed as mean values ± standard error of mean (SEM). Student's t-test or two-tailed t-test was used for comparison of the data between two groups or within each group for each variable. For multigroup comparisons, analysis of variance was

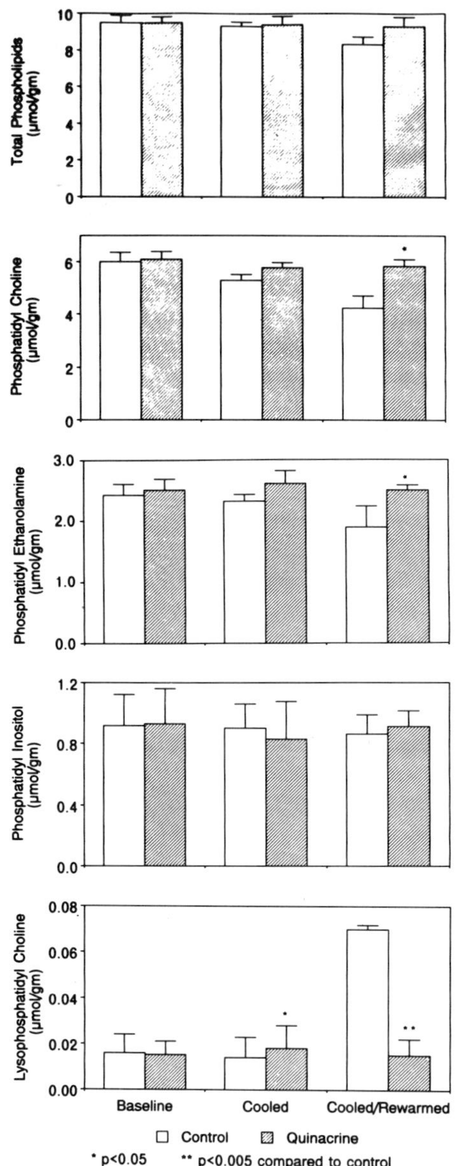

FIGURE 1.   Effects of quinacrine on cooling and rewarming on the phospholipid content on rabbit leg.   Results are expressed as mean ± SEM of 6 rabbits in each group. $^{*}p < 0.05$ and $^{**}p < 0.005$ (compared to untreated control).

used followed by Scheffe's test.   Differences were considered significant when the *p* value was less than 0.05.

RESULTS

To examine whether phospholipid contents of the leg are changed after cooling, total as well as individual phospholipids were measured in the leg which was subjected to cooling and rewarming. No significant differences in any individual or total phospholipid content were seen in any tissues during cooling, except for phosphatidylethanolamine (PE), which was lowered slightly but not significantly (Figure 1). After rewarming, however, all phospholipids except phosphatidylinositol declined compared to baseline values ($p < 0.05$). Quinacrine significantly inhibited the depletion of these phospholipids. Breakdown of phosphatidylcholine (PC) was also accompanied by the accumulation of a significant amount of LPC (Figure 1). Quinacrine was able to reduce LPC accumulation in the leg during cooling and rewarming.

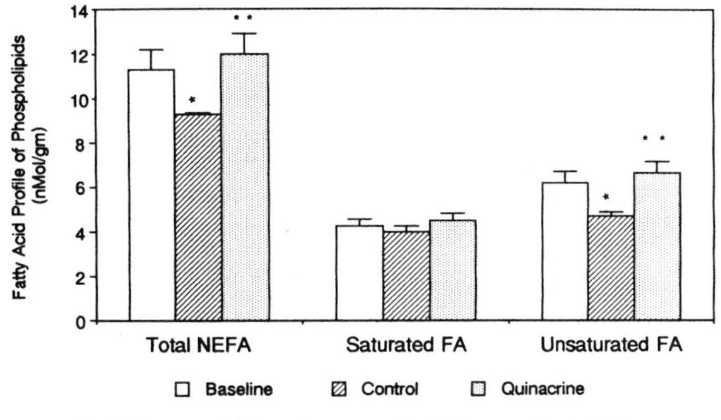

FIGURE 2.   Effects of quinacrine on cooling and rewarming on the total NEFA as well as saturated FA and unsaturated FA content of rabbit leg. Baseline biopsies of legs were obtained before cooling, whereas control and quinacrine biopsies were obtained from the legs which had been subjected to cooling and rewarming. *$p < 0.05$ (compared to baseline), **$p < 0.005$ (compared to untreated control).

249

Since free radicals are known to attack PUFA of membrane phospholipids, we analyzed the fatty acid profile of the total phospholipids. As shown in Figure 2, total NEFA was decreased significantly after cooling and rewarming compared to baseline values. Of these NEFA, loss of unsaturated fatty acid was most significant. Saturated fatty acids did not suffer any loss. Quinacrine was able to preserve the unsaturated fatty acid in the phospholipids.

We also examined the fatty acid composition from 16:0 to 22:6 of the membrane phospholipids. Most of the fatty acids were reduced in the membrane phospholipids, although the reduction of oleate linoleate, arachidonate, and decosahexenoiate was significant (Figure 3). Again, quinacrine preserved these polyunsaturated fatty acids in the membrane phospholipids.

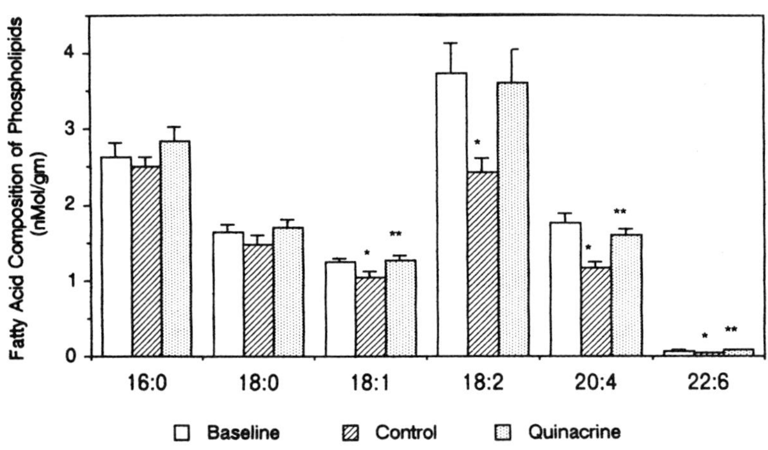

FIGURE 3.   Effects of quinacrine on cooling and rewarming on the fatty acid composition of phospholipids. Baseline biopsies of legs were obtained before cooling, whereas control and quinacrine biopsies were obtained from the legs which had been subjected to cooling and rewarming. $^{*}p < 0.05$ (compared to baseline), $^{**}p < 0.005$ (compared to untreated control).

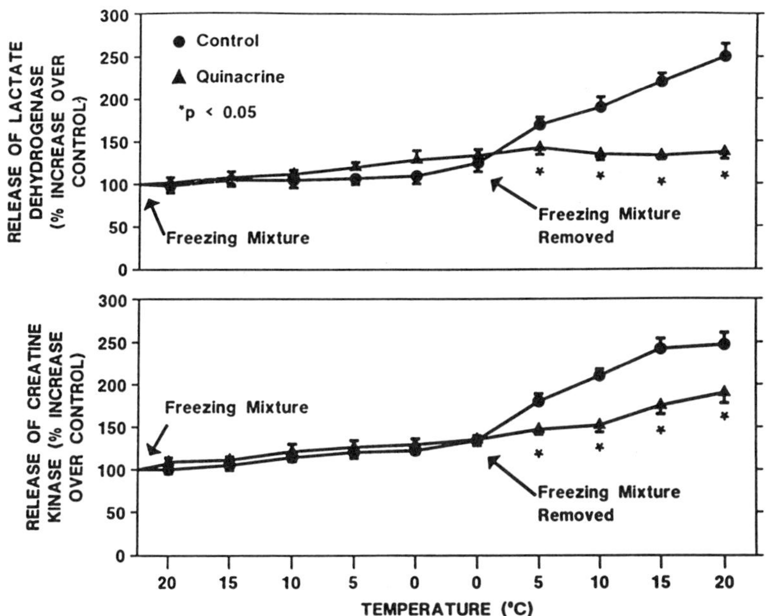

FIGURE 4. Effects of quinacrine on the changes in the level of lactate dehydrogenase (top) and creatine kinase (bottom) in plasma obtained from the femoral artery of rabbit leg during cooling and rewarming. Each point represents the average ± SD of six separate experiments in each group (0 - 0) control; (0 - 0) quinacrine.
*$p < 0.05$ (compared to untreated control).

Cellular injury during cooling and rewarming was monitored by estimating the release of LDH and CK from the tissue. The release of arterial plasma LDH is plotted against temperature during cooling and rewarming of the rabbit leg (Figure 4, top). There was no change in plasma LDH levels during cooling from 303 to 288°K (30°C to 15°C), but below that temperature there was a slight increase in LDH release. A remarkable increase in plasma LDH (about 2-fold) was noted at the end of rewarming, suggesting that tissue damage occurred mostly during the rewarming period. CK, another marker for tissue necrosis, followed a similar pattern (Figure 4, bottom). After cooling, a slight increase in CK activity was noticed, but the differences were not statistically significant. During rewarming, however, CK increased dramatically. At the end of reperfusion, these values were 2.5-fold higher compared to the baseline levels. Treatment with quinacrine reduced the release of both LDH and CK significantly, suggesting protection of the tissue from cold injury.

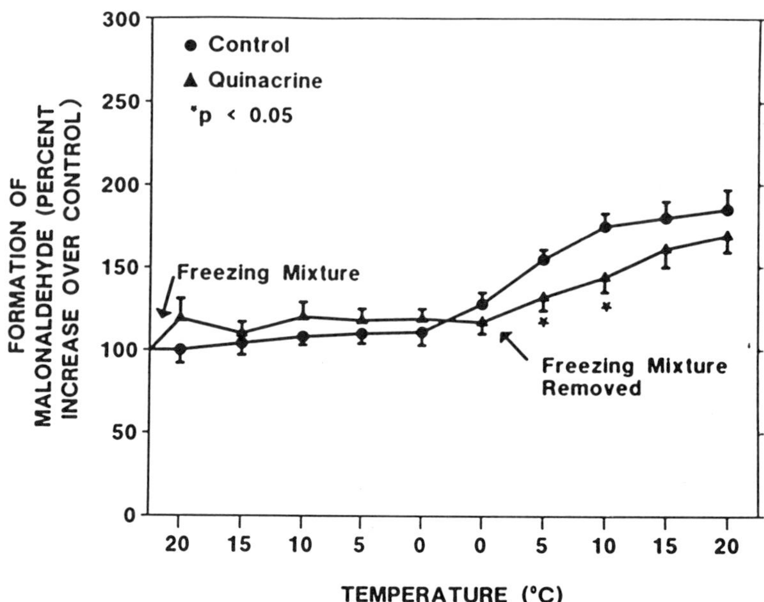

FIGURE 5.   Effects of quinacrine on the changes in the level of malonaldehyde in plasma obtained from the femoral artery of a rabbit leg during cooling and rewarming. Each point represents the average ± SD of six separate experiments in each group. (0 - 0) control; (0 - 0) quinacrine. *p < 0.05 (compared to untreated control).

Malonaldehyde formation, a presumptive marker for lipid peroxidation, remained unchanged during cooling (Figure 5). During rewarming, MDA formation increased significantly and reached a 1.4-fold higher value compared to the control at the end of the rewarming phase. This suggests that lipid peroxidation occurred only during rewarming of the cooled tissue. Quinacrine inhibited the formation of MDA during the rewarming phase.

DISCUSSION

Severe hypothermia occasionally gives rise to tissue injury resulting from vascular stasis.[19,20] Prolonged exposure to cold with or without wet conditions is associated with venous stagnation for multifactorial causes. Increased blood viscosity due to hypothermia in conjunction with hemoconcentration may lead to abnormality in the microvascular system.[21] In addition, during NFCI, the hemoglobin dissociation curve is shifted to the left, thereby preventing release of O_2 to the tissue. This combination of poor perfusion of capillaries and decreased ability to release O_2 induces severe ischemic insult to the affected tissues despite the fact that hypothermia reduces the rate of cellular metabolism significantly. Almost any part of the body, including nerve fibers, muscle, skin, soft tissue and even bone, can be injured. Blood vessel injury is severe and is manifested by severe spasm of small arteries and arterioles. It also increases permeability or leaking of the capillary bed on rewarming of the frozen part. Poor perfusion, anoxia, and flow through cold-injured vessels cause aggregation of red cells, platelets, and polymorphonuclear leukocytes (PMN), leading to thrombosis with tissue infarction (gangrene).

We recently demonstrated that cold exposure is accompanied by severe ischemic insult.[5] Most interestingly, during rewarming the cooled tissue is subjected to additional tissue injury similar to that associated with reperfusion injury.[6-9] The presence of oxygen-derived free radicals was indicated in the cooled-rewarmed tissue, which is believed to play a significant role in the "rewarming injury."

Free radicals, if generated, presumably attack the phospholipid membranes and the lysosomal membranes, causing phospholipid breakdown and release of phospholipases, respectively.[12,22] Phospholipases, especially phospholipase A_2 (PLA_2) and phospholipase C (PLC), can hydrolyze membrane phospholipids.[23] In earlier studies, release of lysosomal enzymes was noticed in hypothermically perfused dog kidneys.[24] Hypothermic perfusion also reduced PC and PE levels in the kidney after 24 h when perfused at 279-283°K (6-10°C).[25] The results of our study indicated that a significant amount of phospholipids, especially PC and PE, were degraded during rewarming of the cooled leg. Breakdown of PC was also associated with the accumulation of LPC. Similar observations were made during the reperfusion of ischemic tissue,[12,23] supporting our previous reports concerning "rewarming injury."[26-28]

Rewarming of the cooled rabbit leg was also accompanied by lipid peroxidation as judged by the increased production of MDA. As mentioned earlier, lipid peroxidation may occur by the direct interaction of lipids and free radicals. Increased lipid peroxidation thus suggests the presence of free radicals, as reported previously[5] and their interaction with PUFA resulting from the breakdown of membrane phospholipids.

In an attempt to inhibit phospholipases and preserve membrane phospholipids, we preperfused the leg with a phospholipase inhibitor. Quinacrine, a known inhibitor of both PLA_2 and PLC,[29] was

able to prevent phospholipid breakdown significantly, as evidenced
by the preservation of PC and PE in the cooled-rewarmed leg.
Prevention of PC breakdown was also reflected in the reduction in
LPC production and NEFA accumulation. LPC is presumably formed from
PC by the action of PLA_2.[12] Accumulated LPC is then hydrolyzed to
form NEFA. Thus, these results indicate the efficacy of quinacrine
in preserving membrane phospholipids during rewarming injury.

In addition to preserving membrane phospholipids, quinacrine
also ameliorated tissue injury occurring during cooling and
rewarming. Release of LDH and CK, presumptive markers of cell death
and tissue necrosis,[18] was reduced by quinacrine treatment. It is
interesting to note that enzyme release did not occur during
cooling, but rather was enhanced during the rewarming phase,
indicating additional tissue injury during rewarming of the cooled
leg. Lipid peroxidation followed a similar pattern. Formation of
MDA increased only during rewarming. Quinacrine reduced MDA
formation significantly.

Lysophosphoglycerides and NEFA are known to affect a number of
intracellular enzymes and membrane functions by their detergent-
like properties.[30,31] It is reasonable to assume that accumulation
of LPC and NEFA would cause tissue injury during cooling and
rewarming. Quinacrine, in addition to inhibiting PLA_2 and PLC, can
bind to the membrane bilayers and modulate lipid packing.[32] All of
these properties of quinacrine may make it a suitable therapeutic
agent for the treatment of cold injury.

The accumulation of NEFA, especially arachidonic acid as seen
in our study, may have further consequences. Arachidonic acid is
a substrate for the cyclooxygenase and lipooxygenase pathways and
may lead to the formation of prostaglandins, thromboxanes, and
leukotrienes. It has been shown that the substances that block the
production of prostaglandins and thromboxanes may be beneficial in
preserving tissue in a frostbite injury.[33] Blister fluid from human
frostbite victims has been shown to contain high levels of throm-
boxane B_2 and prostaglandin $F_{2\alpha}$.[34] The primary step in prostaglandin
synthesis is the release of arachidonic acid from membrane phospholipids. The release of arachidonic acid is directly related to the
activation of phospholipases that can act on various phospholipids.
Thus, these studies in conjunction with our present study clearly
indicate a role of phospholipids in cold injury.

ACKNOWLEDGMENTS

This study was supported in part by a grant from the United
States Department of the Navy. The secretarial assistance of
Mrs. J. D'Aprile is gratefully acknowledged.

LIST OF ABBREVIATIONS AND PNEUMONICS

CK   - creatine kinase
EKG   - electrocardiogram
FA   - fatty acid
FFA   - free fatty acid
LDH   - lactate dehydrogenase
LPC   - lysophosphatidylcholine
MDA   - malonaldehyde
NEFA - nonessential fatty acid
NFCI - nonfreezing cold injury
PC   - phosphatidylcholine
PE   - phosphatidylethanolamine
PI   - phosphatidylinositol
PLA_2 - phospholipase A_2
PLC   - phospholipase C
PMN   - polymorphonuclear leukocytes
PUFA - polyunsaturated fatty acid
SEM   - standard error of mean
TLC   - thin layer chromatography

REFERENCES

1.   H. E. Hanson and R. F. Goldman, Cold injury in man: a review
     of its etiology and discussion of its prediction,
     Milit. Med. 134:1307 (1969).
2.   N. G. Kirby and G. Blackburn, Cold injury, in: "Field Surgery
     Pocketbook," Her Majesty's Stationary Office, London
     (1981).
3.   C. Bangs and M. P. Hamlet, Hypothermia and cold injuries, in:
     "Management of Wilderness and Environmental Emergen-
     cies," P. S. Auerbach and E. C. Geehr, eds., MacMillan
     Publishing Co., New York (1983).
4.   G. F. Purdue and J. L. Hunt, Cold injury: a collective
     review, J. Burn Care Rehabil. 7:331 (1986).
5.   J. Iyengar, A. George, J. Russell and D. K. Das, Generation
     of free radicals during cold injury and rewarming,
     Vasc. Surg. 24:467 (1990).
6.   D. K. Das and R. M. Engelman, Mechanism of free radical
     generation during reperfusion of ischemic myocardium,
     in: "Oxygen Radicals: Systemic Events and Disease
     Process," D. K. Das and W. B. Esmann, eds., S. Karger,
     Basel (1989).
7.   D. K. Das, R. M. Engelman, J. A. Rousou, R. H. Breyer, H.
     Otani and S. Lemeshow, Pathophysiology of superoxide
     radical as potential mediator of reperfusion injury in
     pig heart, Basic Res. Cardiol. 81:155 (1986).
8.   H. Otani, M. R. Prasad, R. M. Jones and D. K. Das, Mechanism
     of membrane phospholipid degradation in ischemic
     reperfused rat hearts, Am. J. Physiol. 257:H252 (1989).

9. D. J. Hearse, Reperfusion of the ischemic myocardium, *J. Mol. Cell. Cardiol.* 9:605 (1977).

10. R. B. Jennings and K. A. Reimer, Lethal myocardial ischemic injury, *Am. J. Pathol.* 102:241 (1981).

11. K. R. Chien, A. Han, A. Sen, L. M. Buja and J. T. Willerson, Accumulation of unesterified arachidonic acid in ischemic canine myocardium, *Circ. Res.* 54:313 (1984).

12. D. K. Das, R. M. Engelman, J. A. Rousou, R. H. Breyer, H. Otani and S. Lemeshow, Role of membrane phospholipids in myocardial injury induced by ischemia and reperfusion, *Am. J. Physiol.* 251:H71 (1986).

13. J. Folch, M. Lees and G. H. Sloan-Stanley, A simple method for the isolation and purification of total lipids from animal tissues, *J. Biol. Chem.* 226:497 (1957).

14. A. M. Gilfillan, A. J. Chu, D. A. Smart and S. A. Rooney, Single plate separation of lung phospholipids including disaturated phosphatidylcholine, *J. Lipid Res.* 24:1651 (1983).

15. G. R. Bartlett, Phosphorus assay in column chromatography, *J. Biol. Chem.* 234:466 (1959).

16. R. Prasad, R. M. Jones, H. S. Young, L. B. Kaplinsky and D. K. Das, Analysis of tissue free fatty acid isolated by aminopropyl bond-phase columns, *J. Chromatogr.* 428:221 (1988).

17. D. K. Das, R. M. Engelman, D. Flansaas, H. Otani, J. Rousou and R. H. Breyer, Development profiles of protective mechanisms of heart against peroxidative injury, *Basic Res. Cardiol.* 82:36 (1987).

18. D. K. Das, D. Flansaas, R. M. Engelman, J. A. Rousou, R. H. Breyer, R. Jones, S. Lemeshow and H. Otani, Age-related development profiles of the antioxidative defense system and the peroxidative status of the pig heart, *Biol. Neonate* 51:156 (1987).

19. B. Albrektsson and P. I. Branemark, Early microvascular reactions to slow and rapid thawing of frozen tissue, *Adv. Microcirc.* 2:37 (1969).

20. D. Kettelkamp, M. Walker and P. Ramsey, Radioactive albumin, red blood cells and sodium as indicators of tissue damage after frostbite, *Cryobiology* 8:79 (1971).

21. J. P. Kulka, Cold injury of the skin: the pathogenic role of microcirculatory impairment, *Arch. Environ. Health* 11:484 (1965).

22. M. R. Prasad and D. K. Das, Effect of oxygen derived free radicals and oxidants on the degradation *in vitro* of membrane phospholipids, *Free Radic. Res. Commun.* 7:381 (1989).

23. H. Otani, R. Prasad, R. M. Jones and D. K. Das, Mechanism of membrane phospholipid degradation in ischemic reperfused rat heart, *Am. J. Physiol.* 257:H252 (1989).

24. G. S. Pavlock, J. H. Southard, J. R. Starling and F. O. Belzer, Lysosomal enzyme release in hypothermically perfused dog kidneys, _Cryobiology_ 21:521 (1984).

25. J. H. Southard, M. S. Ametani, M. F. Lutz and F. O. Belzer, Effects of hypothermic perfusion of kidneys on tissue and mitochondrial phospholipids, _Cryobiology_ 21:20 (1984).

26. D. K. Das, J. C. Russell and R. M. Jones, Reduction of cold injury by superoxide dismutase and catalase, _Free Radic. Res. Commun_. (in press).

27. J. Iyengar, A. George, J. C. Russell and D. K. Das, The effects of an iron chelator on vascular stasis-induced cellular injury evoked by hypothermia, _J. Vasc. Surg_. (in press).

28. D. K. Das, J. Iyengar, R. M. Jones, D. Lu, S. Maity, Protection from nonfreezing cold injury by quinacrine, a phospholipase inhibitor, _Cryobiology_ (in press).

29. H. Kunze, B. Hesse and E. Bohn, Effects of antimalarial drugs on several rat liver lysosomal enzymes involved in phosphatidylethanolamine catabolism, _Biochim. Biophys_. _Acta_ 713:112 (1982).

30. P. B. Corr, R. W. Gross and B. E. Sobel, Amphipathic metabolites and membrane dysfunction in ischemic myocardium, _Circ. Res_. 55:135 (1984).

31. A. M. Katz and F. C. Messineo, Lipid membrane interactions and the pathogenesis of ischemic damage in the myocardium, _Circ. Res_. 48:1 (1981).

32. A. I. Tauber and E. R. Simons, Dissociation of human neutrophil membrane depolarization respiratory burst stimulation, and phospholipid metabolism by quinacrine, _FEBS Lett_. 156:161 (1983).

33. T. J. Raine, M. D. London, L. Goluch, J. P. Heggers and M. C. Robson, Antiprostaglandins and antithromboxanes for treatment of frostbite, _Surg. Forum_ 31:557-559 (1980).

34. M. C. Robson and J. P. Heggers, Evaluation of hand frostbite blister fluid as a clue to pathogenesis. _J. Hand Surg_. 6:43-47 (1981).

SOME CRYOSURGERY-RELATED HEAT TRANSFER PROBLEMS

Peter Hrycak, Martin J. Linden and
Sidney A. Wilchins*

N.J. Institute of Technology
King Blvd., Newark, N.J. 07102
*N.J. Med. School, Univ. of Med. and Dent. of N.J.

ABSTRACT

Cryosurgery represents physically cooling a small
area (in a direct contact with a probe) on a large,
semi-infinite slab, representing here the "body of the
patient." Although done often on an empirical basis,
the process of cryocooling can be handled in a more
rational fashion, according to the accepted principles
of heat transfer. Thus, the greatest extent of cooling
that occurs at the center of the cooled area, is calculated here from the solution of the appropriate boundary-value problems. Similarly, the times required for
the maximum penetration of the freezing temperature are
estimated, with the help of Neumann's solution for
freezing of a semi-infinite slab.

The cooling process may be affected by the internal geometry of the "body." Therefore, the influence
of a major blood vessel that may exist in the neighborhood of the cooled area is considered with its possible adverse effect on the cooling process and the intended beneficial result of the cryosurgical manipu-
lation.

INTRODUCTION

Cryosurgery has been practiced for some time on a
semi-empirical basis, in the treatment of certain types
of benign and malignant lesions. It is indeed inevit-
able that some models of the physical sequences of
events, as they occur in the typical cryosurgical
treatment, would be discussed in a more formal manner.
This sequence of events is namely (a) original cryosurgical temperature penetration of the lesion to the
required depth, (b) estimate of the time required for
the penetration, as determined by the size and geometry
of the cryoprobe, and (d) further effects of the
geometry of the lesion, like the existence, for example, of major blood vessels in the neighborhood of
the lesion, on the treatment.

All the items listed under (a) to (d) above can be
treated mathematically, based on the fundamental principles of heat transfer to a semi-infinite slab[1,2,3]
(representing the body of the subject) cooled at the
surface by the cryoprobe. For an effective treatment
of the problem, the thermophysical properties of the
tissues treated must also be known to some degree.[4,5]

ORIGINAL CRYOSURGICAL PENETRATION

The original cryosurgical penetration, Z, of the
freezing temperature into the subject's body is an
exact solution for the freezing of a homogeneous body,
semi-infinite in extent. The expression for Z,
originally determined by Neumann,[1] is as follows:

$$Z = \nu \, [2k_1(T_f-T_s)t/L]^{1/2} \tag{1}$$

where k_1 = the thermal conductivity of the body being
frozen, T_f-T_s = the temperature differential between
the freezing temperature and the temperature of the
cryoprobe, t = the total time of freezing, and L = the
latent heat for freezing of the tissues treated. The
effect of the sensible heat is expressed by the
coefficient , defined by the formula below

$$\nu = \{[1+a+b) - ((1+a+b)^2-4a)^{1/2}]/2a\}^{1/2} \tag{2}$$

where $a = 1 + \mu/2 + \alpha\mu$ and $b = (2/3)\alpha^2\mu/\delta^2$

The terms occurring in Eq. (2) have the following
physical significance: α = the ratio of the sensible
heat of the liquid zone to that of the frozen zone per

unit volume; μ = the ratio of the sensible heat of the frozen zone to the latent heat per unit volume, and δ^2 = the ratio of the thermal diffusivity of the frozen zone to that of the liquid zone.

Equation (1) determines the depth of freezing temperature penetration into the body of the subject treated. It has no steady state.

ESTIMATE OF THE MAXIMUM FREEZING TEMPERATURE
PENETRATION TIME

While there is no steady-state solution for the one-dimensional freezing process, three-dimensional freezing has a steady-state. The time required for maximum penetration of the freezing temperature, in practically significant geometries, may be estimated from a comparison between Z resulting from Eq. (1) above, and Z_∞ obtained from the steady-state solution for the given geometry, to be derived below. Assuming an exponential approach to steady-state equilibrium, a ratio is formed

$$(Z/Z_\infty)^2 = 1-e^{-t/N} \tag{3}$$

As $t \to \infty$, then also $Z \to Z_\infty$. In Eq. (3), for an approach where $(Z/Z_\infty) = 0.99$, we need a time span, say, equal to t_o. Therefore, by Eq. (3), $t_o/N = 3.91$ is required, whereas for small t (short time, frost penetration), Eq. (3) reduces to $(Z/Z_\infty)^2 = t/N$ with Z from Eq. (1). This yields an estimate of N:

$$N = Z_\infty^2 L/[2\mathcal{V}^2 k_1(T_f-T_s)] \tag{4}$$

On solving Eq. (4) for Z , and using the relation $t = 3.91\ N$, there results the formula for the maximum freezing temperature penetration in three-dimensional geometry

$$Z_\infty = \mathcal{V}[2k_1(T_f-T_s)t/3.91L]^{1/2} \tag{5}$$

By comparison with Eq. (1), it is seem that the time required to reach the depth of freezing equivalent up to 99% of Z_∞ is 3.91 times longer than the "normal" freezing time: $t_o = 3.91\ t$; Z_∞ is to be calculated separately from the solution of the boundary-value problem for the three-dimensional case of freezing temperature penetration.

EFFECT OF NON-HOMOGENEOUS THERMOPHYSICAL PROPERTIES

Known solutions of three-dimensional heat transfer problems in the steady-state, when a small area on the surface of a semi-infinite medium is cooled, normally do not distinguish between the physical properties of the frozen zone (subscript 1) and those of the liquid zone (subscript 2). From the inspection of the exact solution of the problem, we can conclude, however, that at the boundary between the liquid and the frozen zone, there applies the relationship

$$k_1\theta_1 f_1'(0)/f_1(0) - k_2\theta_2 f_2'(0)/(1-f_2(0)) = 0 \qquad (6)$$

where we let $T_f = 0$, $f_1(0)$ and $f_2(0)$ are dimensionless functions of Z_∞, and $\theta_1 = T_f-T_s$, $\theta_2 = T_\infty - T_f$. If $f_1(0) \to f_2(0)$ at the end of the freezing process, then also

$$f_1'(0) \to f_2'(0) \text{ and}$$

$$\lim_{Z \to Z_\infty} [f_1(0), f_2(0)] = f*(0).$$

We can therefore associate $f^*(0)$ with the solution of Eq.(6) and use the expression

$$f^*(0) = k_1(\theta_1)/(k_1\theta_1+k_2\theta_2) \qquad (7)$$

to show the effect of the thermal conductivities, k_1 and k_2, on the maximum depth of frost penetration into an infinite medium. It is seen that if k_1 (frozen region) is larger than k_2 (liquid region), $f^*(0)$ increases, whereas for $k_1/k_2 < 1$, a decrease for $f^*(0)$ would result. In itself, $f^*(0)$ is the maximum depth of freezing temperature penetration at the center of the refrigerated area located at the surface of a semi-infinite medium.

Equation (7) represents the situation at $Z = Z_\infty$ where the temperature ratio $k_1\theta_1/(k_1\theta_1+k_2\theta_2)$ applies exactly when the maximum depth of freezing has actually been reached. With the dimensionless temperature, defined as $\theta = (T-T_\infty)/(T_s-T_\infty)$, the freezing depth Z_∞ becomes, in the steady-state, and in the three-dimensional freezing problems, associated with the temperature

$$\theta_o = k_2\theta_2/(k_1\theta_1+k_2\theta_2). \qquad (7a)$$

The relationship described by Eqs. (7) and (7a) are suggested by the analysis of thawing of permafrost as example, carried out by Porkhayev.[6]

CALCULATION OF THREE-DIMENSIONAL EFFECTS OF FREEZING

Three-dimensional effects associated with the freezing temperature penetration can be described mathematically by a system of partial differential equations that have to satisfy certain conditions at the boundaries of the system, called also "solutions of boundary-value problems." The governing differential equations are linear,[1,2] see Equation (12) below. They represent (in the present case) the situation that results when a small area on the surface of a semi-infinite medium is cooled. "Small" in our case means, for example, a circle of a diameter that, say, represents (1/20) of the depth of the body (of the patient, etc.) to be cooled. This case may still be treated as semi-infinite slab.[2]

Several solutions are available,[1,7] as shown in Figure 1 and Figure 2, that are particularly useful for the present discussion. They differ according to the shape of the cooled area (circular, rectangular, or square) and the temperature, distribution over the area in question. In the simplest case, it is assumed that the dimensionless temperature $\theta = (T-T_\infty)/(T_S-T_\infty)$, is equal to 1.0 over the area, and is zero outside of the area. For a round area of radius R, the solution at the center of area is

$$\theta_0 = 1 - x/(R^2+x^2)^{1/2} \tag{8}$$

and at $r = R$

$$\theta = 0.5 - (2/\pi)^{-1}x[(x/2)^2+R^2]^{-1/2}.$$
$$K\{R[(x/2)^2+R^2]^{-1/2}\} \tag{8a}$$

where K here is the complete elliptical integral of the first kind. Using Eqs. (7) (8) and (8a), the temperature distribution under a disk of diameter 2R can be mapped, Fig. 1.

For a very intensive cooling (physically equivalent to having the area outside the cooled disk with a diameter 2R insulated), the solution for temperature distribution at the center is

$$\theta_0 = (2/\pi)\,\tan^{-1}(R/x) \tag{9}$$

This solution has been actually verified experimentally, for small x/r values, by Ruckli.[2] Similarly,

where x is the generalized distance away from the surface and from the ratio

$$\theta_e = (T_e - T_\infty)/(T_\infty - T_s) = (e \cdot x/M \cdot R)/(T_\infty - T_s) . \quad (14)$$

This equation represents the correction to the temperature distribution calculated by means of Eqs. (7) to (10). The corrected temperature may then be written as

$$\theta_c = \theta_o + \theta_e = k_2(\theta_2 + e \cdot x/M \cdot R)/(k_1\theta_1 + k_2\theta_2) \quad (15)$$

for $x < M \cdot R$. Equation (15) indicates that the extent of freezing at the center of the cooled zone is reduced, to some extent, due to the persistence of T_e given by Eq. (13), in maintaining its temperature in spite of the ongoing cooling process.

EXAMPLE PROBLEMS

Let's consider a hypothetical case of a lesion with a radius $R_e = 0.0254$ m, and a depth of 0.01 m at the center, treated with a cryosurgical contact probe of 0.01 m radius R_{pr}, at 213K. Here, we let $T_f = 273K$, $T_\infty = 310K$, $k_1 = 1.04$ w/mK, $k_2/k_1 = 0.5$, density $\gamma = 600$ kg/m^3, $C_1/\gamma = 2.51$ kJ/kgK, $C_2 = 2 \ C_1$, and $L/\gamma = 333.4$ kJ/kg. The parameters $\alpha = 1.233$, $\mu = 0.375$, and $\delta = 2$ are calculated based on $\theta_2 = 37$ K, $\theta_1 = 60K$. For the effect of the sensible heat, the parameter $\gamma = 0.53875$ is calculated by Eq. (2). This situation corresponds to θ_o, by Eq. (7), equal to $\theta_2/(2\theta_1 + \theta_2) = 0.23557$, that yields, based on Eq. (8), $Z_\infty/R_{pr} = 1.185$, and $Z_\infty = 0.012$ m. The time required for the depth to reach the bottom of the lesion, is

$$t = (0.012/0.539)^2 [333,400 \cdot 600 \cdot 3.91)/(2 \cdot 1.04 \cdot 60 \cdot 3600)$$

$$t = 0.863 \text{ hr}$$

or about 52 min. This analysis includes the three-dimensional effect estimate. Several applications of the probe will be required since the diameter of the lesion is larger than that of the probe.

The thermophysical constants here have been adapted from Barron.[5] For a deeper penetration, a lower temperature of the probe would be required.

In the previous example, the time required for reaching Z_∞ might seem excessive. It is understood to be only an estimate. Considering here the

264

for a parabolic temperature distribution over the disk, (D = 2R), the solution is

$$\theta_O = [(R^2+x^2)^{1/2} - x]^2/R^2 \tag{10}$$

Also of some practical significance is the solution where $0 = 1.0$ applies over an area a x b, and is zero outside of that area:

$$\theta_O = (2/\pi)\tan^{-1}\{ab/[2x(4x^2+a^2+b^2)^{1/2}]\} \tag{11}$$

This would present the maximum effect of cooling with a rectangular cryoprobe. Because of the way 0, the dimensionless temperature, is now defined, the corresponding term for the depth of the maximum freezing temperature, penetration here is

$$\theta_O = 1-f^*(0).$$

where $f^*(0)$ is given by Eq. (7) and, from Eq. (7a)

$$\theta_O = k_2\theta_2/(k_1\theta_1+k_2\theta_2)$$

Other thermal properties, those related to thermal diffusivity, do not appear explicitly in the governing differential equations for the steady-state,

$$\nabla^2 T = 0 \tag{12}$$

but effects of different thermal conductivities will show up in the boundary conditions. As 0_O represents the isotherm that separates the liquid zone from the frozen zone, it may be therefore expected that the right-hand side of Eq. (7a) be a function of k_1 and k_2.

EFFECT OF MAJOR BLOOD VESSELS

Major blood vessels carry blood at a temperature that is slightly higher than the temperature of the body (represented here by the semi-infinite slab). For example $T_e = T_\infty +e$. As a direct effect of circulation, this temperature will persist during the cryological cooling process as long as the blood supply is not interrupted, at location $M \cdot R$ below the surface of the body. M is of the order of 10, and we can write, for the linear effect of the blood vessel on the temperature near the surface of the body,

$$T_e = T_\infty + e \cdot x/M \cdot R \tag{13}$$

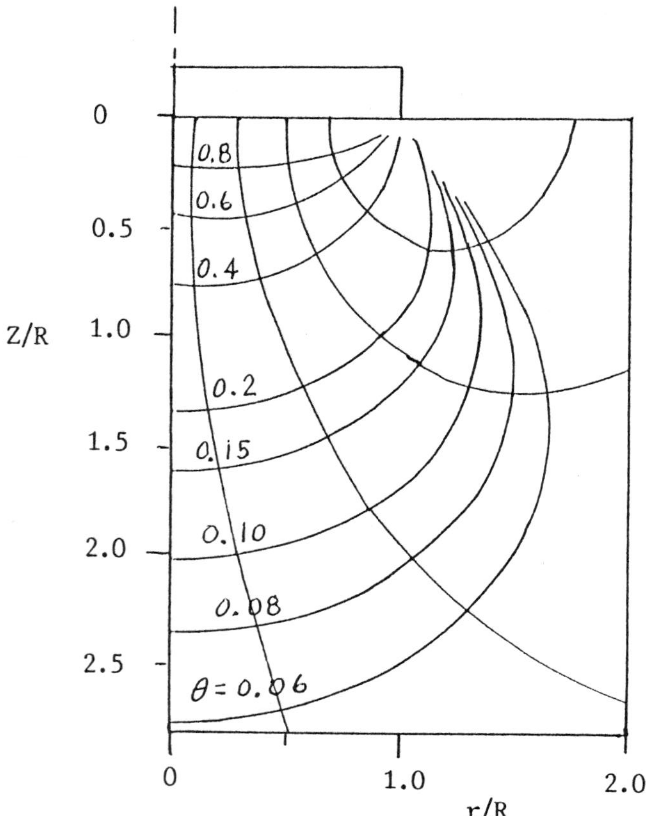

Fig. 1 Geometry of Three-Dimensional Freezing

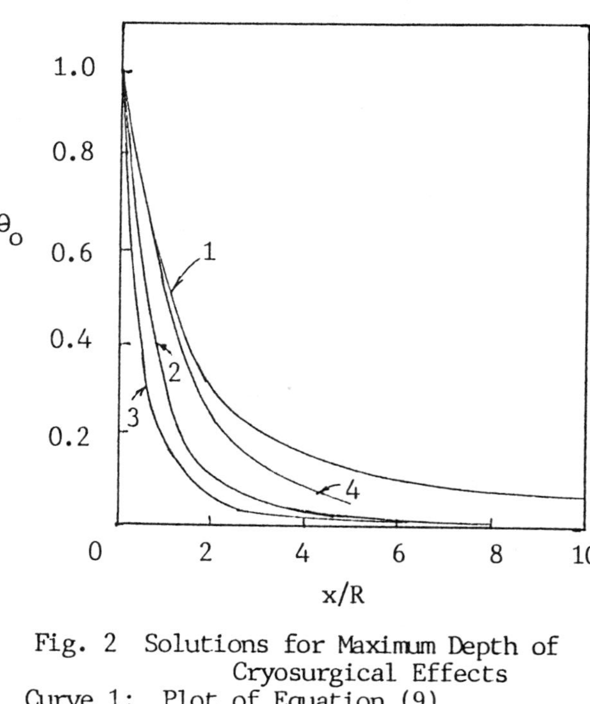

Fig. 2 Solutions for Maximum Depth of
        Cryosurgical Effects
Curve 1: Plot of Equation (9)
Curve 2: Plot of Equation (8)
Curve 3: Plot of Equation (10)
Curve 4: Ruckli's Experimental Results

(hypothetical) case of one-dimensional freezing, the required time would be only 13.2 min. Calculations of one-dimensional freezing, show a reasonable agreement with measurements, if all the needed thermophysical parameters are available.[1,3] And it is obvious that to reach a steady-state in three-dimensions, a lot more time would be required than in one-dimensional freezing. As the freezing process starts as one dimensional, it appears that an improved estimate of the time for the maximum extent of freezing temperature penetration may be based on the average of the one- and three-dimensional times, or 32.5 min.

Now, let's consider the effect of major blood vessels close to the treated area on the results from the example above. In case (a) assume $M = 10$, $x = R$, and $e = 1.5K$. According to Eq. (15), there occurs a -0.4% change in Z_∞; in case (b), assuming the blood vessel is closer to the lesion (with $M = 5$), the change in Z_∞ is -0.7%. In particular cases, one could be even more specific about the decreases of freezing temperature penetration due to the various "heat sources" of which more prominent ones would be large blood vessels.

CONCLUSIONS

A discussion of problems related to contact cryosurgery has been carried out, based on a few standard mathematical models. The time required to achieve a maximum freezing temperature penetration has been estimated. A better knowledge of thermal properties of the bodies to be treated cryosurgically are needed for more exact calculations. Also, effects of the internal geometry of the body, like the existence of heat sources resistant to temperature changes, have been pointed out.

NOMENCLATURE

a,b	linear dimensions of rectangular cryogenic probe, (m)
A	area, (m^2)
C	volumetric specific heat, $(J/m^3\text{-}K)$
k	thermal conductivity, $(W/m\text{-}K)$
L	volumetric latent heat of fusion, (J/m^3)

M   multiple of probe radius to locate the major
             blood versel

N   indicator of three dimensional geometry on
             freezing process

r   coordinate parallel to surface, (m)

R   radius of refrigerated probe, (m)

t   time, (hr)

T   temperature,

x   coordinate perpendicular to surface, (m)

Greek Symbols

α $\theta_2 C_2 / \theta_1 C_1$, dimensionless sensible heat

γ mass density, (kg/m^3)

δ^2 $k_1 C_2 / k_2 C_1$, dimensionless thermal diffusivity

e   temperature excess due to a blood vessel,
             (deg. K)

θ dimensionless temperature

θ_1 $T_f - T_s$, reduced temperature, frozen zone, (K)

θ_2 $T_\infty - T_f$, reduced temperature, liquid zone, (K)

μ $\theta_1 C_1 / L$, reciprocal dimensionless latent heat

ν sensible heat coefficient,
             root of equation (2)

Z   depth of solidification, (m)

Subscripts

1   frozen zone

2   liquid zone

∞ condition far away from surface

o   condition at center of area

e   temperature excess caused by major blood
             versel, also effective radius of cryoprobe

268

f   freezing front

pr   cryosurgical probe

s   condition at the surface

REFERENCES

1.   P. Hrycak, Frost Action Under Refrigerated Areas,
     PhD Thesis, University of Minnesota, Minneapolis,
     1960.

2.   R.F. Ruckli, "Der Frost im Baugrund," Springer,
     Vienna, 1950.

3.   P. Hrycak, Heat Conduction With Solidification in
     a Stratified Medium, _AIChE Journal_, 13:161, 1967.

4.   R. Barron, "Cryogenic Systems," McGraw-Hill, New
     York, 1966.

5.   C. Kellman and I.S. Cooper, Cryogenic Ophthalmic
     Surgery, _Amer.Jour. of Ophth._, 56:731, 1963.

6.   G.V. Porkhayev, Temperature Fields in Foundations,
     Proc., Internat. Conf. NAS-NRC Publ. R87, pp. 285-
     291, 1963.

7.   P. Hrycak, M.J. Levy, and S.A. Wilchins,
     Cryosurgery of Lesions through Contact Surgery
     and Estimates of Penetration Times, _ASME paper_
     75-WA/Bio - 11, 1975.

CRYOBENCH - APPARATUS FOR TESTING

CEBAF CRYOGENIC SUBCOMPONENTS

M. Wiseman, R. Bundy, A. Guerra, J. P. Kelley, T. Lee
W. Schneider, K. Smith, E. Stitts, and H. Whitehead

SURA-CEBAF*
12000 Jefferson Avenue, Newport News, VA 23669

ABSTRACT

The Continuous Electron Beam Accelerator Facility (CEBAF) is a 4 GeV multipass accelerator. At the heart of CEBAF are 338 Superconducting Radio Frequency (SRF) niobium cavities operating at 2.0 K. The cavities are installed into 42 1/4 cryomodules, each having 8 cavities. A cryomodule is the smallest cryogenic unit of the accelerator consisting of a pair of end caps for ingress and egress of the cryogens and, typically, 4 cryounits, each containing a pair of cavities.

During the CEBAF design phase, a number of developmental issues became apparent. These issues include cavity tuners, RF waveguides, rotary feedthroughs, and instrumentation techniques. A modified cryounit, known as the Cryobench, has been designed and assembled to aid in resolving these issues. What follows is a description of the Cryobench, and a report of the initial operating experience. Included in the report are the measured heat loads to the system, results from tuner reliability tests, and an outline of future experiments.

---

*Work performed under U.S. Department of Energy Contract #DE-AC058ER40150.

INTRODUCTION

When completed, the Continuous Electron Beam Accelerator Facility, located in Newport News, Virginia, will be a nuclear physics laboratory used to study quark interactions. The facility's accelerator will produce a 4 GeV electron beam by using 338 superconducting Radio Frequency (SRF) niobium cavities. The cavities are mounted in forty-two and one quarter cryomodules, each containing eight SRF cavities. The cryomodules sustain the cavities with a saturated bath of superfluid helium at 3.14 kilopascals (0.031 Atm) and 2.0 K. The accelerator is housed in an oval tunnel, one and one eighth miles in circumference (Figure 1). The cryomodules are divided between three different areas in the tunnel, the injector, North, and South linear accelerators (linacs). The injector contains two and one quarter cryomodules and delivers electrons to the North linac at relativistic speeds with energies of 45 MeV. The North linac contains twenty cryomodules and boosts the beam energy to 0.4 GeV. After leaving the North linac, the beam is bent through the first arc with dipole magnets and directed into the South linac, which contains the remaining twenty cryomodules. The electrons are then recirculated through the two linacs until they have passed through each five times. The 4 GeV electrons are then separated into three beams and delivered to the three experimental halls.

The cryomodules are divided into four cryounits, with 2 SRF cavities per cryounit (Figure 2). Each cryomodule contains two distinct cryogenic circuits. The primary or 2 K circuit is composed of a Joule-Thomson valve and the four serially connected helium

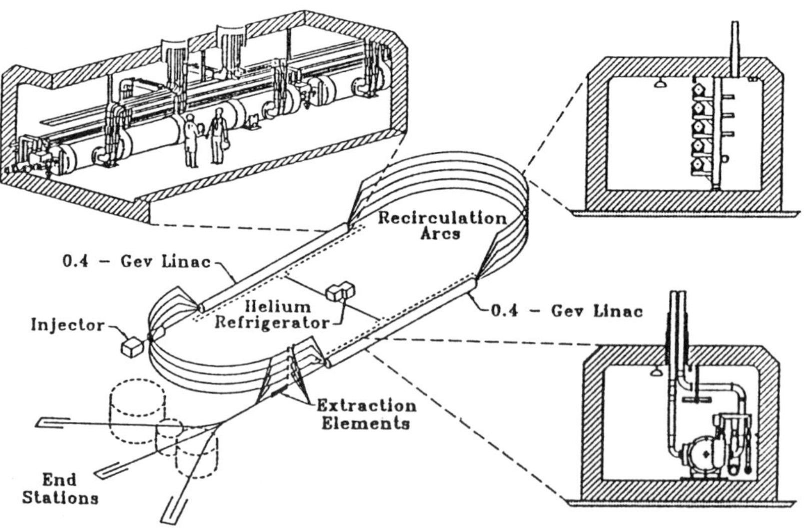

Figure 1. CEBAF Machine Configuration

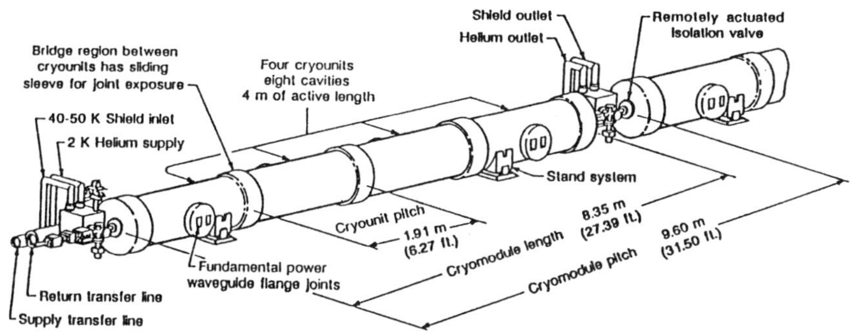

Figure 2.   CEBAF Cryomodule Configuration

vessels which contain the cavities.   A 50 K copper shield is the load for the secondary circuit.   The helium is supplied and removed from the cryomodules via end cans that are mounted axially at the ends of the four series coupled cryounits.

A plan view of a cryounit is shown in Figure 3.   The helium vessel is supported from the vacuum jacket by ten Nitronic 50 support rods (4.7 mm in diameter).   The helium vessel and all 2.0 K systems are wrapped with 24 layers of 0.0254 mm (0.001") thick aluminized Mylar with .0508 mm (0.002") of spun bonded Mylar as a spacer (MLI).   The 50 K copper heat shield is located between the helium vessel and the vacuum vessel.   All 50 K systems are covered with 60 layers of MLI.   RF power at 1497 MHz is supplied through

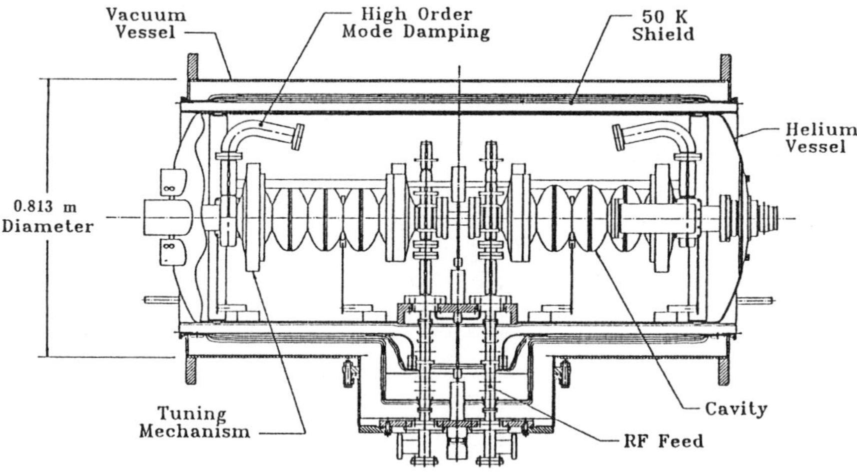

Figure 3.   CEBAF Cryounit Configuration

RF feedthroughs or waveguides on the cryounit's top hat. To reduce the heat load and prevent multipacting, these waveguides are maintained under a maximum vacuum pressure of 13.33 micropascal $(10^{-7}$ torr). This vacuum space is defined by a warm Teflon window on the atmospheric side of RF waveguides and by a Kapton or ceramic window located at the cavity to waveguide flange inside the helium vessel. A cavity's fundamental frequency is maintained at 1497 ± 80 MHz by a mechanical tuner mounted on the first and fifth cavity cells. Torque is supplied to the tuning mechanism by a stepper motor mounted outside the vacuum jacket on the top hat feedthrough flange. The torque is transferred to a tuner with two vacuum rotary feedthroughs. The warm feedthrough is mounted on the top hat and the cold feedthrough is mounted inside the helium vessel. Magnetic shielding (Mu metal) is used on the inside of the vacuum jacket and on the outside of the helium vessel.

The purpose of the magnetic shielding is to limit the earth's magnetic field at the beam line to less than 5×10^{-7} tesla, minimizing flux entrapment by the cavities.

CRYOBENCH DESCRIPTION

The cryobench was designed to allow the performance of long term experiments on several components of the CEBAF cryomodules. For this reason, the cryobench is a single cryounit with two modified end cans. All the components shown in Figure 3 and described for the cryounit are contained in the cryobench with the exception of the SRF cavities. The end cans take cryogens from portable dewars and vent gas to atmosphere (Figure 4). Instrumentation read-outs are mounted on top of the cryobench for local operation and data is displayed and logged by a remote computer, an HP 9000 series Model 300.

A flow schematic for the cryobench is shown in Figure 5. Helium is passed from a 1000 liter dewar through a transfer line and into a phase separator in the supply end can. Liquid from the bottom of the phase separator is then passed into the cryounit's helium vessel. The liquid flow is controlled by the means of an electrically actuated cryogenic valve, a Foxboro Jordan model #1110/AD8110 (LCV20), located between the phase separator and the helium vessel. Boiloff gas then exits the helium vessel and passes through a heater, where it is warmed to room temperature. After passing through the heater, the mass flow is recorded by a Sierra Model #733-16-2 flow transducer (PT12), and the gas vented through a 13.8 kPa gauge pop-off relief valve. In case of a loss of

Figure 4. Cryobench Apparatus During Testing (12/89)

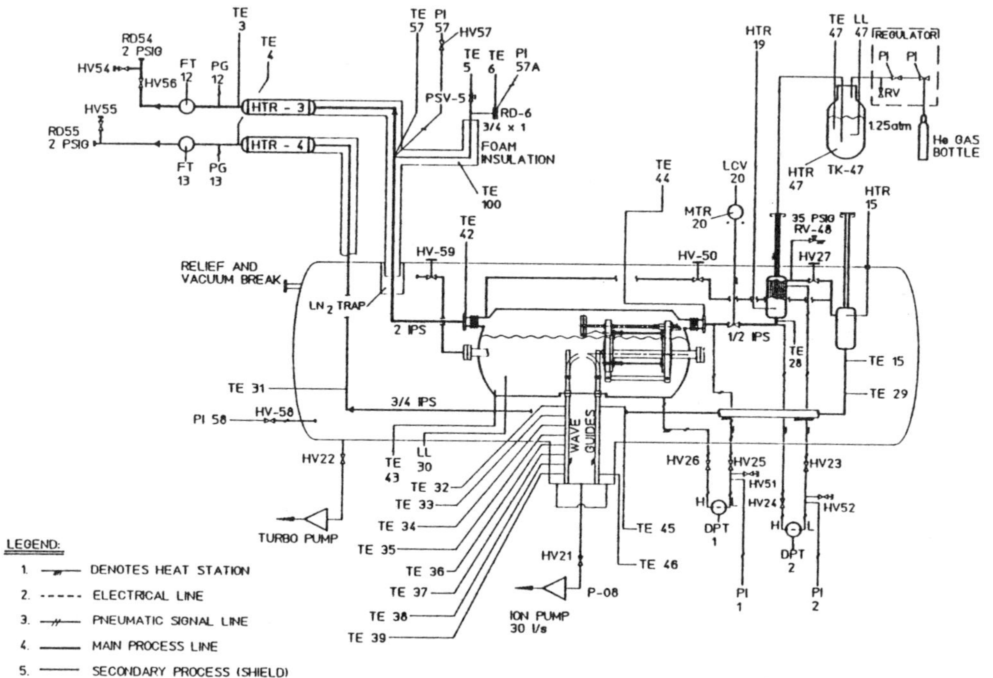

Figure 5. Cryobench Flow Schematic

insulating vacuum, two relief devices are provided on the primary circuit. The first is an Anderson Greenwood Series 80 safety relief valve with a set pressure of 203 kPa absolute. The second is a Fike 0.051 meter (2") reverse acting rupture disc with a set pressure of 304 kPa absolute. Similar relief devices are currently employed on the CEBAF cryomodules.

The shield circuit can be maintained using three different methods. The most economical and most often used method is to use the boiloff gas from the return side of the helium vessel. This gas is routed through a heater mounted inside the supply end. The heater can be used to maintain the shield at a set temperature between 4 and 50 K or deactivated to provide a floating shield. After leaving the heater, the gas cools the shield and exits to the return end can. A heater is used to warm the gas to room temperature. The mass flow is measured by a second Sierra flow transducer (PT13), and then released to atmosphere through another pop-off relief valve set at 13.8 kPa gauge. The second method of operating the shield routes gas from the top of the phase separator into the 4 to 50 K heater and on to the rest of the shield circuit.

A heater is provided at the top of the phase separator to insure that only gas is routed to the shield. This method of operation most closely mimics the actual cryomodule configuration. The third method of operating the shield is provided by a second bayonet on the supply end can. The shield can be isolated from the primary circuit and liquid nitrogen or gaseous helium supplied directly through this second bayonet into the 4 to 50 K heater and the shield circuit. The method of operating the shield is implemented by opening or closing two Cryolab manual cryogenic valves (HV27 and HV50).

Three types of temperature sensors have been used in the cryobench. The helium circuits are primarily monitored with Lake Shore silicon diodes mounted in the stream (TE28, 29, 31, 42-44, and 47). The three heaters on the cryobench are operated by PID controllers which monitor temperature with T-type thermocouples mounted downstream of the heaters (TE3, 4, and 15). RF waveguide temperature profiles are monitored with Cryogenic Linear Temperature Sensors (CLTS), (TE32-39, 45, and 46). Liquid level is monitored in the helium vessel with both a twenty inch AMI level probe (LL20), and a differential pressure gauge (DPT1). Liquid level in the phase separator can also be monitored with a differential pressure gauge (DPT2). Both differential pressure gauges are MKS Type #229HD. Absolute pressures in the helium vessel and phase separator are measured with capacitance manometer pressure gauges

(PT1 and PT2, respectively). For absolute pressure, MKS Type #127A and Vacuum General CML Series B pressure transducers have been used. Pressure is also displayed by dial gauges for both the primary (PE12), and shield circuits (PE13). Additional temperature and pressure gauges are added to match the needs of the experimental setup.

The flow schematic shows several possible experimental setups that can be assembled inside the helium vessel. Included are some possible RF test configurations. Straight RF waveguides inside the helium vessel for 100% reflected RF power studies is one possible experiment while another test uses a 180 degree bend in the waveguide, a configuration which best simulates full beam operations. These tests will measure temperature profiles on the waveguides and heat loads with different experimental setups. Mounting the cavity tuners on springs that have been placed on a two inch 'beam' pipe is another anticipated experimental configuration. The springs simulate the stiffness of the cavities to allow for reliability testing of the tuning mechanisms. This test has been performed and the experimental results are discussed later in this paper. The two inch pipe 'beam' tube can also be used to measure the magnetic fields that are reaching the beamline. This is accomplished by running a flux gate magnetometer through the pipe after cooldown.

COOLDOWN

Since the cryobench experimental set up includes the use of a 1000 liter dewar, the cooldown procedure must include both the dewar and the cryobench. The dewar offers several advantages. First, during cooldown, it acts as a buffer so that the 4 K liquid is not flashed directly into the cryobench. The dewar has also been instrumented with a pressure transducer and a heater, which provides better control of the mass flow during experiments. Another reason for using a 1000 liter dewar is because it contains enough liquid to run an experiment for two to three days, permitting operation with minimal support.

Before beginning cooldown, a warm, dry nitrogen purge is used to remove moisture from the cryobench. Then both the cryobench and 1000 liter dewar undergo several pump and purge cycles with helium to remove nitrogen and other gases from the system. The insulating vacuum on the cryobench must be at 13.33 millipascal or lower before beginning cooldown.

During cooldown, purchased liquid from a 500 liter dewar is passed to the 1000 liter dewar by a small transfer line. A second

transfer line then couples the flashed gas from the 1000 liter dewar
to the phase separator in the cryobench supply end cap. The gas is
then routed through the helium vessel and out through the shield
circuit to atmosphere. This gives the gas the longest possible path
to travel and permits the greatest use of sensible heat.

During the initial part of a cooldown, the 500 liter dewar is
pressurized with a regulated gas bottle. Once the temperatures in
the cryobench are below 70 K, the pressure building circuit provided
with most 500 liter dewars replaces the gas bottle. This provides an
average mass flow to the system of 0.3 g/s at the start of the
cooldown to 1.5 g/s when the system is near liquid helium
temperatures. By using this method, approximately 1000 liters of
liquid helium are sufficient to cool the 1000 liter dewar and the
cryobench to the point where liquid helium starts to accumulate.

Figure 6 shows a typical cooldown curve. The traces show
that transfer from the first 500 liter dewar was started at
approximately 15:00 and lasted until 04:00 the next morning.

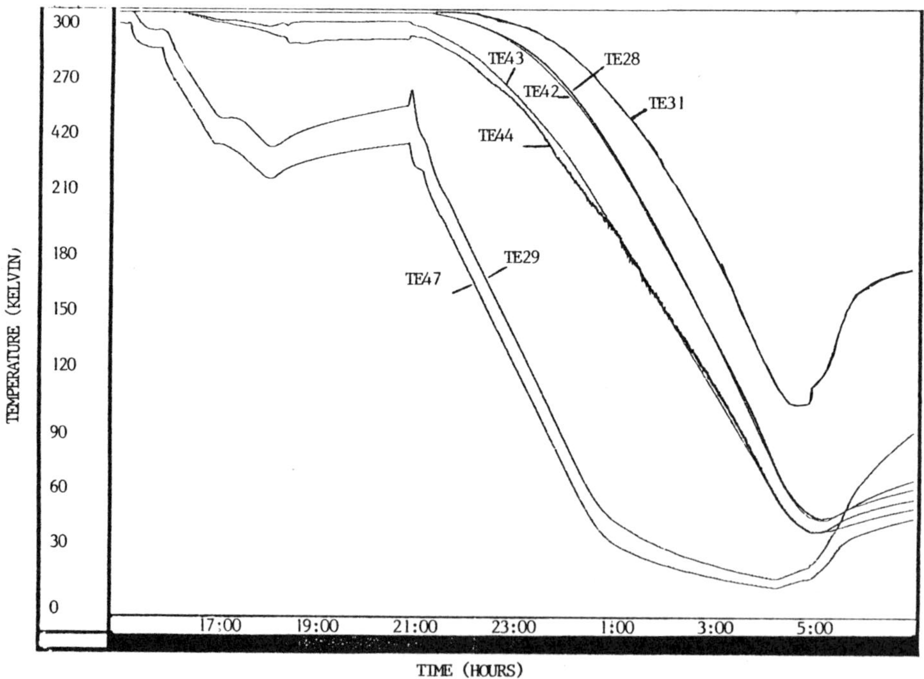

Figure 6. Typical Cool Down Curves

Ideally, transfer from a second 500 liter dewar should have been started at this time. In this case, the second dewar was not started until 09:00 and some of the original cooling was lost. The traces from this cooldown show that the He gas bottle that was used as a pressure builder was depleted around 18:00 and was not replaced until approximately 21:00. Cooldown then continued at a peak rate of nearly 50 K an hour. The curve shows approximately a 5 K differential between the dewar (TE47) and the inlet to the Cryobench (TE29). The helium inlet (TE44) and the helium vessel (TE43) show little temperature difference between the two. The helium vessel outlet (TE42) and shield inlet (TE28) also track together. The shield outlet (TE31) is naturally the last part of the system to complete cooldown.

HEAT LOADS

To date, the cryobench has been used on two separate occasions. For both of these tests, the usual mode of operation involved filling the cryobench with liquid helium during the morning to the 80 to 90 percent level. The fill valve would then be closed, effectively isolating the cryobench. The cryobench then operated essentially as a vapor-shielded storage vessel. This allowed an experiment to run virtually unattended through the rest of the day and night until refilling the next day. Typical liquid losses were twenty to thirty percent of the initial fill level per day.

Heat loads for a vapor shielded storage vessel are:

$$Q_{PRIMARY} = \dot{m}_g h_{fg} = \left[\frac{\dot{m}_{out}}{(1 - \rho_g/\rho_L)} \right] h_{fg}$$

$$Q_{SHIELD} = \dot{m}_{out} \left(h_s - h_1 \right).$$

where $\dot{m}_g$ is the boiloff mass flow rate from the He vessel and $\dot{m}_{out}$ is the measured mass flow[1]. Data was taken by recording the total mass flow through the system at night. The average mass flow is then calculated by dividing the total mass recorded for the evening by the time interval expended. The heat of vaporization (h_{fg}), the gas enthalpy (h_1), the gas density (ρ_g), and the liquid density (ρ_L) are taken from the two phase properties of helium at 115 kPa. The exit gas enthalpy (h_s) is based on the average exit temperature for that evening (TE31) at a pressure of 115 kPa. By using these

equations, and averaging the daily data heat loads for each
experiment, the following heat loads were calculated:

AVERAGE HEAT LOADS (WATTS)

	$Q_{PRIMARY}$	Q_{SHIELD}
EXPERIMENT I (12/09/89 - 12/18/89)	4.55	51.3
EXPERIMENT II (05/16/90) - 5/24/90	6.24	64.4

The reasons for the higher heat loads during the second experiment
are as follows. The fill valve, which isolates the cryobench from the
1000 liter dewar and transfer line, was not sealing completely.
Therefore, part of the recorded mass flow included the boiloff from
the 1000 liter dewar resulting in the higher calculated values.
Additionally, the 300 K layer of MLI had been directly shorted to
the 50 K intercept on the waveguides. This provided an extra heat
load directly to the shield.

CAVITY MECHANICAL TUNER TESTS

One of the first goals for the cryobench was to perform a
reliability test on the mechanical tuners in a 4 K helium bath. In
the accelerator, the tuners will be run daily and must have a usable
life of ten years. This life expectancy translates into 1.25 billion
steps from a stepper motor. One full revolution of the motor shaft
equals 25,000 steps. The tuners are maintained at 2 K and normal
lubrication techniques cannot be used. Instead, all moving parts in
the tuner have been impregnated with dicronite. Dicronite (tungsten
disulfide) is a dry lubrication and does not freeze at cryogenic
temperatures. Little data was available concerning the reliability of
mechanical parts at helium temperatures which necessitated this test.
The torque is transmitted to the tuners through two rotary
feedthroughs. Hence, the reliability of the feedthroughs was also
evaluated.

For these experiments, the tuners were continuously cycled
through a semi-random pattern. This pattern consisted of two basic
parts. In the first part, the feedthroughs were alternatingly jogged
for two or three full rotations (50,000 - 75,000 steps) from the zero
torque mark in both the clockwise and counterclockwise directions.
The second step consisted of turning the feedthroughs approximately
thirty full rotations (750,000 steps) from the zero torque mark in

first the clockwise then counterclockwise directions. Full travel for the tuners is fifty full rotations (1,250,000 steps) in either direction. The number of jogs and the length of full travel was different for each tuner and test. The two completed parts made up one cycle and the number of cycles are recorded and presented in Table 1. Also presented are the number of rotations per cycle as well as the cycle duration. The same two tuners were tested in both experiments. The two cold rotary feedthroughs were also used in both tests. The two cold feedthroughs were modified Varian Model #954-5151 that were purchased by CEBAF in 1988 and are designated as "Original Varian" in Table 1. The "Standard Varian" feedthrough is from a second batch that was purchased in 1989. The "Prototype Varian" incorporates design changes originating from the failure of the two warm "Original Varians" in the first test. One change incorporated in the Prototype was the use of dicronited bearings. The Original and the Standard models were not lubricated.

During the first test, both of the warm feedthroughs failed before reaching the required service life of 50,000 rotations (1.25 billion steps). The data is presented in Table 2 and the torque profile for the top tuner is shown in Figure 7. The torque profile shows that the tuner/feedthroughs initially required a torque of approximately 0.057 N-m (8 in.-oz.) peak to peak. The torque data from the first test show a short break in period followed by smooth operation for about 14 hours. The torque profile exhibits a steady increase from a torque of 0.057 N-m (8 in.-oz.) to 0.57 N-m (80 in.-oz.) peak to peak before a bearing in the rotary feedthrough seized and ended the experiment. Both of the warm feedthroughs failed in this manner during the first test.

The two warm feedthroughs were replaced for the second test while no changes were made to the tuners and the cold feedthroughs. The new Prototype Varian was instrumented and its torque profile is presented in Figure 7. The Prototype also displayed a short break in period followed by smooth operation for approximately 12 hours. The torque profile then changed from 0.057 N-m to 0.092 N-m (8 to 13 in.-oz.) peak to peak for the rest of the experiment. The Prototype Varian did not display the steady degradation apparent during the first test but failed when a vacuum leak developed in the welded bellows. The Standard Varian was not tested to failure due to a slipping drive shaft coupling and the vacuum failure of the Prototype model. The Standard Varian required a torque of 0.18 N-m (26 in.-oz.) peak to peak at the time the test was terminated.

Table 1. Cavity Tuner and Rotary Feedthrough Test Parameters

Test 1	Top Tuner	Warm–Original Varian	317 cycles to failure
		Cold–Original Varian	154.5 rotations/cycle
			11.3 minutes/cycle
	Bottom Tuner	Warm–Original Varian	386 cycles to failure
		Cold–Orginal Varian	103 rotations/cycle
			6.4 minutes/cycle
Test 2	Top Tuner	Warm–Prototype Varian	397 cycles to failure
		Cold–Original Varian	121 rotations/cycle
			10.3 minutes/cycle
	Bottom Tuner	Warm–Standard Varian	387 cycles
		Cold–Original Varian	126 rotations/cycle
			6.6 minutes/cycle

Table 2. Cavity Tuner and Rotary Feedthrough Test Data

		Average Speed (RPM)	Maximum Total Rotations	Percent of 50,000	Maximum # of Steps (Billions)
Top Tuner	Test 1	13.70	48,976.50	97.95	1.22
	Test 2	12.70	52,007.00	104	1.30
	Total	NA	100,983.50	202	2.52
Bottom Tuner	Test 1	16.10	39,758.00	79.52	0.99
	Test 2	19.10	48,762.00	97.50	1.22
	Total	NA	88,520.00	178	2.21

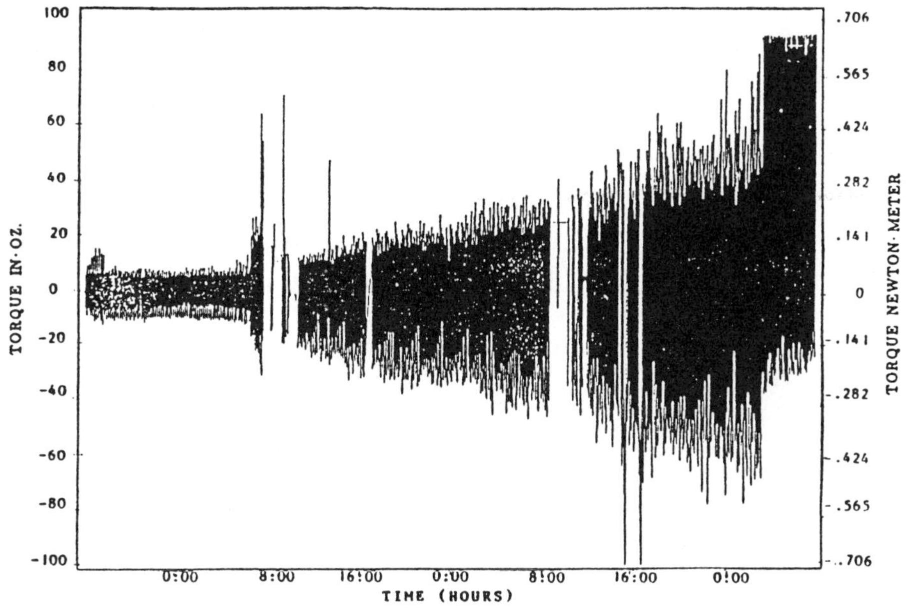

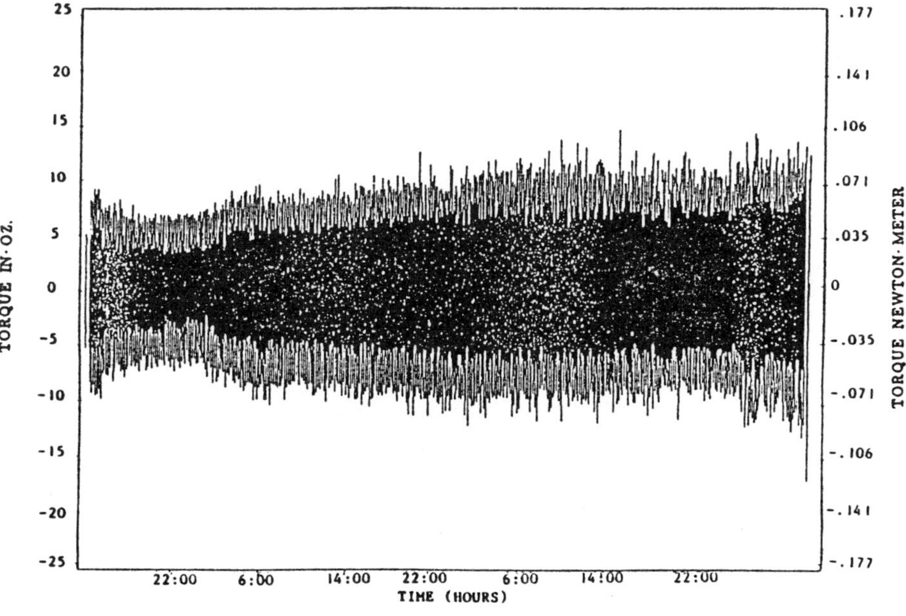

Figure 7.   Tuner Torque Profiles

283

The two cold feedthroughs and the two tuners displayed no noticeable degradation through both experiments and, therefore, easily exceeded their expected life span. Final examination of the feedthroughs and tuners must wait until the helium vessel can be removed from the cryobench.

There were two differences between the cold and warm feedthroughs in both experiments that result from their mounting configuration. The first difference is that the cold feedthroughs were in a 4 K liquid bath while the warm feedthroughs are at room temperature. The second difference is in the way the pressure was distributed across the bellows in each feedthrough. The cold feedthrough has vacuum on the inside of the welded bellows and the warm feedthroughs are presently mounted with the vacuum on the outside of the welded bellows. The difference in the vacuum load changes both the stresses in the welded bellows and the direction of the axial load applied to the drive shaft, a consequence of the differential pressure across the bellows.

FUTURE TESTS

Several more tests are planned for the cryobench. The first of these tests takes advantage of the 2 inch pipe that was used to mount the tuners in the helium vessel. The pipe is open to the insulating vacuum on both sides on the helium vessel and will allow a flux gate magnetometer to be passed through the helium vessel while cold. This test will demonstrate whether the magnetic shielding presently used is sufficient to keep the magnetic field at the beamline to 5×10^{-7} tesla. A higher magnetic field will affect the quality of the electron beam and make it harder to control.

Several tests are also planned using the RF waveguides. Two of the tests will help map both the static and dynamic heat load profiles along the waveguides. This data will be used to help optimize the waveguide's heat intercept by verifying the predictions of an existing mathematical model[2,3]. Other tests with the waveguides will study the high power RF characteristics of the waveguides when coupled with several different configurations in the helium vessel and while using several different window materials. Temperature and RF arc detector monitors that are used as interlocks to safeguard the cavities could also be tested safely to their design limits in the cryobench without endangering the SRF cavities.

Also planned for the cryobench is a series of loss of vacuum experiments to determine if the present safety reliefs have been adequately sized to protect the cavities and helium vessels from over-pressurization. Two of these tests will involve spoiling the insulating vacuum, first to an air leak and then to a helium leak. A loss of vacuum to air is also planned for the waveguide vacuum and for the beamline vacuum.

ACKNOWLEDGEMENTS

The authors would like to thank the SRF secretarial staff, especially Linda Williams, and the SRF technical staff for their time and effort in producing this paper.

BIBLIOGRAPHY

1. L.E. Scott, R.C. Van Meerbeke, B.S. Kirk and G.C. Nubel, Storage, Distribution, and Handling, in: "Technology of Liquid Helium," NBS Mono. 111, R.H. Kropschot, B.W. Birmingham, and D.B. Mann, ed., National Bureau of Standards, Boulder, 1968.

2. P. Brindza, G. Biallas, and L. Phillips, An Optimized Input Waveguide for the CEBAF Superconducting Linac Cavity, IEEE Trans. on Mag., 23:319, 1987.

3. J.P. Kelley and J. Takacs, Thermal Design and Evaluation of the CEBAF Superconducting RF Cavity's Prototype Waveguide, in: "Advances in Cryogenic Engineering, Vol. 35," R.W. Fast, ed., Plenum Press, New York, 1990.

CEBAF CRYOUNIT LOSS OF VACUUM EXPERIMENT*

Mark Wiseman, Richard Bundy, J. Patrick Kelley
and William Schneider

Continuous Electron Beam Accelerator Facility
Newport News, Virginia

ABSTRACT

In order to test the sizing of the safety relief devices provided for the
Continuous Electron Beam Accelerator Facility's 42 1/4 Cryomodules, several loss of vacuum experiments have been planned with a modified 1/4
Cryomodule called the Cryobench. The first of these experiments, presented
here, is a controlled spoiling of the insulating vacuum to air. The most credible vacuum failure was simulated with in a 3.2 mm diameter hole being
opened to the vacuum space yielding an air flow of 0.0033 kg/s and corresponding peak heat fluxes of 749 W/m^2 just before the rupture disc opened.
The maximum calculated helium mass flows of .025 kg/s indicate that the
process relief valve provided on each cryomodule is adequate to handle this
type of vacuum failure. The experimental configuration and recorded data
are presented and discussed.

INTRODUCTION

The Continuous Electron Beam Accelerator Facility (CEBAF) uses 338
superconducting radio frequency (SRF) cavities to produce a 4 GeV electron beam for nuclear physics research. The SRF cavities are housed in the
42 1/4 Cryomodules that are used to construct the three basic sections of
the accelerator: the injector, the North and the South linear accelerators
(LINACs). The Cryomodule is composed of four Cryounits and two end can

---

* The Southeastern Universities Research Association (SURA) operates
the Continuous Electron Beam Accelerator Facility for the United States
Department of Energy under contract DE-AC05-84ER40150.

assemblies. Each Cryounit contains two cavities, which are mounted inside a helium vessel. The helium vessel is surrounded by a copper thermal shield and a vacuum jacket. The end can assemblies provide for the supply and return of helium from the central refrigerator to the helium vessels and thermal shield. Detailed descriptions of the CEBAF accelerator are available elsewhere.[1,2]

Approximately 350 liters of saturated superfluid helium at 2.0 K are stored in each helium vessel. The Cryomodules use multilayer insulation (MLI), thermal shielding, and high insulating vacuums to reduce the heat load to the helium bath. A cross-section of a CEBAF Cryounit is shown in Figure 1. The helium vessels and all 2.0 K circuits are covered by 24 layers of MLI consisting of 0.0254 mm (0.001") thick aluminized Mylar with 0.0508 mm (0.002") of spun bonded mylar as the spacer. The copper shielding is covered by 60 layers of MLI and is maintained between 40 K and 50 K by gaseous helium.

Three separate vacuum systems are contained in each Cryomodule/ Cryounit. The insulating vacuum encompasses all the space between the helium vessels and the vacuum vessel, excluding the Cryomodule's RF feedthroughs or waveguides. The insulating vacuum is maintained at 13 to 1300 μPa (10^{-7} to 10^{-5} torr). The RF waveguides, which are rectangular tubes with sides 25.04 by 134.42 mm (0.986" $\times$ 5.292") respectively, require a separate vacuum system. The waveguide vacuum is defined by a warm Teflon window at room temperatures and pressures, and by a cold Kapton or ceramic window between the waveguide and the fundamental power coupler, which is located inside the helium vessel. To prevent multipacting (the loss of energy from the input RF signal to stray electrons which can cause sputtering) the waveguides require a harder vacuum than does the insulation space, typically 0.13 to 13 μPa (10^{-9} to 10^{-7} torr). The third vacuum system is defined by the SRF cavities and the beam pipe, which traverses the entire accelerator. The beam pipe connects the cavities between helium vessels and between cryomodules. The cavity and beam pipe vacuum is maintained between 0.013 to 0.13 μPa (10^{-10} to 10^{-9} torr). All three vacuum systems may be spoiled by either an air or helium leak.

If any of the vacuum systems are spoiled, there must be adequate safety relief devices mounted to protect the SRF cavities and helium vessels from over pressurization. The cavities are presently rated at 304 kPa (3 atm) differential pressure at room temperature and the helium vessels are rated at 405 kPa (4 atm) differential pressure. Since the yield strength of niobium is inversely proportional to temperature, its cold yield strength allows the two Cryomodule pressure reliefs to be designed to maintain absolute pressures

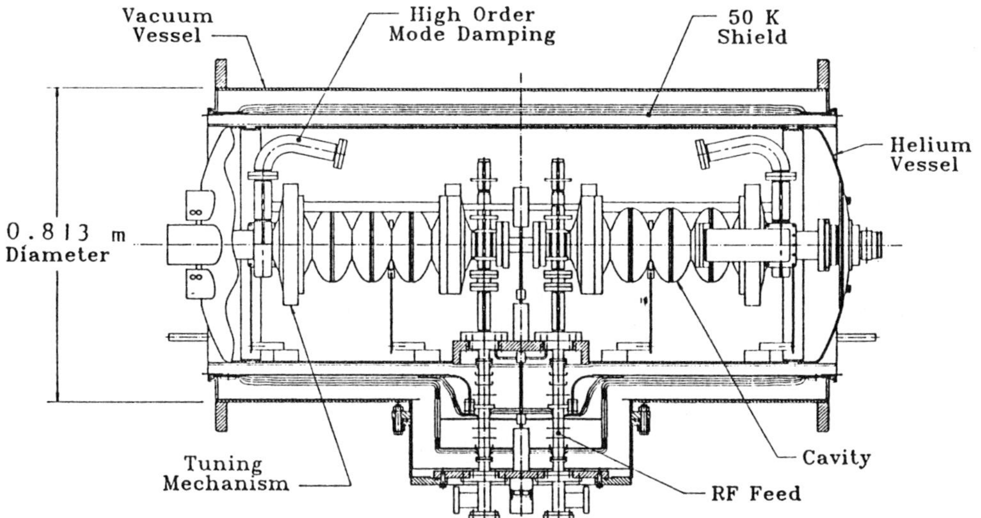

Figure 1. CEBAF Cryounit Configuration

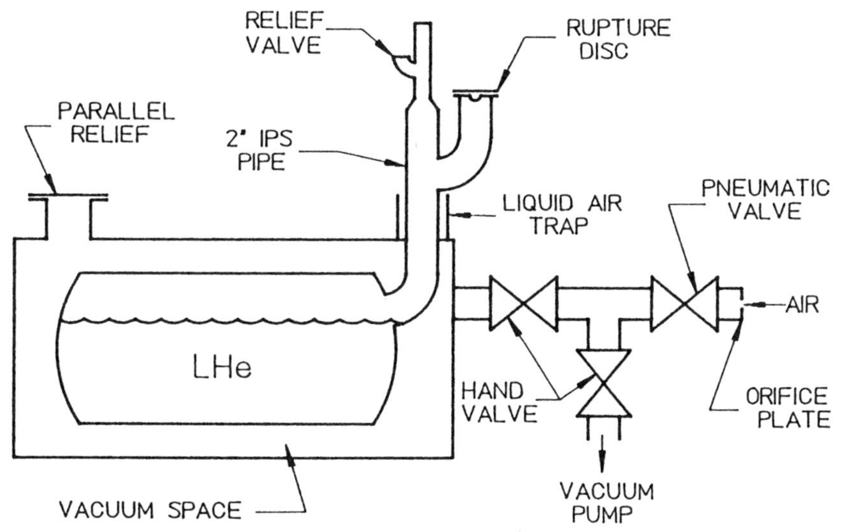

Figure 2. Loss of vacuum experimental setup

below 405 kPa (4 atm). The first relief is an Anderson Greenwood model #
81-B68-8 safety relief valve. This valve has an 0.013 m (0.5") orifice and is
set to relieve at pressures of 203 to 304 kPa (2 to 3 atm) depending on the
condition of the downstream recovery header. The second relief is a 0.051 m
(2") Proquip model # SB10 reverse acting rupture disc with a set pressure of
405 kPa (4 atm). To determine if these reliefs are properly sized, a series of
loss of vacuum experiments will be conducted at CEBAF with the Cryobench
apparatus. The first experiment is a loss of insulating vacuum to air and is
reported here. Planned experiments include a loss of the insulating vacuum
to helium and the loss of vacuum to air in both the waveguide and cavity
vacuums.

EXPERIMENTAL SET UP

The Cryobench is a modified cryounit and was designed as an experimental bench for several of the CEBAF accelerator components[3]. The cryounit
configuration of the cryobench is the same as for the cryomodule and is shown
in Figure 1. Two reliefs are provided on the Cryobench and are similar to
those used for the Cryomodule. The first is the same Anderson Greenwood
relief valve with a desired set pressure of 203 kPa (2 atm) absolute. After
the first experiment, it was discovered that this valve had mistakenly been
set at a pressure of 304 kPa (3 atm). The second relief is a 0.051 m (2") Fike
reverse acting rupture disc set at 304 kPa (3 atm) absolute.

The first experiment to be performed with the Cryobench was a loss of
insulating vacuum to air. The experimental configuration is shown in Figure 2
and a flow schematic is presented in Figure 3. A remotely operated pneumatic
valve was used to open the vacuum space to air and an orifice plate was used
to control the air flow. For this experiment, a .0032 m (1/8") diameter hole
was used for the orifice plate, yielding a sonic mass flow of 3.3 g/s of air into
the vacuum space. The most credible failure of the insulating vacuum space,
a total failure of a rotary feedthroughs welded bellows, determined the size
of the orifice.

Two data logging systems were used for this experiment. The first system consisted of a CAMAC A/D conversion card and an HP 9000 series
Model 300 computer. This system was used to translate and store the data
on a hard disk. Using this method, data can be logged at a speed of just
over one data point per second. In the second system, the raw data from
the CAMAC card was stored directly into computer RAM and written to a
hard disk after the experiment. The raw data was then translated into usable
values at a later date. This system allowed for 33 data points a second to
be recorded until the RAM was filled, about 45 minutes. This method was
used because of the extremely fast response times demonstrated by previous

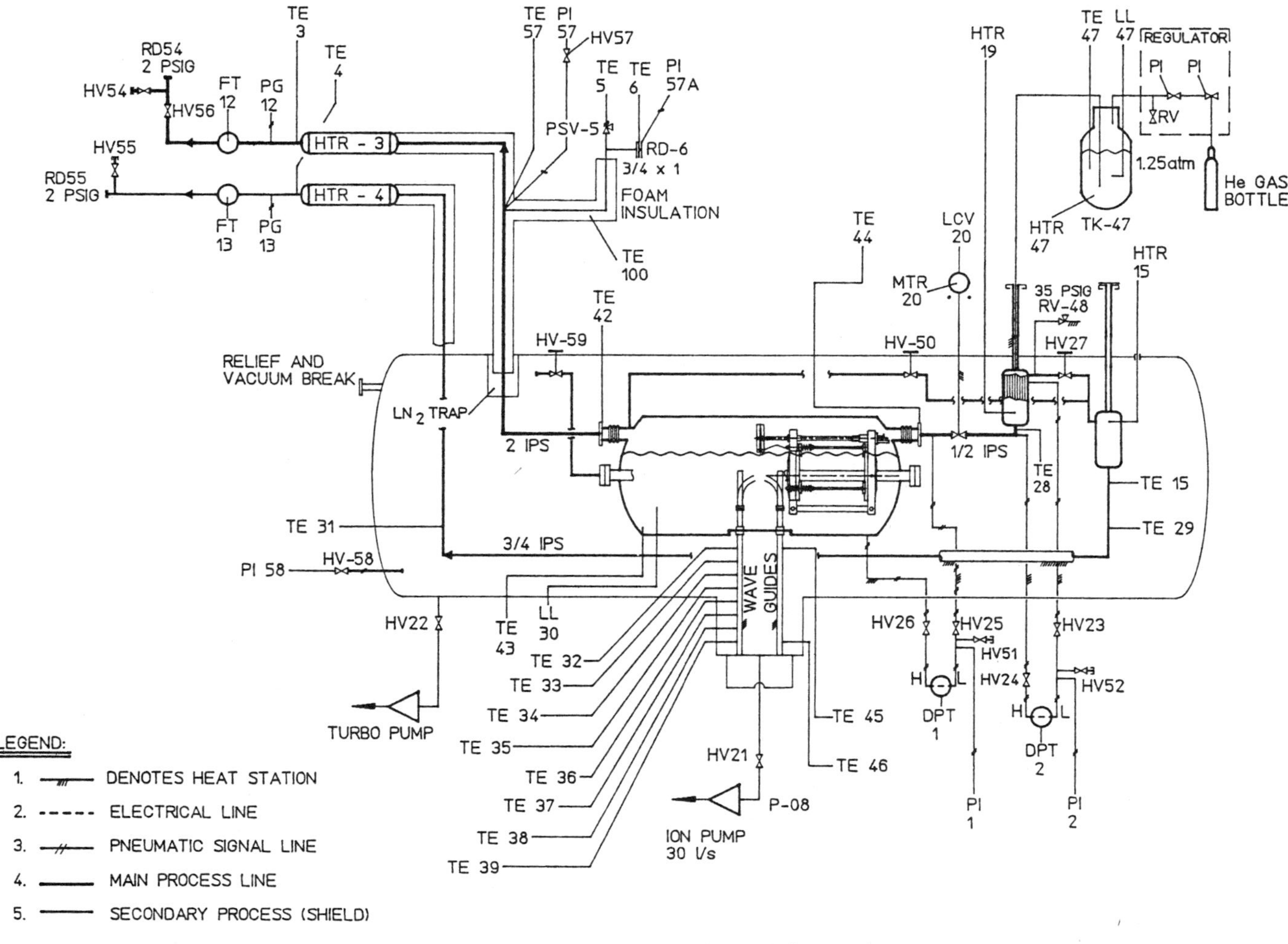

Figure 3. Cryobench Flow Schematic

experiments and was used for the first time during this experiment.[4,5] In
addition to data logging, the venting was videotaped.

The loss of vacuum experiment was performed in a test cave at CEBAF
which is equipped with six oxygen deficiency monitors and a ventilation system to exchange the cave atmosphere every 1 1/2 minutes. The cave was
locked and access restricted during the test. Before starting the experiment,
the Cryobench was filled with liquid helium and the thermal shield was maintained at a stable temperature of 50 K for several hours. The Cryobench was
then isolated from the fill dewar. The thermal shield was isolated from the
helium vessel by closing valves LCV20, HV27, and HV50. The normal boiloff
relief valve (RD54) for the helium vessel was suppressed by closing HV56 so
that the only outlets for the helium were through the safety reliefs: the AGCO
valve (PVS-5) and the rupture disc (RD-6). Pertinent dimensions and initial
conditions for the test are presented in Table 1.

Table 1. Initial Conditions for the Loss of Vacuum
to Air Experiment

<u>Vacuum Space</u>
Volume	1.1 m^3
Vacuum	66.7 × 10^{-6} Pa

<u>Helium Vessel</u>
Total Volume	0.445 m^3
Surface Area	3.48 m^2
Liquid Volume	0.340 m^3 (42.2 kg)
Gas Volume	0.105 m^3 (1.83 kg)
Internal Pressure	105.0 kPa
Liquid Temperature	4.26 K
Gas	2.46 K to 11 K
Shield Temperature	50.0 K

<u>Insulation</u>
24 layers of MLI over the 4 K circuit
60 layers of MLI over the 50 K shield

EXPERIMENTAL RESULTS AND DISCUSSIONS

Pressure curves for the helium vessel (PI1) and a point just upstream of
the relief devices (PI57) are presented in Figures 4 and 5. Figure 4 presents
the pressure characterization for the entire experiment and Figure 5 for the
first three minutes. The pressure downstream of the rupture disc (PI57A)
is also displayed in Figure 4. The pressure inside the helium vessel and the
pressure just upstream of the relief devices show only a small difference,
approximately 2.7 kPa (20 torr), throughout the experiment. Shortly after

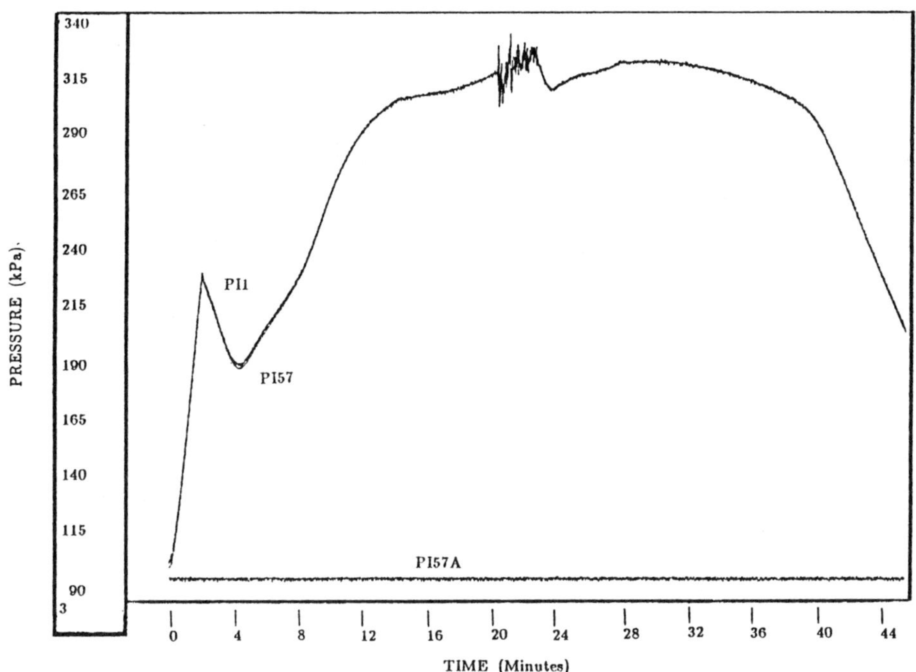

Figure 4.   Pressure Response vs Time

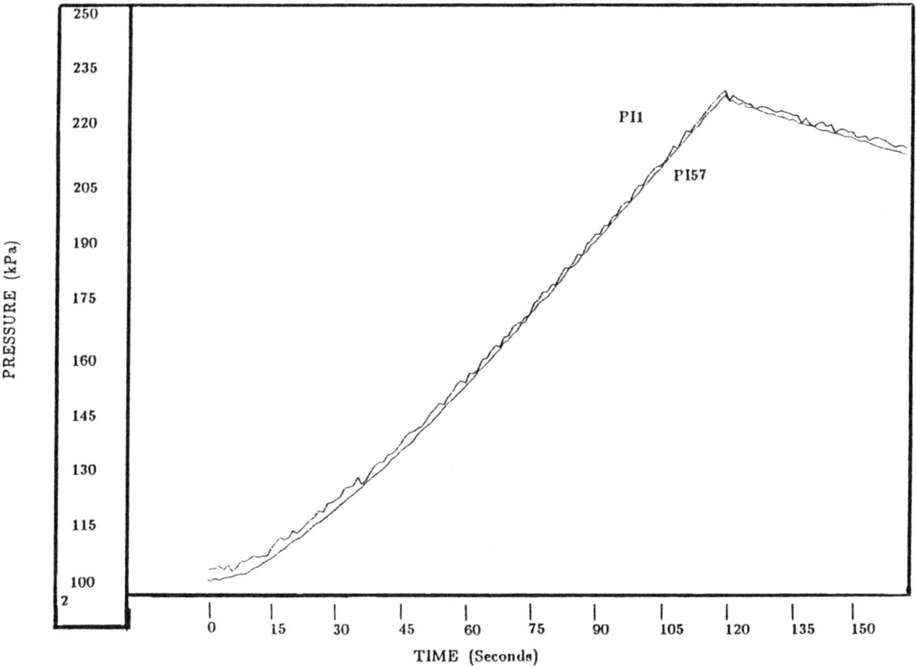

Figure 5.   Initial Pressure Response vs Time

the pneumatic valve was opened, the pressure in the vessel began to rise from an initial value of 105 kPa (790 torr). At approximately 90 s into the experiment the pressure passed the point were the Anderson Greenwood valve was thought to have been set, 204 kPa (1520 torr), and continued to rise. At 109 s and 230.8 kPa (1731 Torr), the Fike burst disc ruptured. This value was much less than the set pressure of 304 kPa (2280 torr). The rupture disc did not open fully, presenting a large restriction to the flow.

After the rupture disc opened, there was a pressure drop to about 184 kPa (1380 torr). The pressure then began a steady rise to approximately 306.7 kPa (2300 torr). The Anderson Greenwood valve began to open at a pressure of 314.7 kPa (2360 torr). The peak pressure recorded was 333.3 kPa (2500 torr). The pneumatic valve was closed 35 minutes into the experiment and the pressure in the helium vessel began to drop thereafter, a consequence of the reduced heat load and the decreasing volume of helium in the vessel.

After 35 minutes, the mass of the air had entered the vacuum space would have a volume, at room temperature, that was five times the volume of the insulating vacuum space. Two hours after closing the pneumatic valve, the pressure in the vacuum space had risen to one atmosphere.

Temperature profiles for the experiment are given in Figures 6 and 7. The pressure in the helium vessel (PI1) is plotted in both figures to provide reference points. Note that the data in Figures 4 and 5 were fast data logged on the second computer system. The pressure data in Figures 4 and 5 are, therefore, more detailed than in Figures 6 and 7 and have different time references on the X-axes due to the different clocks in the computers. Notice in Figure 6 that the temperatures at the top of the helium vessel, TE44 and TE42, show short term temperature rises just after the burst disc ruptured and when the Anderson Greenwood valve opened. These two occurrences caused warm stratified gases above the diodes to mix resulting in the short term temperature rises. Also of interest in Figure 6 is the difference in temperatures between the supply side of the helium vessel (TE44) and the return side (TE42) which is where the vacuum leak occurred. The temperature on the return side (TE42) also increased faster than the temperature at the exhaust (TE57) after the pneumatic valve was closed. By this time, the LN2 trap surrounding the length of the exhaust pipe was full of liquid air and the exhaust pipe was covered with frozen air which cooled the exhaust gas. Additionally, diode TE42 may have been in stratified gas, or been influenced by conduction from frozen air on the return side both causing the greater temperature rise.

Figure 7 displays the response of Cryogenic Linear Temperature Sensors (CLTS) that were positioned along the waveguides during the experiment.

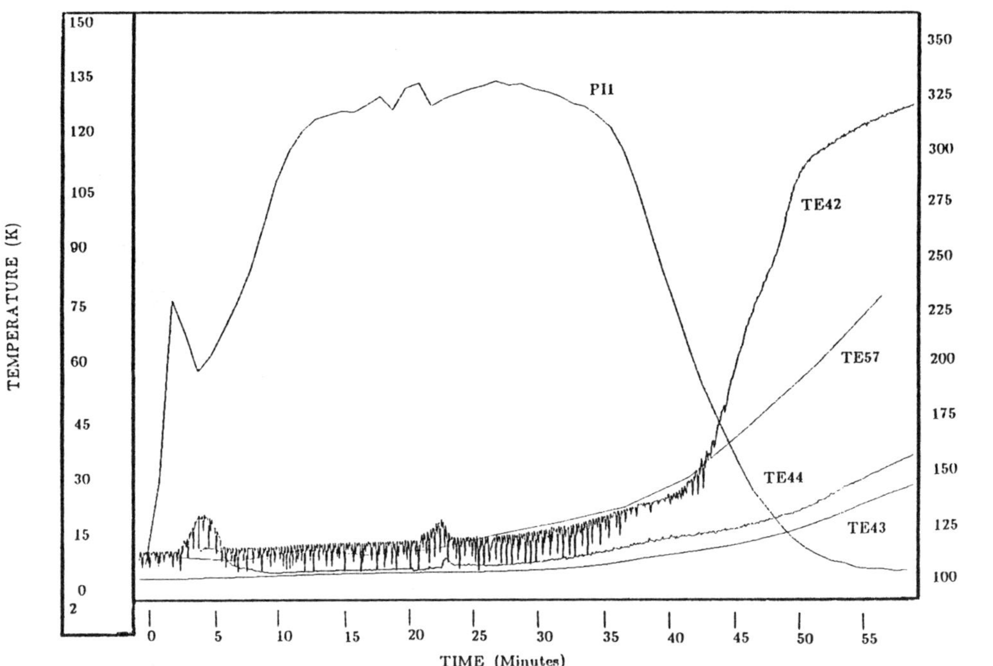

Figure 6. Helium Vessel Temperatures vs Time

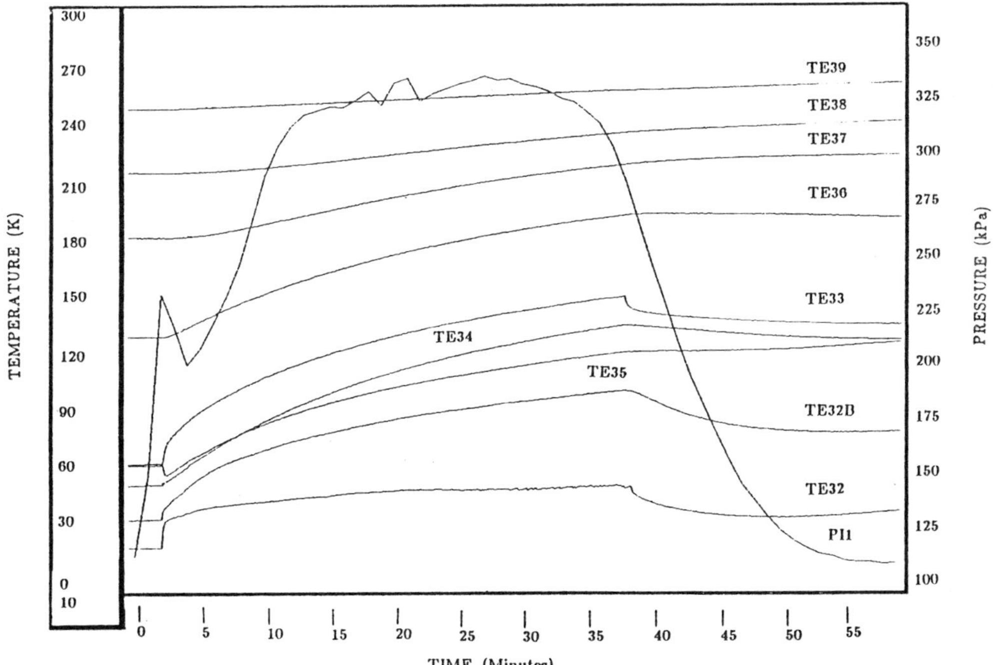

Figure 7. Waveguide Temperatures vs Time

TE33 is mounted on the waveguide thermal intercept and TE34 is mounted on the 50 K thermal shield. There is nearly a two minute delay between when the pneumatic valve was opened and when the CLTS registered any response. When the temperatures did change, there was a fast response in the curves closest to the helium vessel (TE32 and TE32B), followed by a steady temperature rise. TE35, which is 12.7 mm (1/2") from the warm side of the thermal intercept, shows a temperature drop at this point. This is attributed to either liquid runoff from the shield or from air freezing on the waveguide at this point. There is also a knee in many of the curves at the point were the pneumatic valve was closed, a consequence of the reduced heat load.

Thermodynamically, the helium in the vessel during the first 109 s of the experiment may be treated as a closed system. Since the vessel volume is constant and no mass is transferred, the specific volume and hence, the system density are constant during this time period. A Pressure - Internal Energy plot for helium displays lines of constant specific volume (Figure 8). The specific volume for this system is:

$$v = \frac{V_T}{m_L + m_g} = \frac{V_T}{\rho_L V_L + \rho_g V_g} = 10.12 \frac{cm^3}{g} \tag{1}$$

The volumes of the liquid and gas, V_L and V_g respectively, are determined using the initial liquid level probe reading and the knowledge of its position in the vessel. The densities are obtained from the initial pressure and the assumption of saturated conditions in the vessel. Considering the constant specific volume line of 10.12 cm^3/g, we see that initially, at $t = 0$ s and 105 kPa (1.04 atm), the helium is two phased, since the line is below the liquid-vapor dome. As pressure is increased, the quality of the system decreases, reaching zero when the constant specific volume line intersects the liquid-vapor dome to the left of the critical point. This occurs at 90 s. Crossing the liquid-vapor dome, the helium state is initially a compressed liquid, but when it exceeds the critical pressure at 107 s, it becomes a supercritical fluid.

It is of interest to estimate the instantaneous heat load to the helium. The net energy entering a system when no work is done on the system is

$$Q = \Delta U = m_T \Delta u \tag{2}$$

Dividing by the time difference between initial and final states and taking the limit as t approaches 0 we have

$$\frac{dQ}{dt} = m_T \frac{du}{dt}. \tag{3}$$

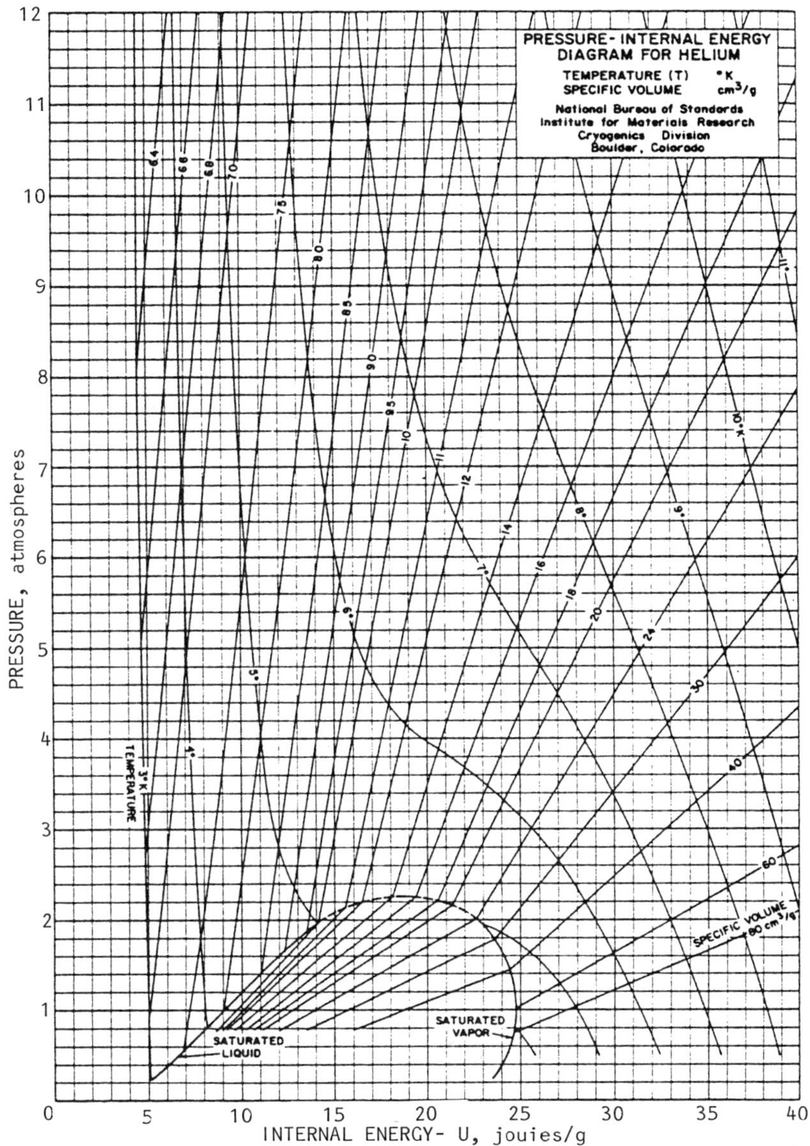

Figure 8. Pressure-Internal Energy Diagram for Helium[6]

This is the instantaneous heat load to the helium. Dividing by the internal surface area of the vessel, we obtain the heat flux to the helium,

$$q = \frac{m_T}{A} \frac{du}{dt}. \tag{4}$$

To apply this equation, we first perform a least squares fit of the pressure data obtaining

$$P = (*)Pa \quad \text{where } (*) \text{ for the proper time interval is given as}$$

$$(105,400 - 87.46t + 40.34t^2 - 0.3729t^3) \qquad 1 \le t \le 3s$$
$$(105,800 - 281.6t + 62.26t^2 - 1.067t^3) \qquad 3 \le t \le 21s$$
$$(102,000 + 543.3t + 9.866t^2 - 0.03772t^3) \qquad 21 \le t \le 89s$$
$$(196,000 - 2141t + 35.20t^2 - 0.1171t^3) \qquad 89 \le t \le 109s \tag{5}$$

Using the pressure and specific volume, values for the internal energy are obtained for specific time values. Again using a least squares fit, we obtain the following time dependent expressions for internal energy

$$u = (**)J/g \quad \text{where } (**) \text{ for the proper time interval is given as}$$

$$(9.646 - 0.001t + 0.002t^2 - 4.186 \times 10^{-5}t^3$$
$$+ 4.504 \times 10^{-7}t^4 - 1.835 \times 10^{-9}t^5) \qquad t \le 10s$$
$$(9.583 + 0.018t + 5.651 \times 10^{-4}t^2$$
$$- 2.703 \times 10^{-6}t^3) \qquad 10 \le t \le 35s$$
$$(9.456 + 0.029t + 3.145 \times 10^{-4}t^2$$
$$- 1.081 \times 10^{-6}t^3) \qquad 35 \le t \le 90s$$
$$(0.01213t + 12.75) \qquad t \ge 90s \tag{6}$$

Hence

$$q = \left(\frac{m_T}{A}\right)(***)\frac{J}{m^2} \quad \text{where } (***) \text{ for the proper time interval is given as}$$

$$(-0.001 + 0.004t - 1.256 \times 10^{-4}t^2$$
$$+ 1.802 \times 10^{-6}t^3 - 9.175 \times 10^{-9}t^4) \qquad t \le 10s$$
$$(0.018 + 1.130 \times 10^{-3}t - 8.109t^2) \qquad 10 \le t \le 35s$$
$$(0.029 + 6.290 \times 10^{-4}t - 3.243 \times 10^{-6}t^2) \qquad 35 \le t \le 90s$$
$$(0.01213) \qquad t \ge 90s \tag{7}$$

These equations are plotted on Figure 9. The heat flux to the helium rises sharply at first and appears to be approaching a constant value of around 749 W/m^2 when the flux decreases drastically at 90 s, and remains constant. The discontinuity occurs because the transfer of heat from a surface to a bath of compressed helium liquid is governed by natural convection, an inefficient heat transfer mechanism when compared to boiling heat transfer, which occurs while the bath state is below the liquid-vapor dome. Hence, the heat load from the incoming ambient air, since it is no longer being absorbed efficiently by the helium, drastically increases the temperature of the helium vessel and other materials within the vacuum vessel. This is shown in Figure 7, which is a plot of the temperatures recorded by the CLTS mounted on one of the waveguides. There is little change in the readings of the CLTS during the first two minutes of the experiment during which pool boiling occurs in the vessel. Then, there is a rapid increase in the temperature of the CLTS near the helium vessel followed by a slower, steady temperature rise. The rapid rise in temperature corresponds to the period of time when natural convection is the heat transfer mechanism to the helium and the additional heat that is no longer absorbed by the helium heats the helium vessel and other material in the vacuum vessel. It is expected that as the temperature of the helium vessel increases, heat transferred to the helium would increase, approaching the flux values seen previous to the transition to a compressed liquid.

The maximum heat that can be removed from the incoming air results in a heat flux of 427 W/m^2. Figure 9 shows that we are initially far below this value, a consequence of the thermal lag caused by the freezing of the air on cold masses other than the helium vessel. As time goes on, the heat flux increases far above this value. However, as t increases, the insulation vacuum deteriorates, permitting gas conduction and convection to contribute to the heat load to the helium vessel.

To aid in sizing the relief devices, an estimate of the mass flowrate through the burst disc was performed. To make the estimate, the open area of the burst disc was required. This was obtained by removing the burst disc from the Cryobench and installing it at the bottom of a water tank. Timing the passage of a known volume of water through the disc, the pressure upstream of the disc is calculated using the energy equation for a laminar incompressible fluid flow and the fact that the pressures at the top of the water tank and downstream of the disc are atmospheric. The details of this calculation may be found in any undergraduate fluid mechanics text and are not presented here. The averaged mass flowrate for 3 trials was 0.0307 kg/s of room temperature water. Using the calculated upstream pressure, the known downstream pressure and the orifice plate equations[7]:

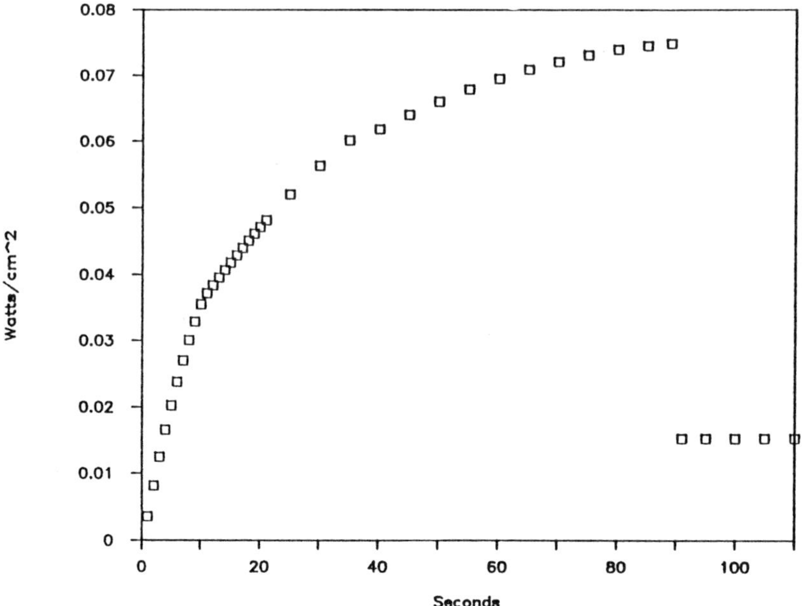

Figure 9.   Heat Flux vs Time

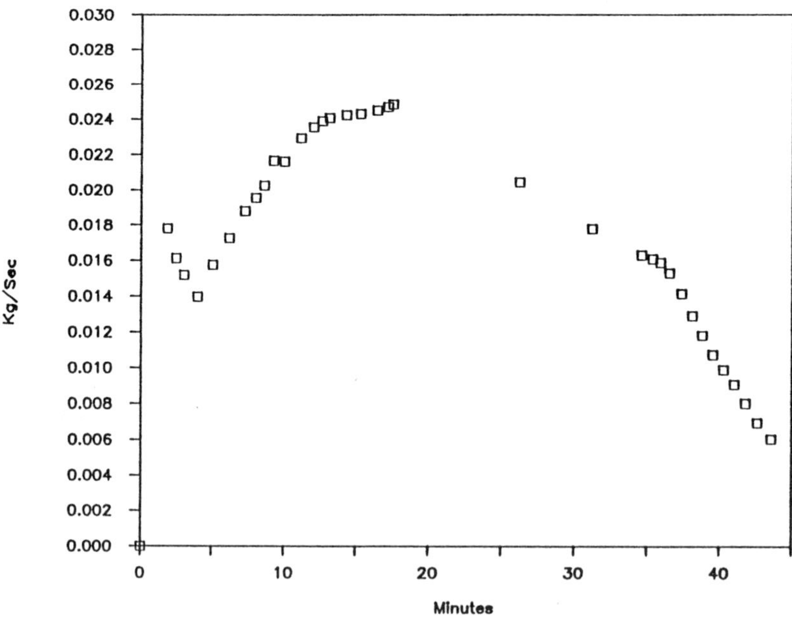

Figure 10.   Calculated Mass Flow Through the Burst Disc

300

$$\dot{m} = A_o C Y \left(\frac{2\rho g_c \Delta P}{1 - \beta^4} \right)^{1/2} \qquad (8)$$

where

$$Y = 1 - (0.41 + 0.35\beta^4)\frac{\left(1 - \frac{P_u}{P_D}\right)}{\kappa}$$
$$C = 0.61 \qquad \text{(See Reference 8)} \qquad (9)$$

$$\beta^2 = \frac{A_o}{A_u} \qquad (10)$$

the area of the burst disc may be found iteratively by estimating the area and calculating the corresponding mass flowrate. The burst disc's venting area was found to be 0.196 cm^2 (0.0303 in^2), which agreed with the physical evidence.

With the open area of the burst disc known, the mass flowrate through the disc during the loss of vacuum experiment can be estimated. Using the orifice plate, equation (8), a downstream pressure of 101.325 kPa (1 atm), and the raw data for the pressure and temperature upstream of the disc, the mass flowrate is calculated using helium properties. The results are presented in Figure 10. Note that these results do not include the mass flowrate through the relief valve which was judged not to be significant by studying the video tape recording of the behavior of the relief devices during the experiment. Numerically integrating the data used for the curve in Figure 10 (Trapezoidal Rule), a total mass of 47 kg is obtained. Although this value is greater than the mass initially estimated for the system (44 kg), it is well within the margin of error for the experiment. Since the relief valve can handle 0.128 g/s at these temperatures and pressures, we can conclude that, if properly set, the relief valve on the Cryomodule can easily handle this loss of vacuum scenario.

CONCLUSIONS

The most credible loss of vacuum accident to air for a Cryomodule was simulated as a 3.2 mm diameter hole was opened between air and the insulating vacuum space. A rapid rise in the heat flux to the helium occurred, peaking 90 s into the experiment at 749 W/m^2, just before the helium bath crossed the liquid-vapor dome. The highest heat flux that can occur solely from the complete cryopumping of air in the helium vessel is 427 W/m^2. The difference between these two values is a consequence of the conduction and convection of air resulting from the rapid spoiling of the insulating vacuum.

After crossing the liquid-vapor dome, the bath was a compressed liquid and the measured heat flux dropped dramatically as convection became the prime heat transfer mechanism to the helium instead of boiling heat transfer. At this point, temperature increases in the rest of the cold mass were recorded as the metal began absorbing the remaining system heat load.

From this experiment, it is evident that the process relief valve provided on the Cryomodules could handle this loss of vacuum scenario and that the rupture disc should not have been needed. Peak calculated mass flows of 0.025 kg/s are well under the 0.128 kg/s that the process relief can vent. Future experiments will be needed to investigate other vacuum failures.

ACKNOWLEDGEMENTS

The authors would like to acknowledge the support and assistance of the rest of the CEBAF SRF technical staff for their efforts in putting this experiment together, Karen Nolker and the Thaumaturgical computer group for their work on the fast data logging system employed, and Darrel Thompson for his work that helped explain some of the data presented in this paper. In addition, we acknowledge Julie Lilley who graciously typed the manuscript.

NOMENCLATURE

A	surface area of helium vessel, 3.48 m²
A_o	pipe area of burst disc, .196 cm² (.0303 in²)
A_u	pipe area upstream of burst disc, .051 m (2")
C	discharge coefficient
g_c	1.0 kg m/N s²
κ	ratio of the specific heat
m_g	initial mass of the gas, 1826 g
m_T	total mass of helium in the system, 43940 g
$\dot{m}$	mass flowrate through an orifice/burst disc, kg/s
P	pressure, Pa
P_D	pressure downstream of the burst disc, 1,013 kPa (1 atm)
P_u	pressure upstream of the burst disc, kPa
Q	net energy entering the system, J
q	instantenous heat flux to the helium, J/m²
t	time, s
U	internal energy, J
u	specific internal energy, J/g
V_g	initial volume of gas, 104,900 cm³
V_L	initial volume of liquid, 339,700 cm³
V_T	total volume of the system, 444,600 cm³
v	specific volume of the system, 10.12 cm³/g

Y	orifice expansion factor
β	$(A_o/A_u)^{1/2}$
ΔP	change in pressure
ρ	density of fluid flowing through the burst disc
ρ_g	initial density of the gas, 0.0174 g/cm^3
ρ_L	initial density of the liquid, 0.124 g/cm^3

REFERENCES

1. H.A. Grunder, J.J. Bisognano, W.I. Diamond, B.K. Hartline, C.W. Leemann, J. Mougey, R.M. Sundelin, The Continuous Electron Beam Accelerator Facility, <u>Proceedings of the 1987 IEEE Particle Accelerator Conference</u>, ed. E.R. Lindstrom and L.S. Taylor, (1987).

2. R.M. Sundelin, RF Superconductivity at CEBAF, <u>Proceedings of the 4th Workshop on RF Conductivity</u>, ed.Y. Kojima, National Laboratory for High Energy Physics, Japan (1990).

3. M. Wiseman, R. Bundy, A. Guerra, J.P. Kelley, T. Lee, W. Schneider, K. Smith, E. Stitts, H. Whitehead, Cryobench Apparatus for Testing CEBAF Cryogenic Subcomponents, presented at the CRYO '90 Conference, Binghampton, NY (1990).

4. G. Cauallame, I. Gonine, D. Güsewell, and R. Stierlien, Safety Tests with the LEP Superconducting Cavity, CERN (1989).

5. V.D. Bartenev, V.I. Datskov, Yu. A. Shishou, and A.G. Zel'dovich, Study of the Process of Insulation Vacuum Failure in Helium Cryostat, "Cryogenics", Vol. 26, (1986).

6. D.B. Mann, Properties of Helium, in: "Technology of Liquid Helium," NBS Mono. 111, R.H. Kropshot, B.W. Birmingham, and D.B. Mann, ed., National Bureau of Standards, Boulder (1968)

7. B.M. Coulter, Jr., "Compressible Flow Manual", Fluid Research Publishing, Houston (1984).

8. Flow of Fluids, Technical Paper No. 410, Crane Co., New York (1969).

Novel Approaches for Attaining High Accelerating Fields in Superconducting Cavities

Q. S. Shu, J. Graber, W. Hartung, J. Kirchgessner,
D. Moffat, R. Noer[*], H. Padamsee, D. Rubin, and J. Sears[+]

Laboratory of Nuclear Studies, Cornell University, Ithaca, NY 14853

ABSTRACT

Present-day superconducting (SC) radio-frequency (rf) cavity structures used in particle accelerators provide accelerating fields (E_{acc}) up to 10 MV/m. Field emission is the most serious obstacle to reaching the higher fields called for in future applications. We have used heat treatment (up to 1500°C), along with high-power processing of cavities and temperature mapping, to suppress field emission and analyze emitter properties. In 27 fired cavities, we have raised the average E_{acc} to 26 MV/m from the 14 MV/m obtained with chemical treatment (CT) alone; the highest E_{acc} reached is 30 MV/m. Non-accelerating cavities have also been made to investigate the highest rf field SC Nb can support; 145 MV/m has been reached. A 6-cell cavity has been constructed in an effort to extend our achievements from single-cell test cavities toward the accelerating structures planned for TESLA (a TeV e^-e^+ linear collider); preliminary measurements with CT only reached E_{acc} = 17 MV/m. The conceptual design of a B-factory cavity is also briefly discussed.

INTRODUCTION

Superconducting (SC) radio frequency cavities have already come to play an important role in particle accelerators around the world. Present-day SC cavity structures provide accelerating fields (E_{acc}) up to 10 MV/m. However, for many reasons, future large scale applications need higher fields — e.g., 30 MV/m. Field

---

[*] Permanent address: Carleton College, Northfield, MN 55057.
[+] Work supported by the National Science Foundation, with supplementary support from the US-Japan collaboration.

emission is the most serious obstacle to reaching these higher fields. In this paper, the focus is on the success of our efforts to suppress field emission in single-cell test cavities by using high-temperature heat treatment, and to identify sources of emission. We will also present preliminary results for chemical treatment of a 6-cell SC accelerating structure, and briefly discuss the conceptual design of a high-field B-factory SC cavity.

APPLICATIONS OF SC CAVITIES

Current Applications

As a result of their great advantages over conventional structures, SC cavities have been widely and rapidly applied in beam acceleration systems around the world[1]. The following is a list of six laboratories that have tested and operated SC cavity structures in cryomodules:

Argonne: SC heavy-ion linac ($.007 < \beta < .06$).
CEBAF: Will make 338 5-cell cavities to be used with a continuous beam. Several cavities have been tested (6-10 MV/m).
CERN: Four 4-cell cavities tested in a beam for 2000 hrs.
Cornell: Two 5-cell cavities tested in the CESR storage ring.
DESY: Sixteen 4-cell cavities for HERA.
KEK: Thirty-two 5-cell cavities increase TRISTAN's energy from 28.5 to 33 GeV; 3000 hrs in the beam line.

All these cavities, designed for accelerators and tested and operated with particle beams, are fabricated from Nb with chemically-etched surfaces (Nb-coated Cu and Nb_3Sn are under study). They provide accelerating fields (E_{acc}) between 5 and 10 MV/m, as shown in Fig. 1.

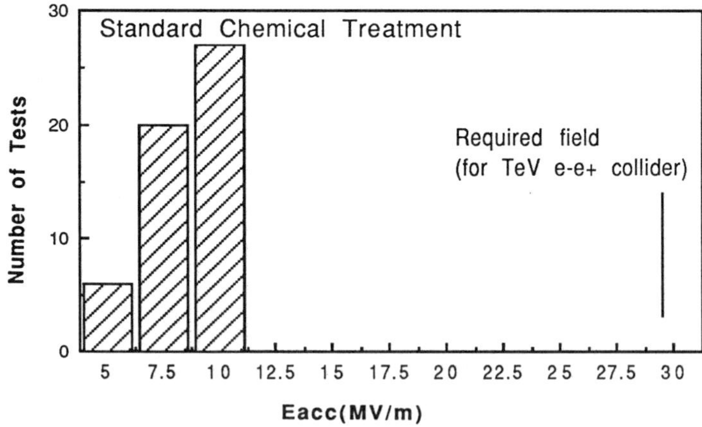

Fig. 1. Highest accelerating fields attained in chemically etched superconducting Nb cavities installed in operating accelerators.

<u>The Challenges</u>

In addition to the current applications of SC cavities listed above, there are several other applications planned and proposed: at Cornell (the B-factory facility to increase the luminosity of CESR by a factor of 100), Los Alamos (LAMPF and PILAC), Italy (LISA), and Saclay (the MACSE continuous-beam facility). TESLA, the TeV-Energy e^-e^+ Superconducting Linear Accelerator currently under discussion, is the largest potential application of SC cavities[2].

Many of these applications need yet higher E_{acc} than has been achieved in current applications. For example, PILAC and TESLA need $E_{acc} > 15$ and 30 MV/m, respectively. The theoretical capability of a Nb cavity is as high as $E_{acc} = 50$ MV/m (corresponding to a surface electric field (E_{pk}) = 100 MV/m), set by the critical magnetic field of $H_{rf} \sim 2000$ Gauss. Approaching this potential would have a profound impact on the cost of future accelerators.

STRATEGY TO INCREASE CAVITY FIELDS

Over the years, three obstacles have retarded the attainment of higher accelerating fields. Multipacting was overcome with the use of spherical or elliptical cavity shapes. Thermal breakdown was eliminated through the use of high purity (high RRR) Nb material. Currently, field emission (FE) is the limiting obstacle.

Field emitted electrons are accelerated to hundreds of keV by the cavity fields; they then strike the cavity walls, causing heating and the emission of bremsstrahlung X rays. Sources of field emission include particles of dust and other foreign matter lying on the cavity surface, and impurity inclusions in the Nb surface, with gases condensed on either of the above often increasing their emission. Accordingly, strategies to increase cavity fields are now directed toward eliminating or deactivating such FE sources. Such strategies have included high-temperature heat treatment, high-power rf processing, and the use of diagnostic tools to better understand the mechanism causing this emission.

HEAT TREATEMENT OF SC CAVITIES

<u>Intermediate Temperature Results</u>

Substantial progress was made by heat treatment (HT) of cavities at 1200°C-1350°C in a UHV (Ultra High Vacuum) furnace for 2-4 hours[3]. The vacuum in the furnace (measured above the heat shields — see Fig. 2) was typically between 3×10^{-7} and 1×10^{-8} torr during the firing. The average E_{pk} reached in these cavities increased to 40 MV/m from the 28 MV/m reached in chemically treated (CT) cavities.

One of the problems we faced with these HTs[4,5] was that the RRR of high purity Nb dropped due to absorption of oxygen into the bulk from the residual gases in the furnace. Final bulk RRR values for HT cavities fell to between 130

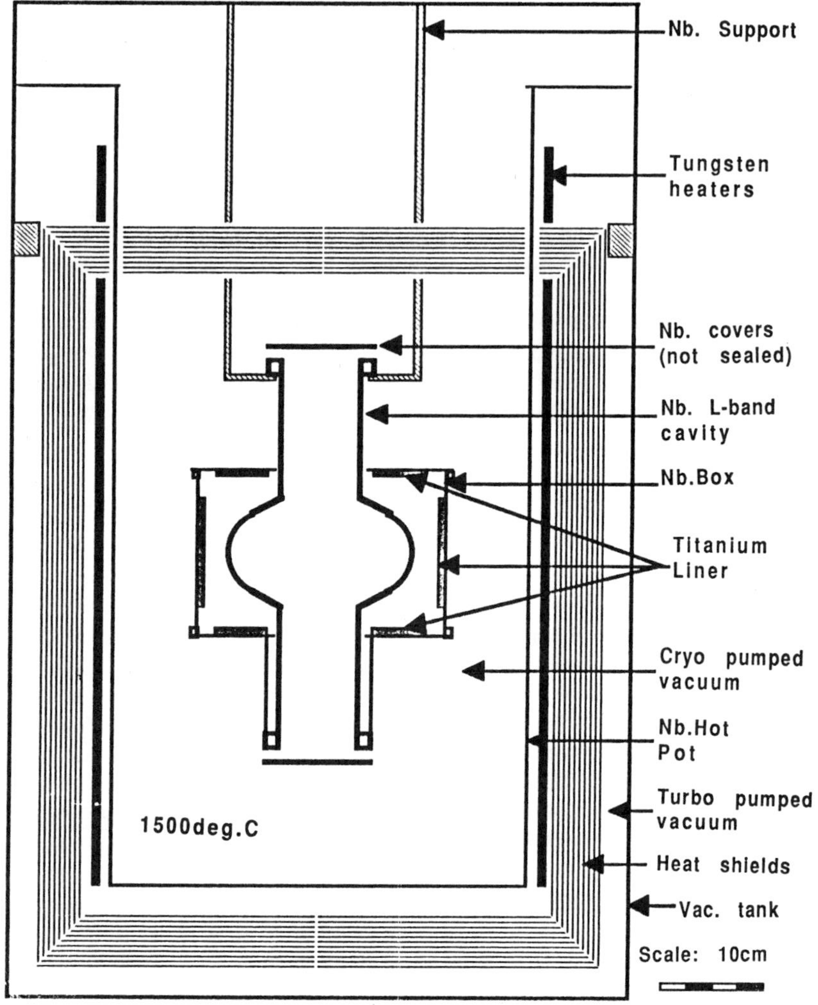

Fig. 2. Heat-treatment furnace with cavity and Ti box.

and 260, increasing the possibility of breakdowns. To minimize this effect, we initially restricted the heating time and temperatures.

<u>Higher Temperature Results</u>

In order to explore whether HT at higher temperatures might be even more effective, we developed the protection technique[3] shown in Fig. 2. A Nb box completely surrounds the cavity cell and short segments of the beam tube at both ends of the cavity. On the inside, the box is lined with Ti sheets. During heat treatment, the high furnace temperatures ensure that a coating of Ti is evaporated onto the outer wall of the cavity. (The vapor pressure of Ti at 1350°C is 2×10^{-5} torr.) Oxygen diffusing into the cavity wall from the residual gas is

308

removed by solid state gettering at the Nb-Ti interface[6]. In our furnace the vacuum is sufficiently good that the net effect is oxygen loss from the cavity wall. The RRR of the cavity increases if its pre-HT performance was limited by bulk oxygen and stays constant if the cavity has already been completely depleted of oxygen by previous solid-state gettering cycles.

Using the above procedure, 10 cavities were fired at 1500°C for 4 hours. With rf processing alone, it was possible to reach an average E_{pk} of 48 MV/m, with 61 MV/m as the best value. He processing raised the average to 52 MV/m. While He processing still plays a role in raising the field for higher-temperature treated cavities, its effect is diminished over that for intermediate temperature treatment.

Fig. 3(a) shows the overall progress in increasing superconducting cavity fields, comparing heat treatments with standard chemical treatments. Fig. 3(b) shows a comparison between intermediate- and higher-temperature treated cavities.

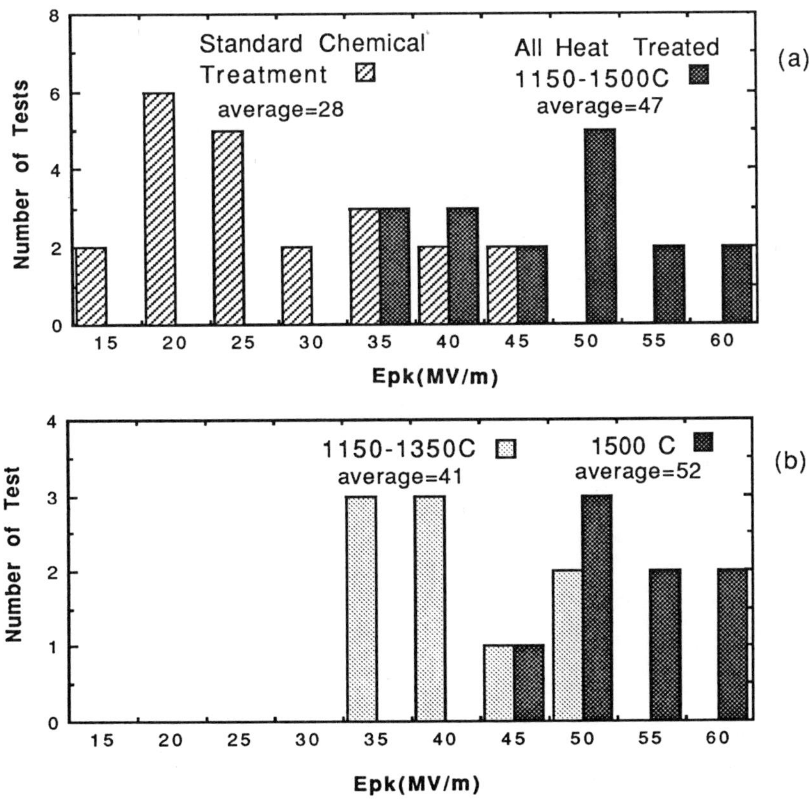

Fig. 3. Highest peak surface fields (E_{pk}) attained in Cornell test cavities. (a) Cavities with chemical etching alone compared with heat treated cavities. (b) Intermediate-temperature vs higher-temperature treatments.

Fig. 4 is a comparison of the overall rf performance among chemically treated (a), intermediate-temperature treated (b), and higher-temperature treated (c) cavities. It can be seen that chemically treated cavities often have Q values lower than 1×10^9 before reaching 20 MV/m, while most of the cavities treated at 1100°C-1350°C have Q values greater than 1×10^9 after reaching 40 MV/m, and the higher-temperature treated cavities even pass 50 MV/m while maintaining Q values greater than 1×10^9.

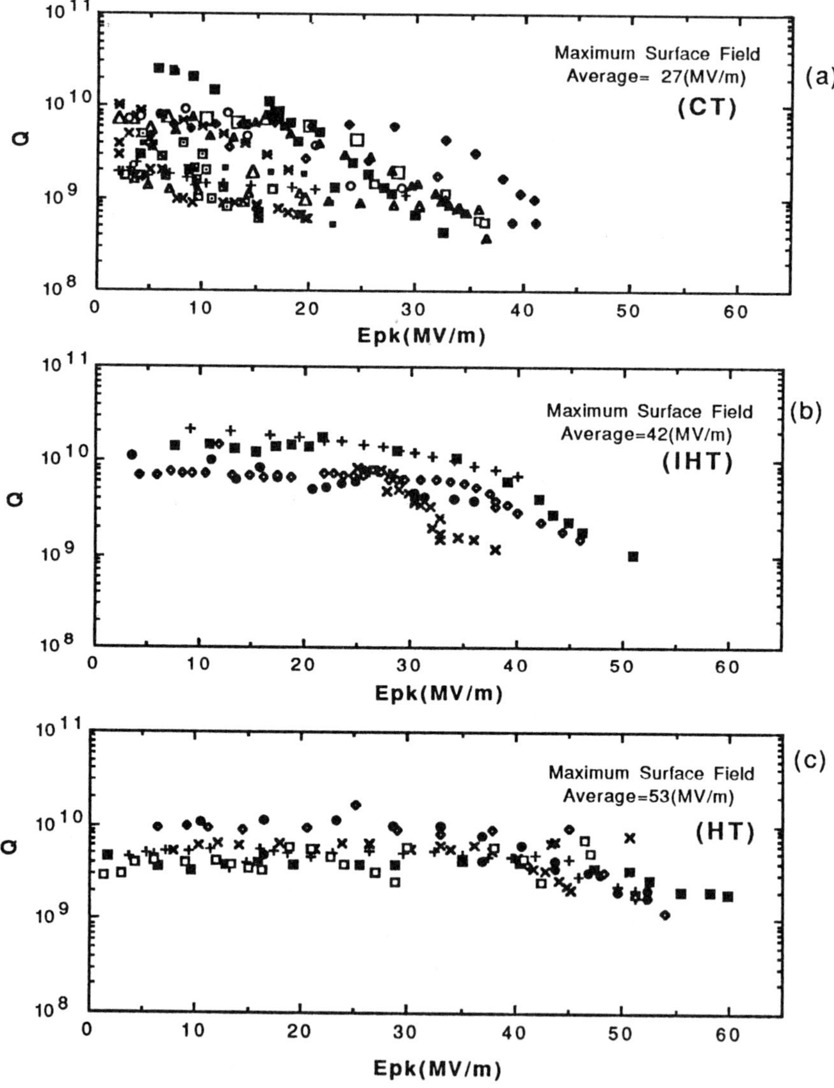

Fig. 4. Overall rf performance of cavities with (a) chemical treatment only, (b) intermediate-temperature heat treatment, and (c) higher-temperature heat treatment.

Comparison of Field Emission

Our temperature-mapping diagnostic system[7] allows us to examine the effects of field emission in detail. Figure 5 shows maps comparing three cavities that were treated with both CT and HT. Each map was taken at a field of 30 MV/m. The three in the left column correspond to CT alone; those in the right column are for the same cavities after heat treatment and show drastically reduced emission.

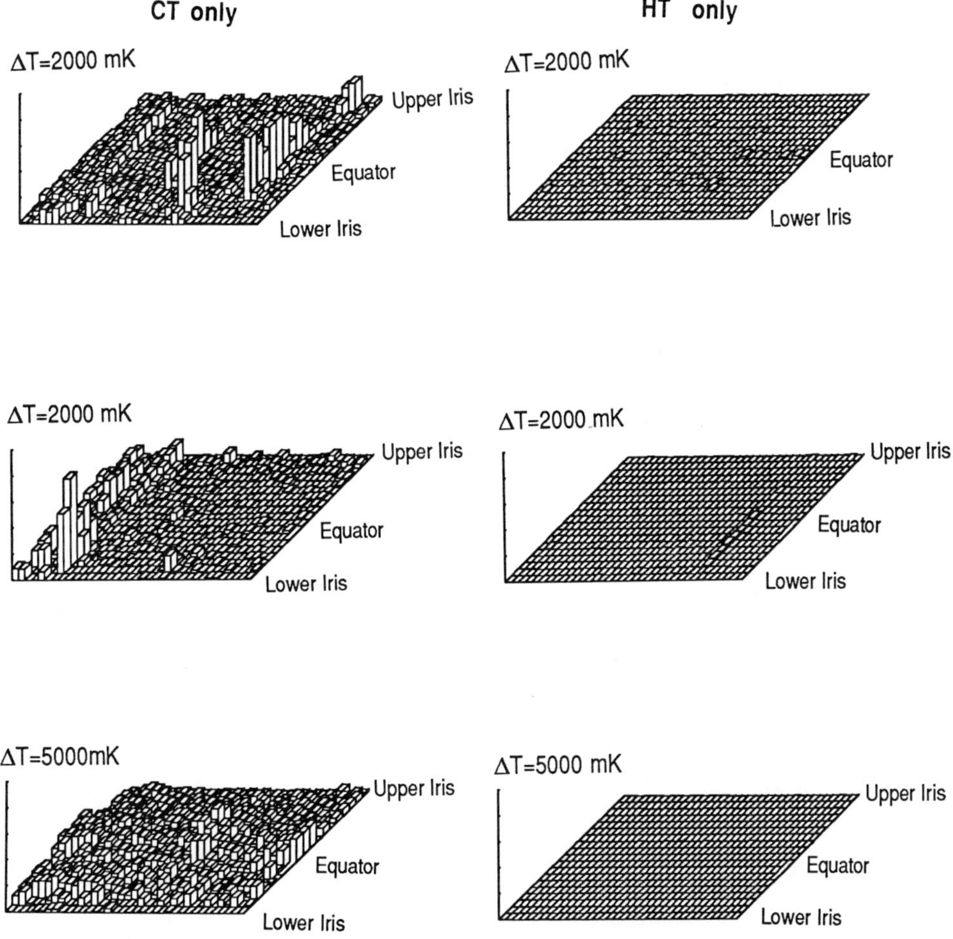

Fig. 5. Temperature maps for three different cavities, with chemical treatment only (left) and after heat treatment (right). Horizontal axis is cavity longitude; a row of 19 thermometer resistors is placed every 10°. Vertical axis is temperature difference between thermometer and helium bath. All maps were taken at E_{pk} = 30 MV/m.

Figure 6 shows a comparison of the emitter densities[8] for CT and HT cavity surfaces. It is evident that, at the same field level, CT cavities have much higher emitter densities than HT cavities.

We can also examine the properties of individual emitters in CT and HT cavities[8,9]. As is well known, an emitter may be characterized by two properties: the field enhancement factor, β, and the equivalent emitter area, A. For this discussion, we use the intercept of the Fowler-Nordheim plot in place of A, since this intercept is proportional to A and can be directly found from the plot. Figure 7 plots this intercept vs β, showing that the distribution for CT cavities

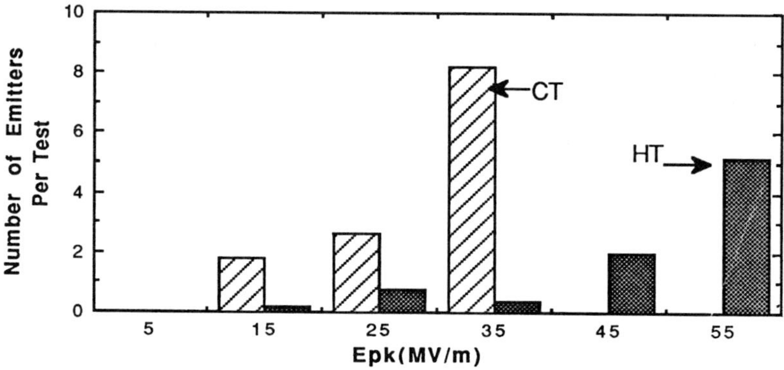

Fig. 6. Emitter densities for cavities with chemical treatment alone (CT) and with heat treatment (HT).

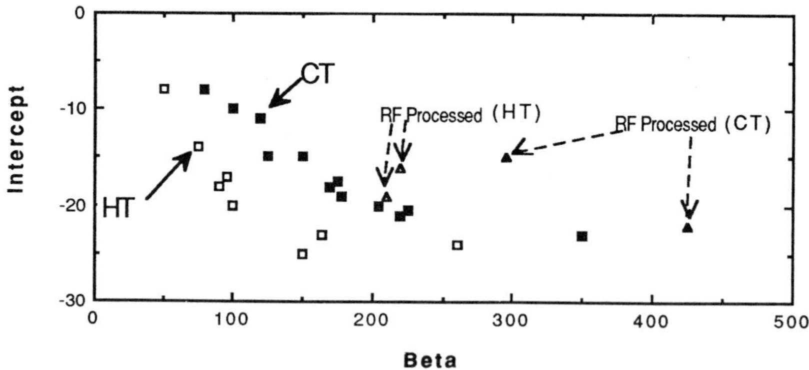

Fig. 7. Intercept of Fowler-Nordheim plot (proportional to the emitter area) vs field enhancement factor, β, for individual emitters. Solid squares:- Chemical treatment only; open squares:- heat treatment; triangles:- after rf processing.

312

lies in the upper right (higher emission) region, while that for HT cavities lies toward the lower left. The stronger emitters were processed out first in both CT and HT cavities.

EXPERIMENTS TO IDENTIFY EMITTER SOURCES

After we found that condensed gases activated emission associated with potential emitters[10], a systematic study was conducted to identify the source of our emitters, since field emission continues to block the way to higher accelerating field, even after HT of cavities. Such knowledge would make treatment and processing more effective. Our strategy has been to intentionally expose well processed and characterized rf cavity "reference" surfaces to various media used in cavity surface preparation[7]. After exposure tests, we carefully study both the new emitters and the resulting rf performance degradation. These tests have indicated that:

(1) Clean air and clean menthanol do not introduce new emitters onto a surface.
(2) Water may degrade a surface, but performance can usually be recovered by He processing. More tests are needed to verify this.
(3) Chemical etching agents are a problem, reducing the Q and E_{pk}.

HIGH PEAK POWER PROCESSING

The benefits of rf processing at a fixed power level diminish after a short period, but further gains are possible as the rf power increases. Therefore, a 3-GHz

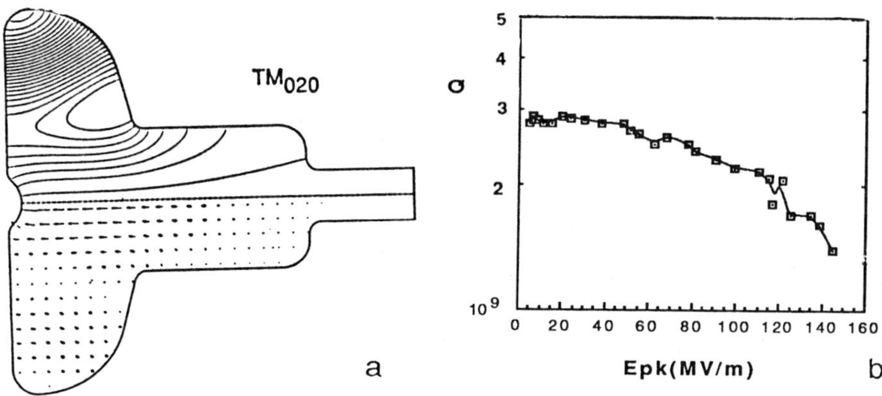

Fig. 8. Mushroom cavity. (a) Electric field (bottom) and emitted electron trajectories (top) for TM_{020} mode. The maximum E is at the center of the dimple; only electrons emitted from the dimple center will arrive at a current probe placed on the axis at right. (b) Overall rf performance for the test attaining the highest E_{pk}.

pulsed high power facility was installed, with a peak power of 200 kW, a maximum pulse width of 2.5 ms and a 1-Hz repetition rate. Preliminary results show a reduction of field emission after high peak power processing[11,12].

SPECIAL CAVITIES DESIGNED TO ATTAIN HIGH E_{pk}

Specially designed non-accelerating "mushroom" cavities[13] have been developed to investigate the maximum rf electric field a Nb surface can support. As shown in Fig 8, the cavity has a very high E_{pk} over only a small area at the center of the dimple. (We use the TM_{020} mode.) In addition the surface magnetic field $(H_{pk})/E_{pk}$ is much less than in normal cavities. This design makes it possible to attain a very high E_{pk} without reaching an H_{pk} limitation while maintaining a low probability of activating FE. The highest surface electric field reached is 145 MV/m with $Q = 3 \times 10^9$ at 1.5K.

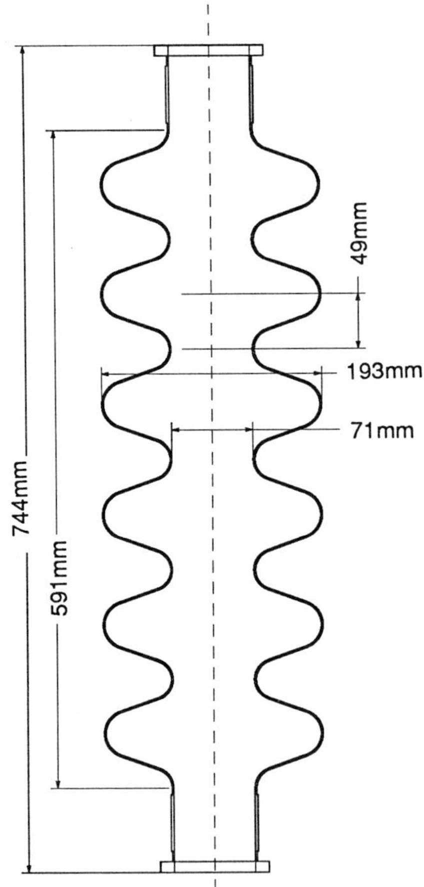

Fig. 9. Six-cell TESLA prototype cavity.

DEVELOPMENT OF SC CAVITY STRUCTURES

One of our major goals is to transfer to real accelerating structures our success in achieving high accelerating gradients in test cavities. To this end, we have chosen two different structures: a prototype ten-cell SC cavity for TESLA, and single cell cavities for a B-factory.

TESLA Cavities

If one wishes to build TESLA, two requirements have to be met: E_{acc} must be increased to 30 MV/m (corresponding to an E_{pk} of about 60 MV/m) and the structure costs must be reduced to an acceptable level. Three considerations will help meet these requirements:

(1) Reduce E_{pk}/E_{acc} from 2.5 to 2 without hurting propagation of higher order modes.
(2) Heat-treat the multicell cavities to reach higher fields.
(3) Use 10 cells per module to reduce costs.

In view of the limited size of our UHV furnace, the first multicell TESLA cavity we have constructed (Figure 9)[14] has six cells. Its E_{pk}/E_{acc} is 2.1. This cavity has been chemically treated and rf tested. The E_{acc} reached in the first test was 17 MV/m with $Q = 4 \times 10^9$ at 1.5K.

B-Factory Cavity

To study the B meson, it is necessary to increase the luminosity of the Cornell CESR beam by a factor of 100. Reaching this goal will present a great challenge for the SC cavity design[15], as can be seen from the following table:

A comparison of the proposed B-factory cavity with the cavities used today

	Today	B-factory
Operating field	5 MV/m	7-10 MV/m
Maximum beam current	30 mA	Several A
Fundamental coupler	100 kW/coupler	0.5 MW/coupler
HOM coupler	10-100 W/coupler	10 kW/coupler
$Q_{external}$ (HOM)	Several 1000	< 100

Figure 10 is a comparison of the cell shapes of 500-MHz cavities used in the Cornell CESR storage ring, in the CERN LEP storage ring, and proposed for the Cornell B-factory. The most apparent difference between them is that the B-factory cavity has the largest beam tube diameter. Figure 11 shows a conceptual design for a SC B-factory cavity.

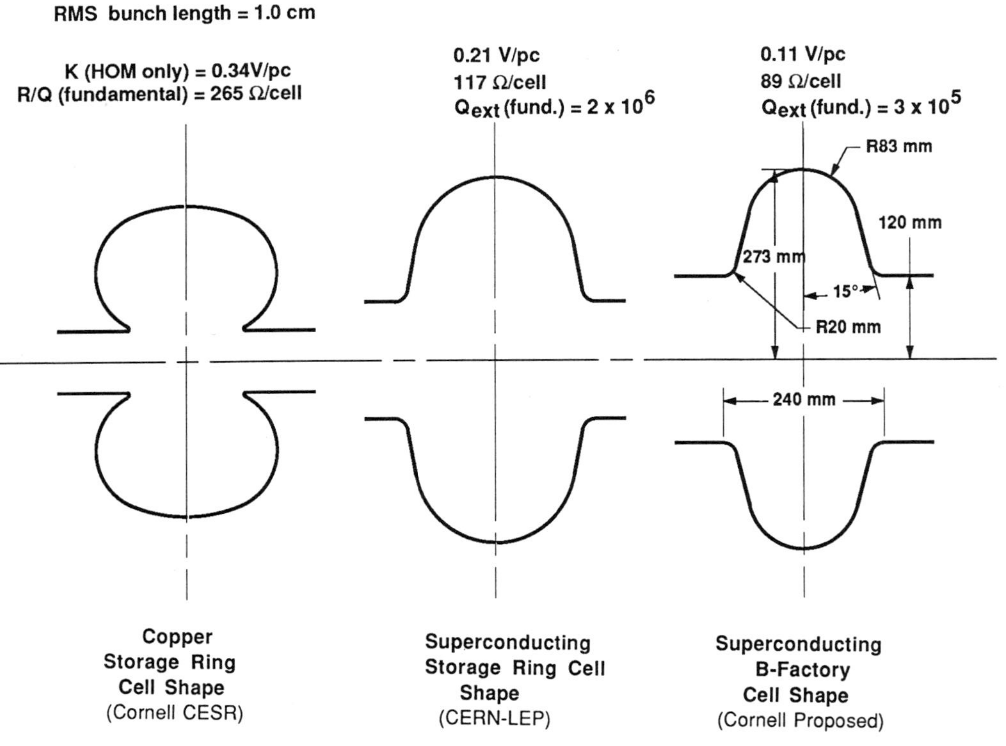

Fig. 10. Comparison of cell shapes for 500-MHz cavities.

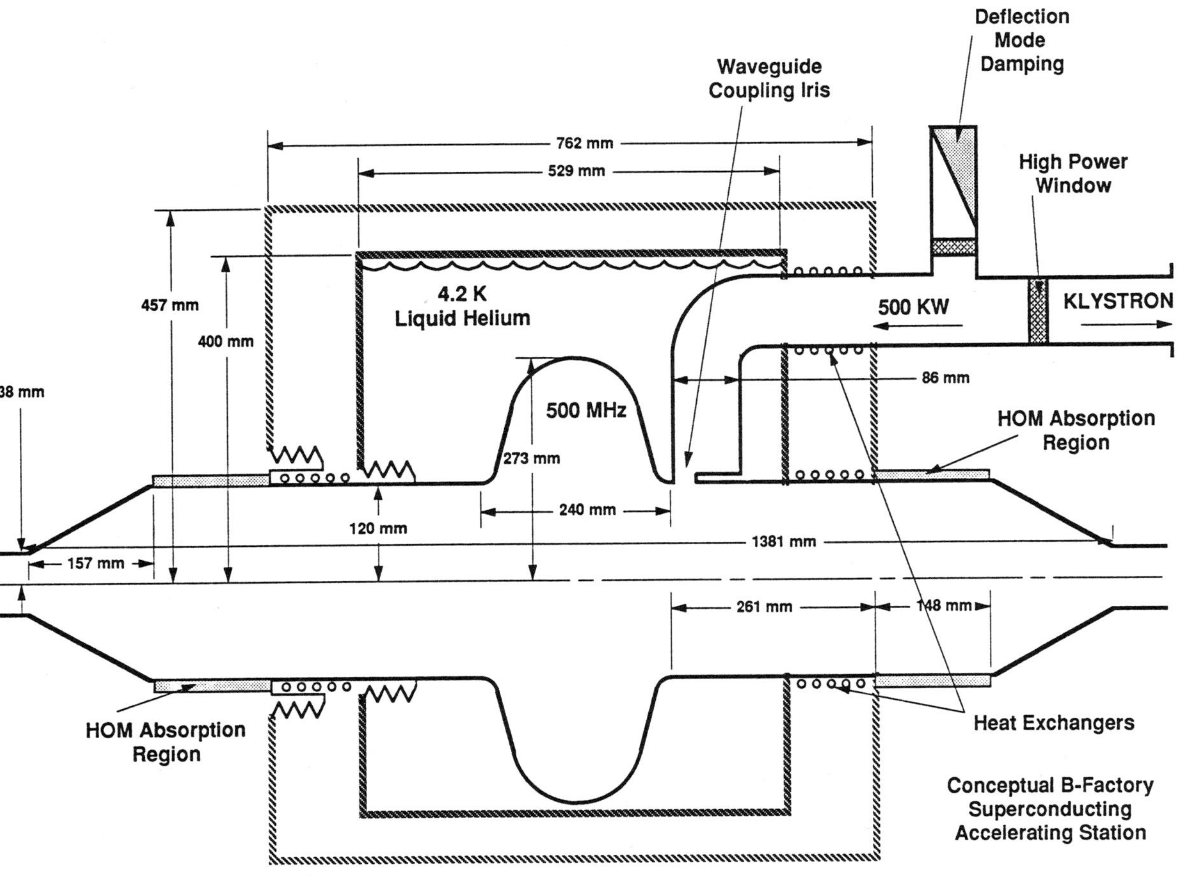

Fig. 11. Conceptual design for proposed superconducting B-factory cavity.

ACKNOWLEDGEMENTS

Sincere thanks are given to the Cornell technicians and machine shop personnel for their indispensable support. A. Leibovich and D. Chandler made a significant contribution to preparation of this paper. The authors express their appreciation for many fruitful discussions with colleagues at Argonne, CEBAF, CERN, Cornell, DESY, KEK, SLAC and the Universities of Geneva and Wuppertal.

REFERENCES

1. Proc. 4th Workshop on RF Superconductivity, KEK, Tsukuba, Japan, Aug. 14-18, 1989.

2. D. Rubin, J. Graber, W. Hartung, J. Kirchgessner, H. Padamsee, J. Sears, Q. S. Shu, Superconducting RF Linear Collider, Proc. 1989 Particle Accelerator Conference, Chicago IL.

3. Q. S. Shu, K. Gendreau, W. Hartung, J. Kirchgessner, D. Moffat, R. Noer, H. Padamsee, D. Rubin, J. Sears, R & D in Progress to Overcome Field Emission in Superconducting Accelerator Cavities, Proc. 1989 Particle Accelerator Conference, Chicago IL.

4. H. Padamsee, Does UHV Annealing above 1100°C as Final Surface Treatment Reduce Field Emission Loading in Superconducting Cavities?, Proc. 1988 Linac Conference, Williamsburg VA.

5. Q. S. Shu, W. Hartung, J. Kirchgessner, D. Moffat, H. Padamsee, D. Rubin, J. Sears, A Study of Influence of Heat Treatment on Field Emission in Superconducting RF Cavities, Nuc. Instr. and Methods in Phys. A278, 329-338 (1989).

6. P. Kneisel, Use of Ti Solid State Gettering Process for Improvement of Performance of Superconducting RF Cavities, J. Less Common Metals 139, 179-188 (1988).

7. H. Padamsee, W. Hartung, R. Noer, C. Reece, Q.S. Shu, RF Field Emission in Superconducting Cavities, Proc. 3rd Workshop on RF Superconductivity, Argonne IL, 251-273 (1987).

8. H. Padamsee, J. Kirchgessner, D. Moffat , R. Noer, D. Rubin, J. Sears, Q.S. Shu, New Results on RF and DC Field Emission, Proc. 4th Workshop on RF Superconductivity, KEK, Tsukuba, Japan, 207-248 (1989).

9. Q. S. Shu, T. Flynn, K. Gendreau, W. Hartung, J. Kirchgessner, D. Moffat, R. Noer, H. Padamsee, S. Rubel, D. Rubin, J. Sears, The Effects of Exposing Well Processed RF Surfaces of Superconducting Cavities to Various Media, Proc. 4th Workshop on RF Superconductivity, KEK, Tsukuba, Japan, 539-569 (1989).

10. Q. S. Shu, K. Gendreau, W. Hartung, J. Kirchgessner, D. Moffat, R. Noer, H. Padamsee, D. Rubin, J.Sears, Influence of Condensed Gases on Field Emission and Performance of Superconducting RF Cavities, IEEE Trans. Mag. 25, 1868-1872 (1989).

11. J. Kirchgessner, Superconducting RF Activities at Cornell University, Proc. 4th Workshop on RF Superconductivity, KEK, Tsukuba, Japan, 37-52 (1989).

12.   J. Graber, High Peak Power Processing of Superconducting Cavities, Proc. 4th Workshop on RF Superconductivity, KEK, Tsukuba, Japan, 529-538 (1989).

13.   D. Moffat, J. Kirchgessner, H. Padamsee, D. Rubin, J.Sears, Q.S. Shu, Superconducting Niobium RF Cavities Designed to Attain High Surface Electric Field, Proc. 4th Workshop on RF Superconductivity, KEK, Tsukuba, Japan, 445-466 (1989).

14.   H. Padamsee, W. Hartung, J. Kirchgessner, D. Moffat, D. Rubin, D. Saraniti, Q.S. Shu, A New Shape Candidate for a Multi-Cell Superconducting Cavity for TESLA, CLNS: 90/985, Cornell University (1990).

15.   H. Padamsee, W. Hartung, J. Kirchgessner, D. Moffat, D. Rubin, D. Saraniti, Y. Samed, J. Sears, Q.S. Shu, Superconducting Cavities for a B-factory — Interim Progress Report, CLNS: 90/978, Cornell University (1990).